Topics in Applied Physics   Volume 6

Topics in Applied Physics Founded by Helmut K. V. Lotsch

Picture Processing
and
Digital Filtering

Edited by T.S. Huang

With Contributions by
H. C. Andrews   F. C. Billingsley
J. G. Fiasconaro   B. R. Frieden   T. S. Huang
R. R. Read   J. L. Shanks   S. Treitel

Second Corrected and Updated Edition

With 113 Figures

Springer-Verlag Berlin Heidelberg GmbH 1979

Professor THOMAS S. HUANG, PhD

Purdue University, School of Electrical Engineering
West Lafayette, IN 47907, USA

ISBN 978-3-540-09339-8   ISBN 978-3-540-35246-4 (eBook)
DOI 10.1007/978-3-540-35246-4

Library of Congress Cataloging in Publication Data. Main entry under title: Picture processing and digital filtering. (Topics in applied physics; v. 6). Includes bibliographical references and index. 1. Image processing. 2. Digital filters. (Mathematics) I. Huang, Thomas S., 1936–. II. Andrews, Harry C. TA1632.P53 1979   621.38'0414   79-4526

© by Springer-Verlag Berlin Heidelberg 1975 and 1979
Originally published by Springer-Verlag Berlin Heidelberg New York in 1979

2153/3130-543210

Preface to the Second Edition

The warm reception accorded to our original volume made us decide to put out this updated paperback edition so that the book can be more accessible to graduate students.

This paperback edition is essentially identical to the original hardcover one except for the addition of a new chapter (Chapter 7) which reviews the recent advances in two-dimensional transforms and filters, and image restoration and enhancement. One hundred additional references have been evaluated and cited. A few typographic errors in the original edition were corrected.

Because of limitation of space, we can do little more in Chapter 7 than point the reader to the literature. We hope that in-depth treatments of some of the important recent results in picture processing and digital filtering will appear in future volumes of the Springer physics program.

West Lafayette, Indiana
December 1978   THOMAS S. HUANG

Preface to the First Edition

In every scientific and engineering endeavor, we encounter signal processing. Many signals are multi-dimensional, i. e., they are functions of several variables. Examples include medical and industrial radiographs, electron-micrographs, radar and sonar maps, seismic data, television images, and satellite (such as ERTS) photographs. The purpose of processing could be: signal generation and display, quality enhancement, information extraction, pattern recognition, efficient coding for transmission or storage, etc. Some of the useful and exciting applications of multi-dimensional signal processing are: character recognition, enhancement of satellite pictures of the moon and Mars, mapping of earth resources from ERTS photographs, and transaxial tomography.

Signal processing can be done either digitally or analogly. However, digital techniques are by far the more flexible. It is because of the rapid progress in digital technology that many multi-dimensional signal processing tasks have become feasible.

In the present book, we bring to the reader in-depth treatment of selected topics in the digital processing of two-dimensional signals (i.e., pictures or images): Chapters 2–4 are on two-dimensional transforms and filters, Chapter 5 is on image enhancement and restoration, and Chapter 6 is on the noise problem in digital signal processing hardware, especially scanners. These chapters are tutorial in nature, yet they bring the reader to the very forefront of current research. We envision that this book will be useful either as a reference book for working scientists and engineers or a supplemetary text book in courses on digital signal processing, image processing, and digital filtering.

We are grateful to ARPA for supporting the editing of this book as well as the work described in the introductory chapter (under contract no. MDA 703-74-0098). We would also like to thank Professor Dr. ADOLF LOHMANN, Technical University of Erlangen, who first suggested the need for a book such as this.

West Lafayette, Indiana
April 1975

THOMAS S. HUANG

Contents

5. Image Enhancement and Restoration. By B.R. FRIEDEN
(With 17 Figures)

6. Noise Considerations in Digital Image Processing Hardware.
By F.C. BILLINGSLEY (With 16 Figures)

Contributors

ANDREWS, HARRY C.

Department of Electrical Engineering and Computer Sciences Program, Image Processing Institute, University of Southern California, Los Angeles, CA 90007, USA

BILLINGSLEY, FREDERIC C.

Jet Propulsion Laboratory, California Institute of Technology, Pasadena, CA 91103, USA

FIASCONARO, JAMES G.

M. I. T. Lincoln Laboratory, Lexington, MA 02173, USA

FRIEDEN, B. ROY

Optical Sciences Center, University of Arizona, Tucson, AZ 85721, USA

HUANG, THOMAS S.

Purdue University, School of Electrical Engineering West Lafayette, IN 47907, USA

READ, RANDOL R.
SHANKS, JOHN L.
TREITEL, SVEN

Amoco Production Company, Research Center, Tulsa, OK 74102, USA

1. Introduction

T. S. HUANG

With 12 Figures

1.1. What is Picture Processing?

In a broad sense, picture or image processing means the manipulation of multi-dimensional signals (which are functions of several variables). Examples of multi-dimensional signals include television images, reconnaissance photographs, medical X-ray pictures, electronmicrographs of molecules, radar and sonar maps, and seismic data.

The purpose of processing these multi-dimensional signals is manifold. However, in most cases it falls into one of the following four categories: enhancement, efficient coding, pattern recognition, and computer graphics.

In many cases, we wish to process a signal to enhance its quality. A prominent example was the work done at the Jet Propulsion Laboratory, California Institue of Technology, in the enhancement of pictures of the Moon and Mars. We have all seen the before and after pictures in the newspaper. The improvement of the processed pictures over the unprocessed ones was truly amazing. When we take the point of view that we are undoing the degradations suffered by the pictures, we talk of image restoration.

When the transmission or storage of a signal requires excessive channel or storage capacity, we would like to code it more efficiently to reduce the requirement. One example is the videophone. With conventional transmission methods, the channel capacity it requires is several hundred times that of a voice channel. Without efficient coding, it is hardly viable economically. Another example is the storage of X-ray pictures in hospitals. Since the number of such pictures is tremendous, their storage takes up much space and their retrieval is almost impossible. Application of efficient coding is badly needed to remedy the situation.

We use the term "pattern recognition" loosely to mean the detection and extraction of patterns or other information from signals. As such, it covers a broad spectrum, ranging from simply calculating the average value of a signal to Newton's recognition of the natural pattern which we call the law of gravity. The area of image pattern recognition which has received the most attention is character recognition. Numerous commercial machines are available for recognizing printed characters, especially single-font. However, the recognition of handwritten characters is still considered as a difficult task. Another area which has recently become rather active is the use of automatic pattern recognition to aid medical diagnosis. Useful schemes are now beginning to emerge. We might mention another emerging area: the automatic mapping of earth resources from satellite photographs.

Computer graphics is concerned with the input and output of pictures into and out of computers and the related programming and data structure problems. One intriguing area of computer graphics is the display of three-dimensional objects. Perhaps, by using holography, someday we could have a truly three-dimensional display.

1.2. Outline of Book and General References

We present in this book in-depth treatment of selected topics in digital picture processing. Specifically: image restoration is stressed, efficient coding is discussed only briefly (in Chapter 2), and pictorial pattern recognition and computer graphics not at all.

In doing picture processing, we make use of many mathematical techniques, some linear and some nonlinear. However, only linear techniques (especially unitary transforms and linear shift-invariant or spatially-invariant operations) can be treated in a general and systematic way. There has been no practically useful general theory of nonlinear operations, and we have to treat each one individually as it comes up in particular applications.

In Chapters 2–4, linear techniques useful in picture processing are presented. Specifically, they treat, respectively: Two-dimensional transforms, nonrecursive filters, and recursive filters. Chapter 5 discusses image restoration, including many nonlinear techniques. Finally, Chapter 6 considers digital image processing hardware with emphasis on the noise problem.

Readers uninitiated in picture processing might wish to read ROSENFELD [1.1] and HUANG et al. [1.2] to gain perspective on the field. They might also wish to skim through the two special issues of the Proceedings of IEEE, one on digital picture processing [1.3], and the other on digital pattern recognition [1.4]. Readers interested in digging more deeply into special areas in picture processing will find ROSENFELD's literature survey papers [1.5–7] most useful.

1.3. Two-Dimensional Transforms

We shall devote the rest of this Introduction to comments on the chapters in the book. Some of the comments might be best appreciated after the reader would have read the appropriate chapters.

In Chapter 2, ANDREWS discusses two-dimensional transforms in the general framework of outer-product expansions. For background and for fast computational algorithms, the reader is referred to ANDREW'S book [1.8]. The most interesting part of Chapter 2 is perhaps the applications of singular value decomposition (SVD) to picture processing, especially image restoration. As described in the chapter, SVD is closely related to matrix pseudoinverses. For an insightful treatment of the topic, see LANCZOS [1.9].

In image restoration, the use of SVD is particularly effective in combatting the noise which is inherent in all degraded images. Using the notations of Chapter 2, we model a nonseparable linear spatially-variant degradation by

$$g = [H]f + n,\tag{1.1}$$

where f and g are column matrices containing the samples from the original object and the degraded image, respectively. The numbers of elements in g and f need not be equal. The rectangular matrix $[H]$ is derived from the impulse response of the degradating system. And n is a column matrix containing noise samples. The noise may, for example, be due to the detector. The problem is: Given g and $[H]$, estimate f.

A good estimate is

$$\hat{f} = [H]^+ g,\tag{1.2}$$

where $[H]^+$ is the Moore-Penrose pseudoinverse of $[H]$. The nice thing about the pseudoinverse is that it always exists so that we do not have to worry about whether the set of linear equations represented by (1.1) has a solution or whether the solution is unique. In fact, $\hat{f}$ is the minimum-norm least-square solution to (1.1) when $n = 0$. In the presence of noise, we have

$$\hat{f} = [H]^+ [H]f + [H]^+ n,\tag{1.3}$$

where the first term on the right-hand side of the equation is the minimum-norm least-square estimate in the absence of noise, and the second term represents the contribution due to noise. Unfortunately,

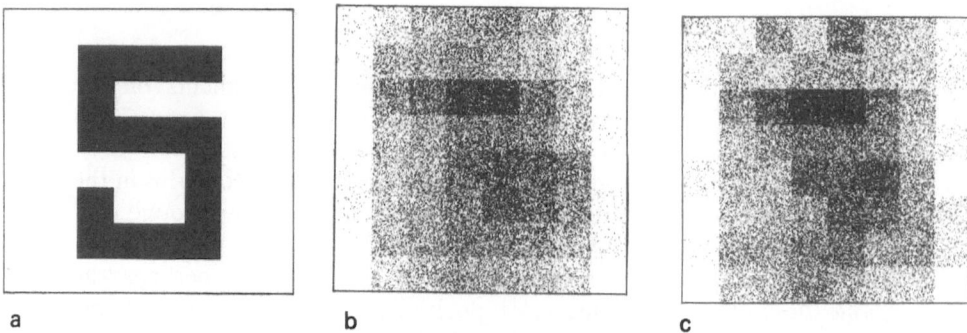

a   b   c

Fig. 1.1 (a) Original. (b) Smeared image with additive Gaussian noise (mean = 0, standard deviation = 0.1). (c) Smeared image with additive Gaussian noise (mean = 0, standard deviation = 0.5)

in many cases, the noise effect dominates and the signal part of (1.3) may be totally obscured.

The use of SVD in calculating the pseudoinverse remedies this situation. As mentioned in passing in Chapter 2, it is possible to derive a generalized Wiener filter. However, the effectiveness of SVD can be demonstrated simply by noting that we can trade off between the amount of noise and the signal quality by choosing the number of terms we use in the SVD of the pseudoinverse. Using the SVD of $[H]^+$, (1.3) becomes

$$\hat{f} = \sum_{i=1}^{R} \lambda_i^{-1/2} V_i U_i^t \{[H]f\} + \sum_{i=1}^{R} \lambda_i^{-1/2} V_i U_i^t n, \qquad (1.4)$$

where U_i and V_i are eigenvectors of $[H][H]^t$ and $[H]^t[H]$, respectively, and λ_i the eigenvalues of either. Generally, each term in the first summation has more or less comparable magnitudes, while the magnitudes of the terms in the second summation increase as $1/\sqrt{\lambda_i}$ (λ_i are in the order of decreasing magnitudes). When we use more and more terms in the summations in (1.4), the first summation becomes closer and closer to the original object, but the signal-to-noise ratio (the ratio of the first summation to the second summation) becomes smaller and smaller. What we would like to do is to achieve a reasonable balance between the two effects. One possibility is to stop at the term where the noise magnitude becomes comparable to the signal magnitude. A better alternative is to look at the result after adding in each new term and stop at the visually-best restoration.

A computer-simulation example is shown in Figs. 1.1–3. The original is a character "5", sampled with 8×8 points. The 8×8 matrix representing the digitized original is shown in Fig. 1.1a. Each point inside

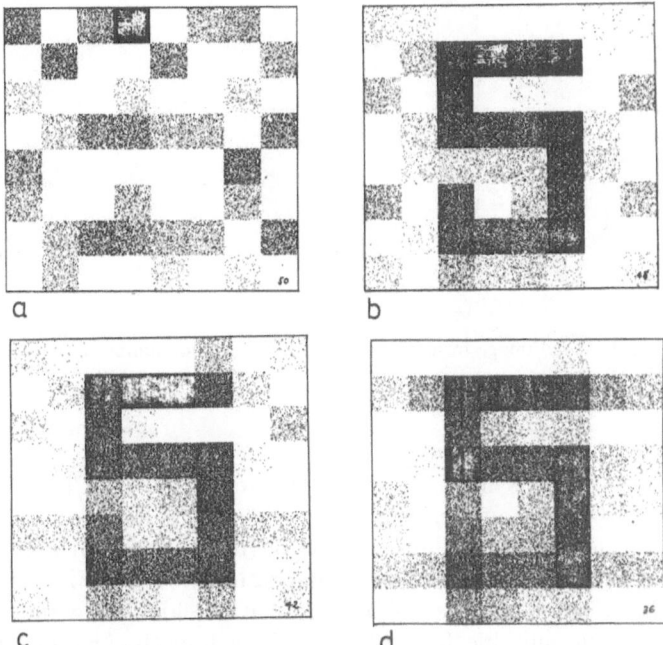

Fig. 1.2a—d. Restored image from Fig. 1.1b using SVD. The number of terms used are: (a) 50, (b) 48, (c) 42, (d) 36

the character was given a value 7; each point outside, a value 0. We blurred this picture by replacing each point by the average of nine points located in the 3×3 block centered around the point in question. Then zero-mean Gaussian random noise was added to it. Two degraded images are shown in Figs. 1.1b and c, the noise standard deviations in these images being 0.1 and 0.5, respectively.

The restoration was done using the equation

$$\hat{f} = \sum_{i=1}^{P} \lambda_i^{-1/2} V_i U_i^t g .$$ (1.5)

For each degraded image, we tried $P = 1, 2, ..., 64$ and looked at all the 64 restorations. Some of the selected restored images are shown in Figs. 1.2 and 1.3. These images were plotted on an electrostatic plotter using dot density modulation. Each point of the 8×8 point image is represented by a square block. An estimation of the 64 eigenvalues of $[H][H]^t$ revealed that 15 of them are practically zero (they are much smaller than the others). Therefore, one obviously should not use

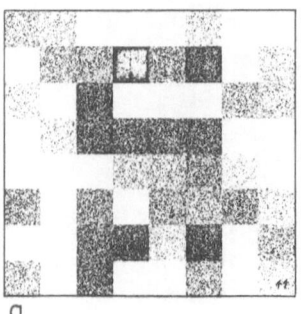 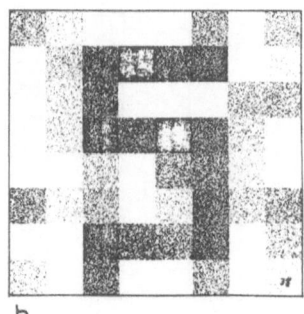 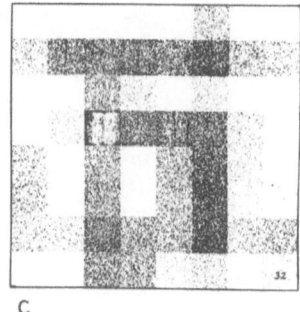

a   b   c

Fig. 1.3a—c. Restored image from Fig. 1.1c using SVD. The number of terms used are: (a) 44, (b) 38, (c) 32

more than 49 terms in (1.5). This is demonstrated in Fig. 1.2a. It is important to note, however, that when the noise is large, one may want to use even fewer terms. Thus, for the degraded image in Fig. 1.1c, 38 terms seemed to give the best (visually) restoration, see Fig. 1.3.

We have seen that the SVD approach of calculating the pseudo-inverse is quite suitable for restoring noisy linearly-degraded images. However, there is one major drawback; viz., even for moderately-sized images, we have to find the eigenvectors and the eigenvalues of very large matrices. For example, for a 100×100 point image, the matrix will be 10000×10000. As discussed in Chapter 2, if the degrading impulse response is separable, then we can simplify the problem considerably. But what can we do in the nonseparable case? One possibility is suggested in Chapter 2: to approximate the nonseparable impulse response by a sum of separable ones. An alternative approach is the following. Usually, the spatial extent of the degrading impulse response is much smaller than that of the picture. Therefore, we can reduce the matrix size by dividing the degraded image into smaller subpictures (the size of each one is still much larger than that of the degrading impulse response) and restore each one separately. The problem one will encounter is: how do we treat the border effect? The points in each subpicture which are near the border are dependent on points in the neighbouring subpictures; therefore, in theory, we cannot treat each sub-picture independent of the others. A mathematical solution to this problem is not available at this time. However, some related work [1.10] indicates that it might invoke Wiener-Hopf techniques. From a practical point of view, we can use this subdivision method by choosing overlapping subpictures, doing the restoration of each separately, and throwing away the borders.

1.4. Two-Dimensional Digital Filters

Two-dimensional (linear spatially-invariant) digital filters find applications in many areas of picture processing, including image enhancement, restoration of linearly degraded images, pattern detection by
matched filtering, and edge extraction. Chapter 3 (FIASCONARO) and
Chapter 4 (READ et al.) of this book present surveys of design techniques for two-dimensional nonrecursive and recursive digital filters,
respectively.

Although the design techniques for one-dimensional digital filters
are relatively well developed [1.11, 12], their extension to two dimensions has been wrought with difficulties. In the case of nonrecursive
filters, we simply encounter the curse of dimensionality. But in the
case of recursive filters, we face an additional curse: a polynomial
in two variables cannot, in general, be factored into a product of first-
order factors. This makes the stability test in two dimensions extremely
cumbersome. Also, it implies that a general two-dimensional recursive
filter cannot be realized as a combination of low-order filters to reduce
the effect of quantization and roundoff noise. It is perhaps fair to say
that there is at present no good general design techniques for two-
dimensional recursive filters. We note that the reason for using recursive filters is that potentially they require less computation time
than nonrecursive filters. We should go through the trouble of designing
a recursive filter only if this potential can be realized.

In the past few years, research in two-dimensional digital filters
has been mushrooming. Most of this research has been in recursive
filters. Therefore, although Chapter 3 covers almost all known techniques
in two-dimensional nonrecursive filter design, Chapter 4 does not include
some of the most recent works in recursive filter design. Among the
omissions, we would like to bring the reader's attention to MARIA and
FAHMY [1.13] and SID-AHMED and JULIEN [1.14].

These authors took the attitude that in order to keep the effects of
quantization and roundoff errors manageable it is imperative that a
recursive filter be synthesized as a combination of low-order sections.
Therefore, they used cascades of second-order sections and determine
the filter coefficients by mathematical optimization techniques. To save
computation time, unconstrained minimization techniques were used.
In the case of MARIA and FAHMY, stability was checked after each
iteration of the minimization procedure, and if the filter was unstable,
the step size was reduced to achieve stability. In the case of SID-AHMED
and JULIEN, stability was ensured by suitable change of variables. Since
a general two-variable rational function cannot be expressed as a product of second-order factors, the filters designed by this approach are

Fig. 1.4. Horizontally smeared image with additive noise

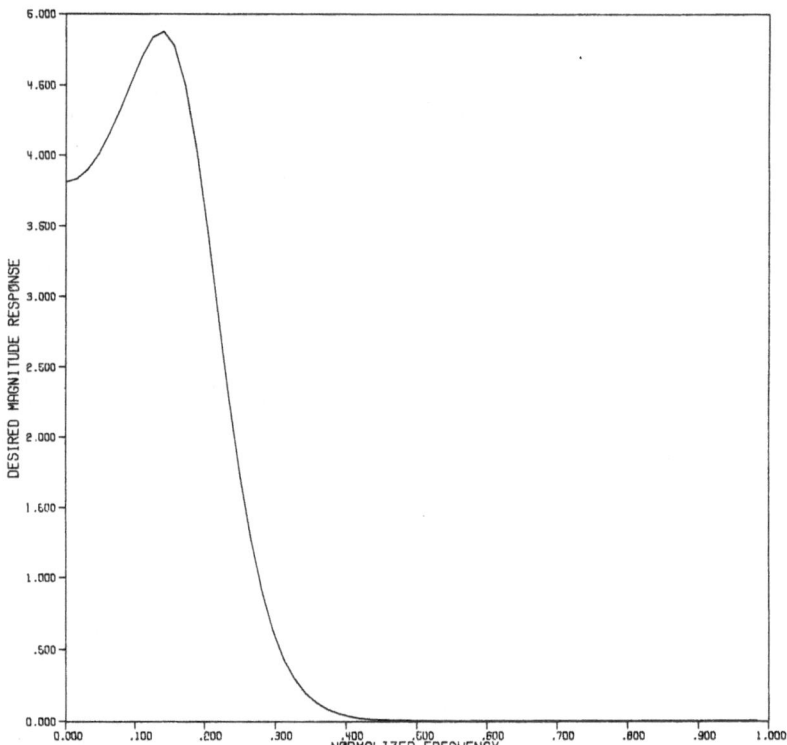

Fig. 1.5. Magnitude of frequency response of least-square inverse filter

only suboptimum. However, they might be good enough for the applications at hand.

Before we leave the subject of two-dimensional digital filters, we would like to emphasize two interrelated points which are often over-

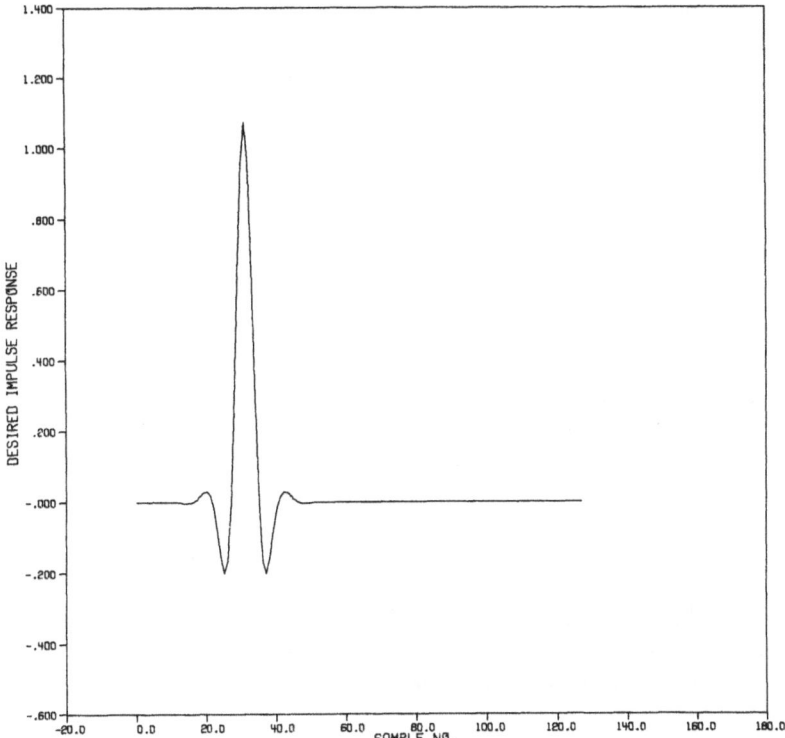

Fig. 1.6. Impulse response of least-square inverse filter

looked: (i) The phase of the Fourier transform of a picture is usually more important than the magnitude. In fact, if we take the inverse transform of the magnitude (setting the phase to zero), we get a blob bearing no resemblance to the original picture; while if we take the inverse transform of the phase (setting the magnitude to a constant). We can still get a likeness of the original picture. (ii) In many image processing applications (e.g., matched-filtering and restoring images degraded by linear variable-velocity motion), the desired filters have nonlinear phases.

The implication of point (ii) is that more research should be directed toward design techniques which allow phase as well as magnitude specifications. For example, it would be worthwhile to study the possibility of extending the linear programming method of designing nonrecursive filters (one- as well as two-dimensional) to include phase specifications.

The implication of point (i) is that in the case of designing recursive filters to process images, even if the desired frequency response

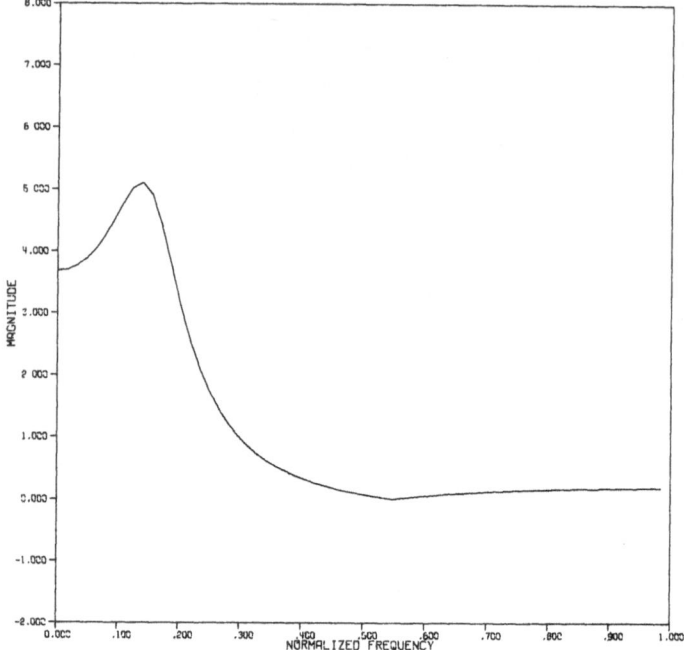

Fig. 1.7. Magnitude of frequency response of designed recursive filter

has linear phase, we should still include it in the specification rather than just specify the magnitude and hope that the phase will turn out to be almost linear—the latter is what we usually do in designing one-dimensional recursive filters. It is true that if the whole image to be filtered is available, then we can achieve zero phase by applying the same recursive filter twice starting from opposite corners of the image. However, this solution is not applicable if we are interested in real-time filtering.

To illustrate the implication of point (i) we offer the following simple example. An image of a letter "T" was sampled with 128×128 points. Each point inside the letter was given a value 5, while each point outside the letter value 0. This image was linearly smeared in the horizontal direction with a Gaussian weighting function (standard deviation $= 3$ points). Then white Gaussian noise with a standard deviation of 0.25 was added, resulting in the noisy blurred image shown in Fig. 1.4.

A least-square inverse filter [1.15] for the linear Gaussian smear was calculated, using a raised-cosine for the signal spectrum. The magnitude of the frequency response of this inverse filter is shown in

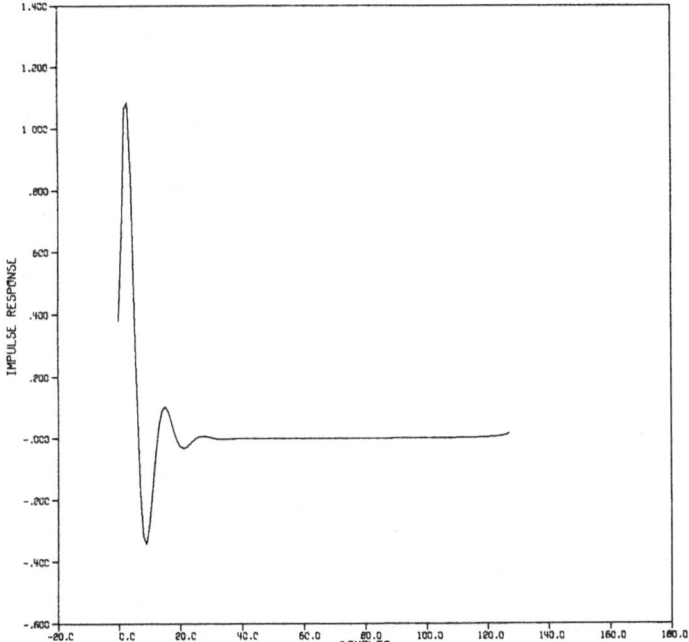

Fig. 1.8. Impulse response of designed recursive filter

Fig. 1.9. Restored image from Fig. 1.4,
using the designed recursive filter

Fig. 1.5. The phase angle is zero. The impulse response of the inverse filter is shown (delayed) in Fig. 1.6.

To restore the image by recursive filtering, we designed a fourth-order (4 poles and 4 zeros) recursive filter, the magnitude of whose frequency response approximates that of the inverse filter (Fig. 1.5). A technique similar to DECZKY's [1.16], was used in the design. The magnitude of the frequency response of the recursive filter is shown in Fig. 1.7, which is almost identical to that of the ideal inverse filter (Fig. 1.5). However, the impulse response of the recursive filter (Fig. 1.8) is quite different from that of the ideal inverse filter (Fig. 1.6). As a result, the application of this recursive filter to the degraded image in Fig. 1.4 yielded the image shown in Fig. 1.9 which contains spurious patterns.

1.5. Image Enhancement and Restoration

In image restoration, we strive to compensate for the degradations introduced into the picture by the imaging process. Typical degradations include camera motion, lens aberrations, low-pass characteristics of electro-optical systems, and atmospheric turbulence. In image enhancement, in addition to or instead of compensating for image degradations, we try to put the image into a form which is more suitable for either human viewing or further machine processing. Examples include edge sharpening, and false-color display.

In Chapter 5, FRIEDEN presents a survey of image restoration techniques. Here, we give some complementary discussions on the following topics: iterative methods, methods based on mathematical programming, linear spatially-varying degradations, and recursive estimation.

In Chapter 5, the iterative methods of VAN CITTERT and JANSSON are discussed. We describe now a different iterative method, the projection method [1.17], which has certain advantages. We represent a monochromatic image by a function of two variables, where the two independent variables are the spatial coordinates and the value of the function is the brightness of the image at a particular point. Then the problem of image restoration can be stated mathematically as follows. Given a degraded image

$$g(x, y) = D[f(x, y)] + n(x, y), \tag{1.6}$$

where $f(x, y)$ is the original, D a degrading operator, and $n(x, y)$ noise, we wish to find a good estimate of $f(x, y)$.

One general approach to attacking the image restoration problem is to try to solve (1.6) by iterative methods. At the kth iteration, the guess solution $f^{(k)}(x, y)$ is determined by $f^{(k-1)}(x, y)$, the guess at the pre-

vious iteration, and $g(x, y) - D[f^{(k-1)}(x, y)]$. A good initial guess might
be

$$f^{(0)}(x, y) = g(x, y).\tag{1.7}$$

Assume the operator D is linear. Then the digitized version of (1.6), neglecting the noise, is

$$\begin{aligned}
a_{11} f_1 + a_{12} f_2 + \cdots + a_{1N} f_N &= g_1 \\
a_{21} f_1 + a_{22} f_2 + \cdots + a_{2N} f_N &= g_2 \\
&\vdots \\
a_{M1} f_1 + a_{M2} f_2 + \cdots + a_{MN} f_N &= g_M,
\end{aligned}\tag{1.8}$$

where f_i and g_j are samples from $f(x, y)$ and $g(x, y)$, respectively, and a_{ij} are constants. The numbers of samples from f and g are N and M, respectively.

The projection method can be best described from a geometrical point of view. We consider $f = (f_1, f_2, ..., f_N)$ as a vector or point in an N-space. Then each equation in (1.8) represents a hyperplane. Let our initial guess solution be $f^{(0)} = (f_1^{(0)}, f_2^{(0)}, ..., f_N^{(0)})$. The next guess solution $f^{(1)}$ is the projection of $f^{(0)}$ on the hyperplane $a_{11} f_1 + \cdots + a_{1N} f_N = g_1$, i.e.,

$$f^{(1)} = f^{(0)} - \frac{(f^{(0)} \cdot a_1 - g_1)}{a_1 \cdot a_1} a_1,\tag{1.9}$$

where $a_1 = (a_{11}, a_{12}, ..., a_{1N})$, and "$\cdot$" denotes the usual dot product. Then we take the projection of $f^{(1)}$ on the hyperplane $a_{21} f_1 + \cdots + a_{2N} f_N = g_2$, and call it $f^{(2)}$,..., until we get $f^{(M)}$ which satisfies the last equation of (1.8). This completes our first cycle of iteration. We then start from the first equation of (1.8) again: We take the projection of $f^{(M)}$ on $a_{11} f_1 + \cdots + a_{1N} f_N = g_1$, and call it $f^{(M+1)}$, and then take the projection of $f^{(M+1)}$ on $a_{21} f_1 + \cdots + a_{2N} f_N = g_2$, ..., until we get to the last equation of (1.8), thus completing our second cycle of iteration. It can be shown that if we continue our iteration in this way, the vector sequence $f^{(0)}, f^{(M)}, f^{(2M)}, f^{(3M)},...$ always converges for any given, N, M, and a_{ij},

$$\lim_{k \to \infty} f^{(kM)} = f.\tag{1.10}$$

It can also be shown that if (1.8) has a unique solution, then f is equal to that solution; and that if those equations have infinitely many solutions, then f is the solution which minimizes

$$|f - f^{(0)}|^2 = (f_1 - f_1^{(0)})^2 + (f_2 - f_2^{(0)})^2 + \cdots (f_N - f_N^{(0)})^2 . \qquad (1.11)$$

Therefore, in the case of infinitely many solutions, we can hope to get a good solution if we start with a good guess $f^{(0)}$.

In applying the projection method, many types of a priori information on the image can be taken advantage of. Specifically, the fact that $f_i \geqq 0$ can be made use of by setting any negative components of $f^{(k)}$ to zero before taking the next projection. And if we know that the original image is confined in a certain region, then we can set to zero those a_{ij}'s in (1.8) that correspond to f_i's lying outside the region.

Some computer-simulation results are shown in Figs. 1.10–12. An image of a numeral "5" was sampled to 64×64 points. Each point in the numeral was given a value 7, while each point outside the numeral was given a value 0. This image was blurred by linear smearing, and the projection method was used to restore the original in the presence of noise.

In the images shown in Figs. 1.10–12, each point was quantized to 16 levels (0–15), the levels 10–15 being represented by the letters A–F. The original image, shown in Fig. 1.10 was smeared vertically over 10 points. Then noise was added to the blurred image resulting in the noisy blurred image shown in Fig. 1.11. The noise was independent of the signal, white, and had a uniform distribution with a standard deviation of 0.5. The projection method was applied to this noisy blurred image. The results after 5 iterations is shown in Fig. 1.12. We note that even at this relatively low signal-to-noise ratio, the result of applying the projection method is surprisingly good. Our initial guess solution was $f^{(0)} = 0$.

Although in the above example the degradation was spatially-invariant and one-dimensional, the projection method can obviously be applied to any linear degradations (generally spatially-varying and two-dimensional). If the image contains $N = n \times n$ points and if the impulse response of the degrading operator contains at most $L = l \times l$ points, then one cycle of iteration requires approximately $2LN$ multiplications and as many additions. For example, if an image contains $N = 200 \times 200$ points and the degrading impulse response contains at most $L = 10 \times 10$ points, then one cycle of iteration requires approximately 8×10^6 multiplications and as many additions. If the computer time for one multiplication and one addition is 1 μsec, then the computer time for one cycle of iteration will be approximately 8 sec.

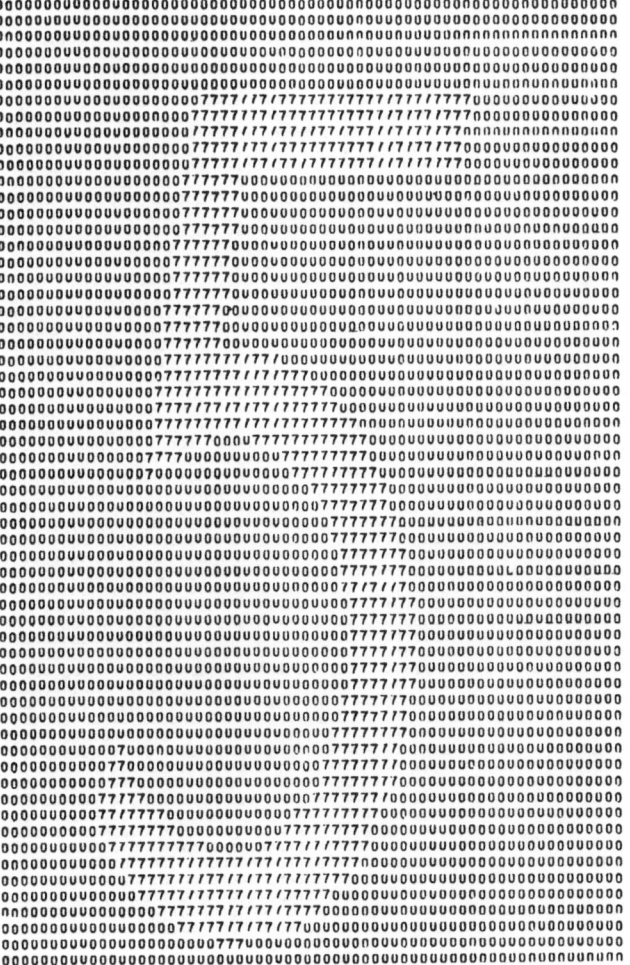

Fig. 1.10. Original image

An alternative approach to solving (1.8) (where g_i are contaminated by noise) under the constraints $f_i \geqq 0$ (and other equality and inequality constraints) is to formulate the problem in the framework of mathematical programming. For example, we can try to find f to minimize

$$F(f) = \sum_{i=1}^{M} \left(\sum_{j=1}^{N} a_{ij} f_j - g_i \right)^2 \qquad (1.12)$$

```
00000011000000011101123342234234222333232323333131001000000001100
1000001010101000000023334L433344434435433434334441100010000001000
00000102000100110004334542344424443324342332544001010000021000000
11000001101010020101155554331553443334434243342444310101100000110000
00000000000010091024665643434345335434343444343100000001000010000
000000000010010000104677675344344553344325344433341101000200000000
00000000001010000013676A644432444353653434442343320111110210000200
01000020000001000124777864343353333332334143323200001000001000100
00000000000000100001567A75222111333122102233237231010000100100001
10000000011110010004567685422102111312001121100201110000010120000
01000010201001001478777420100111012011201121000000000010100010000
00000000110010000015686A63010000011001010101000100000000011001000
10000011000011010347777531000001010010000100001001000010010000100010
101000000000000000367766610002100000000010001001010110100000001100000
01000000000000010137777A521211110001100100010000001100000000010001
000001000010000148677641212523110010010000010000100000000110000
01001000011000112576684443232312110000010100000000010000010010
0000100200000011257886652432433312100000110010010100001011111
101010001000000015778a74045543343321100001000000000000001000100
000000100000000100036776544444443333331310100000010000000001011
100101000000011013666653344445656434233101200000101000000012000000
100100001000001234655344444555566443321112000101000010210000000011
10000010010011003454442444455445566442210211101000002100001002
1010006000000000134524433333545564543422000001000000000010100000
000100100000002113332431212423443646666422000101000000010000000000
0100000000000001323232311412133535567542210000000001001100000000
010100010000111027210220101211123446767652221100100000000000010100
11001000001010111101100001102333456544431010000010000000101000
00100010000101000110100201111000202247777531010000000101001010000
01100020000000030020000110100101233685476524220000100000210010100
00000000100000000100000010000111125778475430000000010010101001100
0000010101021000001000010000000024687765440001101000000010000
10010000000000113000001000010000037567744110010110100000010110001
0000100010100000010000100111000113000013a8778531001010010011001011000
100100000001000000000000000100010137767740001000010000000001101
000201011000000010010001111010100210039476764100010000010000000000
1000010000000001003000121000001000001378768744001000011000010001000
00012000000002100000001101100001000027676765100010011000010010200
00000000010012001210030000000000002477764540000000101010001110100
101110000001121010200031000000000101025676674400010010000000000000
001000000012222110001010010000000124677785610000000001000010002100
010100010020253110011001000001201201257867754100020000100000000100
000000000033422011100000010110112346665553211000100010010010000001
1002000003136412120111000000012124358765442110000001001000000000
00001100003366343122100011120121347686545200100011001000000000
01000001001554455222221101233345478665321100000010000100000000100
001100000373665443323422134435567665433300010000100000000110
020000000234545656444334224344556466543100010001100010000110100
000110000034654544444333332557556654321101100000011001100010010
0100100024344244454446243455555554310100001101000001001010000
011001000003234474433435534546466445322000010000000000001001000
000010010011223404554545442455543421100001001000002010000001000
100010000002234345444544534543234010000010000010201010000001
000000100100122132444344444334233111000020000000000000001000100001
10000210111001212132432342223343222101000100000011010000001000
00000000010100011122234212232111100001100000001001000100001000
00000000000000001012212221122221100000001000010010010000000000
10000111000010001110112100102110100000000100001000000000000000
01100010100110100002000020000000000100010001000010010000010101
011201000003000100010001010101100010000110010000100000010010001
```

under the constraints

$$f_i \geqq 0; \quad i = 1, 2, ..., N.$$ (1.13)

This is a quadratic programming problem. For details, the reader is referred to RUST and BURRUS [1.18] and MASCARENHAS [1.19].

The technique of MCADAM [1.20] mentioned in passing at the end of Chapter 5 is similar (but not identical) to mathematical programming. MCADAM attempted to find solutions to (1.8) under inequality con-

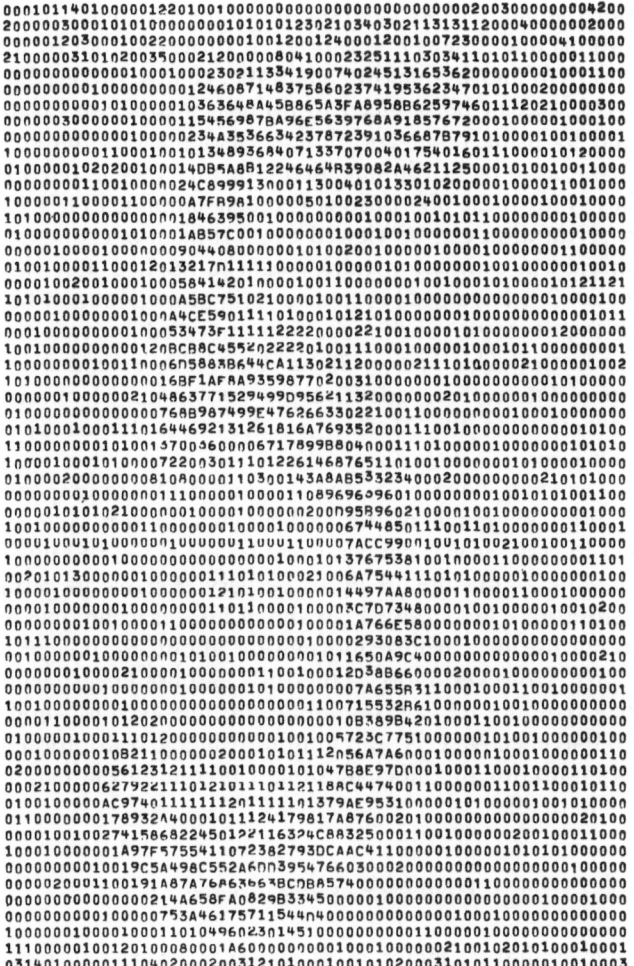

Fig. 1.12. Restored image from Fig. 1.11 using the projection method (5 cycles of iteration)

straints on f_i, g_i, and a_{ij}. There are generally infinitely many solutions. For the linear spatially-invariant degradation case, MCADAM devised computationally efficient algorithms for finding a solution.

Many of the techniques to be described in Chapter 5 as well as the projection method and the mathematical programming approach discussed above and the SVD technique of Chapter 2 can be applied to general linear spatially-varying (LSV) degradations. All these methods, however, require excessive computation. When we consider specific LSV degradations, sometimes we can take shortcuts. One interesting example is degradations due to lens aberrations. It turns out that by using polar coordinates (r, θ) and by making a suitable change of variable in r, the coma aberration can be reduced to a linear spatially-invariant system [1.21]. And by using polar coordinates, astigmatism and curvature of field are reduced to be LSI in one direction, and LSV in the other. Thus, compensation for these aberrations is much simplified [1.22].

Image restoration can be viewed as a problem of estimation. Therefore techniques in estimation theory can be applied. Recently, several authors [1.23–25] have used recursive estimators (Kalman-bucy filters) to reduce image noise. The image was modeled by two-dimensional Markov processes. This approach can be readily extended to apply to restoration of linearly degraded images contaminated by noise. Some ways of generalizing the Markov model were studied by JAIN and ANGEL [1.26].

It was stated in Chapter 5, and we would like to emphasize again here that although much of the earlier work in restoring images degraded by LSI systems was concentrated on the use of LSI inverse filters, such inverse filters are not likely to be very effective in getting good-quality restorations. Generally, LSV and nonlinear techniques (especially those making use of a priori information about the image) are much more effective. However, most of the LSV and nonlinear techniques we know today require excessive computation time even for moderate-size images. To find effective LSV and nonlinear image restoration technique that are computationally efficient should be one of the major goals of researchers in the field.

1.6. Digital Image Processing Hardware

The ultimate limiting factor in most image processing task is noise. In Chapter 6, BILLINGSLEY considers the noise problem, concentrating on the noise analysis of scanners. The scanner converts an analog image into digital data, and is essential whether the processing is going to be done on a general-purpose computer or by special-purpose hardware. The chapter contains many practical and useful results which the researchers in digital image processing will find invaluable.

The reader's attention is called to the rapid progress in surface-wave and charge-coupled devices which can perform one-dimensional convolution extremely fast [1.27]. These devices can also be used to do both image sensing and processing.

It is interesting to note that hardware development has a strong impact on processing algorithms. Thus, in order to take advantage of surface-wave and charge-coupled devices, it was proposed [1.28] that the real-time calculation of two-dimensional Fourier transforms be reduced to that of one-dimensional convolutions.

1.7. Other Areas

Although the concepts and techniques discussed in this book are useful in all areas of image processing, our emphasis is on image enhancement and restoration. For the other areas: we refer the readers to [1.29, 30] for efficient picture coding, [1.31, 32] for pictorial pattern recognition, and [1.33, 34] for computer graphics.

Acknowledgement. This work was supported by ARPA contract MDA 903-74 C-0098. The computer simulation experiments were performed by P. NARENDRA, J. BURNETT, and D. BARKER.

References

1.1. A. ROSENFELD: *Picture Processing by Computer* (Academic Press, New York, 1969).
1.2. T. S. HUANG, W. F. SCHREIBER, O. J. TRETIAK: Proc. IEEE **59**, 1588 (1971).
1.3. Proc. IEEE, Special issue on Digital Picture Processing (July, 1972).
1.4. Proc. IEEE, Special issue on Digital Pattern recognition (October, 1972).
1.5. A. ROSENFELD: Computing Survey **5**, 81 (1973).
1.6. A. ROSENFELD: Computer Graphics and Image Processing **1**, 394 (1972).
1.7. A. ROSENFELD: Computer Graphics and Image Processing **3**, 178 (1974).
1.8. H. ANDREWS: *Computer Techniques in Image Processing* (Academic Press, New York, 1970).
1.9. C. LANCZOS: *Linear Differential Operators* (Van Nostrand, London, 1961), Chap. 3.
1.10. G. M. ROBBINS, T. S. HUANG: "Inverse filtering for linearly shift-variant imaging systems"; Proc. Symp. "Bildverarbeitung und interaktive Systeme", DLR-Mitt. 73–11, pp. 40–49, (DFVLR-Forschungszentrum, Oberpfaffenhofen, West Germany, December 1971).
1.11. B. GOLD, C. RADER: *Digital Processing of Signals* (McGraw-Hill, New York, 1969).
1.12. L. RABINER, C. RADER (eds.): *Digital Signal Processing* (IEEE Press, New York, 1972).
1.13. G. A. MARIA, M. M. FAHMY: IEEE Trans. Acoustics, Speech, Signal Proc. ASSP-**22**, 15 (1974).
1.14. M. A. SID-AHMED, G. A. JULIEN: "Frequency domain design of a class of stable two-dimensional recursive filters", preprint, University of Windsor, Windsor, Ontario, Canada (June, 1974).
1.15. C. W. HELSTROM: Opt. Soc. Am. **57**, 297 (1967).
1.16. A. G. DECZKY: IEEE Trans. Audio Electroacoustics AU-**20**, 257 (1972).
1.17. K. TANABA: Numerical Mathematics **17**, 203 (1971).
1.18. B. W. RUST, W. R. BURRUS: *Mathematical Programming and the Numerical Solution of Linear Equations* (Americal Elsevier, New York, 1972).

1.19. N.D.MASCARENHAS: "Digital image restoration under a regression model—the unconstrained, linear equality and inequality constrained approches", USCIPI Rep. 520, Image Processing Institute, University of Southern California, Los Angeles, Calif. (January, 1974).

1.20. D.P.McADAM: Opt. Soc. Am. **60**, 1617 (1970).

1.21. G.M.ROBBINS, T.S.HUANG: Proc. IEEE **60**, 862 (1972).

1.22. A.A.SAWCHUK, M.J.PEYROVIAN: "Space-variant restoration of astigmatism and curvature of field", USCIPI Rep. 530, Image Processing Institute, University of Southern California, Los Angeles, Calif. (March, 1974), pp. 68–74.

1.23. A.HABIBI: Proc. IEEE **60**, 878 (1972).

1.24. N.E.NAHI: Proc. IEEE **60**, 872 (1972).

1.25. A.C.KAK, D.PANDA: "Statistical modeling and recursive filtering of images", Rep. TR-EE 74–22, Purdue University, West Lafayette, Indiana (June, 1974), pp. 35–40.

1.26. A.K.JAIN, E.ANGEL: IEEE Trans. Computers C-**32**, 470 (1974).

1.27. T.S.HUANG, O.J.TRETIAK (eds.): Proc. of Seminar on Signal Processing, NEREM, Boston, Mass. (1973).

1.28. J.M.SPEISER, H.J.WHITEHOUSE: A two-dimensional discrete Fourier transform architecture"; NUC TN 1221 Naval Undersea Center, San Diego, Calif. (October, 1973).

1.29. T.S.HUANG, O.J.TRETIAK (eds.): *Picture Bandwidth Compression* (Gordon and Breach, New York, 1972).

1.30. T.S.HUANG: "Bandwidth Compression of Optical Images", Progress in Optics, Vol. 10 (North-Holland Publishing Co., Amsterdam, 1972) pp. 1–44.

1.31. R.DUDA, P.HART: *Pattern Classification and Scene Analysis* (J. Wiley and Sons, New York, 1973).

1.32 T.COVER: IEEE Trans. Inf.Theory IT-**19**, 827 (1973).

1.33. R.J.PANKHURST: In *Advances in Information Systems Science*, Vol. 3, ed. by J. T. Tou (Plenum Press, New York, 1970), pp. 215–282.

1.34. M.D.PRINCE: *Interactive Graphics for Computer-Aided Design* (Addison-Wesley, Reading, Mass., 1971).

Further References with Titles

T.S.HUANG: *Image Processing* (to be published by North-Holland, Amsterdam, 1980).

T.S.HUANG, D.BARKER, S.BERGER: Iterative image restoration. Appl. Opt. **14**, No. 5, 1165–1168 (1975).

T.S.HUANG, J.BURNETT, A.DECZKY: The importance of phase in image processing filters. IEEE Trans. Acoustics, Speech, and Signal Proc. ASSP-**23**, No. 6, 529–542 (1975).

T.S.HUANG, P.NARENDRA: Image restoration by singular value decomposition. Appl. Opt. **14**, No. 9, 2213–2216 (1975).

Special issue on Digital Signal Processing Proc. IEEE **63**, No. 4 (1975).

Special issue Digital Filtering and Image Proc. IEEE Trans. on Circuits and Systems CAS-**22**, No. 3 (1975).

G.A.JULIEN, M.A.SID-AHMED: Stability constraints used in computeraided design of recursive digital filters. IEEE Trans. Acoustics, Speech and Signal Proc. ASSP-**22**, No. 2, 153–158 (1974).

A.V.OPPENHEIM, R.SCHAFER: *Digital Signal Processing*. (Prentice-Hall, Englewood Cliffs, NJ., 1975).

L.RABINER, B.GOLD: *Theory and Applications of Digital Signal Processing*. (Prentice-Hall, Englewood Cliffs, NJ., 1975).

A.ROSENFELD, A.C.KAK: *Digital Picture Processing* (Academic Press, New York, 1976).

2. Two-Dimensional Transforms

H. C. ANDREWS

With 18 Figures

This chapter is designed to provide the reader with an overview of two-dimensional transforms and their uses in digital picture processing by general purpose computers. The notation is selected such that casual familiarity with matrix algebra should be sufficient to grasp the basic principles of image coding and restoration in conjunction with the use of two-dimensional image transforms. The chapter is presented in a sequence of mathematical development followed by image coding examples from both the spatial and transform domain viewpoints. Object restoration is discussed in some detail in which numerical techniques in matrix algebra are utilized to avoid singular restorations. The observant reader will note that the assumption of separability becomes a key factor in many of the formulations set forth in the chapter. The chapter contains both tutorial material as well as quite recent object restoration procedures, and it is the author's hope that the common framework of two-dimensional transforms and matrix quadratic forms successfully provide the reader with a smooth continuum in transition between the more tutorial material and newer processing results.

2.1. Introduction

2.1.1. Motivation—Coding, Restoration, Feature Extraction

The popularity of digital picture processing has risen quite drastically over the past few years with major emphasis being placed on the aspects of image coding, image restoration, and image feature extraction. Image coding usually refers to the attempt at transmitting pictures over digital communication networks in as efficient a manner as is possible. This usually implies seeking a reduction in the number of bits describing an image such that fewer bits need be transmitted over the communication system. Examples of such systems might include nationwide computer and time sharing networks for tele-conferencing, image sharing, medical consultation, multispectral satellite imaging, space craft probes, remote piloted vehicles, facsimile transmission of images ranging from fingerprints to text, and even image transmission over existing telephone networks.

Historically speaking, the subject of digital image coding has been a center of research interest for longer than that of digital image restoration, and as such has probably reached a more advanced state-of-the-art. However, recently interest has been directed toward those images which are degraded in quality from their respective original objects and there is motivation for restoring the images to a better estimate of those objects. Examples here might include focus correction, motion blur reconstruction, imaging system restoration such as aberration correction, depth of field and geometric distortion correction, as well as film tilt, other non-optimum imaging conditions, and noise filtering and removal. Usually restoration techniques are conditioned upon some *a priori* knowledge or model of the degradation phenomena involved in the imaging circumstances. The motivation for such processes becomes self evident when one realizes the tremendous emphasis man puts on his visual senses for survival. Considering the relative success achieved in one-dimensional (usually time) signal processing, it is to be expected that far greater strides could be made in the visual two-dimensional realm of signal processing.

The areas of space imagery, biomedical imagery, industrial radiographs, photoreconnaissance images, television, forward looking infrared (FLIR), side looking radar (SLR), and several multispectral or other esoteric forms of mapping scenes or objects onto a two-dimensional format are all likely candidates for digital image processing. Yet many non-natural images are also subject to digital processing techniques. By non-natural images one might refer to two-dimensional formats for general data presentation for more efficient human consumption. Thus range-range rate planes, range-time planes, voiceprints, sonagrams, etc. may also find themselves subject to general two-dimensional enhancement and restoration techniques.

For the sake of semantics we will define restoration to be the reconstruction of an image toward an object (original) by inversion of some degradation phenomena. Restoration techniques require some form of knowledge concerning the degradation phenomena if an attempt at inversion of that phenomena is to be made. This knowledge may come in the form of analytic models, statistical models or other *a priori* information coupled with the knowledge (or assumption) of some physical system which provided the imaging process in the first place. Thus considerable emphasis must be placed on sources and their models of degradation.

Feature extraction in the context of digital picture processing refers to the selection of specific attributes descriptive of some aspect of the picture for subsequent pattern recognition, classification, decision or interpretation. Texture, shape, statistical parameters, etc. all might qualify as relevant features. For certain applications relevant features are often obtained from a two-dimensional transformation of the original image into a space where the new coordinates are more statistically uncorrelated than in the original picture representation. These more uncorrelated spaces then provide more efficient features for ultimate classification purposes due to the fact that uncorrelated attributes require fewer dimensions for description than do correlated parameters.

2.1.2. Linear Models

Normally most analysis models for image processing systems are approximated by linear assumptions. These linear imaging models then give a clue as to the restorations necessary for degradation inversion as well as coded image transmission system under the linear approximation.

For notational convenience, the following format will be established for consistency and ease of understanding (Fig. 2.1):

$f(\xi, \eta)$ will be the object,
$g(x, y)$ will be the image,
$h(x, y, \xi, \eta)$ will be known as the impulse response or point spread function (PSF) if the imaging system is linear.

Emerging from the potpourri of methods in use for digital image processing are a set of models in which attempts are made for the restoration of images by the effective inversion of degradation phenomena through which the object itself was imaged. Let $g(x, y)$ be the image of the object $f(\xi, \eta)$ which has been degraded by the linear operator $h(x, y, \xi, \eta)$ such that

$$g(x, y) = \int\limits_{-\infty}^{\infty}\!\!\int f(\xi, \eta)\, h(x, y, \xi, \eta)\, d\xi\, d\eta. \tag{2.1}$$

The system degradation, $h(\cdots)$, is known as the impulse response or point spread function and is physically likened to the output of the system when the input is a delta function or point source of light. If, as the point source explores the object plane, the form of the impulse

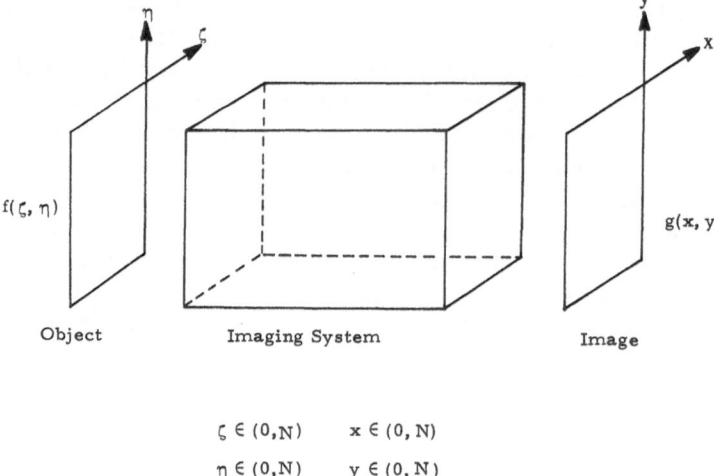

$$\zeta \in (0,N) \qquad x \in (0,N)$$
$$\eta \in (0,N) \qquad y \in (0,N)$$

Fig. 2.1. A Linear imaging system model

response remains fixed except for position in the image plane, then the system is said to be spatially invariant, i.e., a spatially invariant point spread function (SIPSF) exists. If this is not the case, then a spatially variant point spread function system results (SVPSF). In this case (2.1) holds, and in the SIPSF case

$$g(x, y) = \int\int_{-\infty}^{\infty} f(\xi, \eta)\, h(x - \xi, y - \eta)\, d\xi\, d\eta \,. \tag{2.2}$$

Most researchers have been satisfied with the model of (2.2), with variations such as additive noise, multiplicative noise, etc. Fourier techniques work well in attempting to obtain $f(\xi, \eta)$ from $g(x, y)$ through the inversion of $h(\cdots)$ of (2.2) due to the Fourier-convolution relationship. Thus, in the absence of noise

$$G_f(u, v) = H_f(u, v)\, F_f(u, v), \tag{2.3}$$

where G_f, H_f and F_f are the Fourier transforms of g, h, f, and the determination of $F_f(u, v)$ simply requires the inversion of H_f, if it exists. If the SVPSF model is used, the Fourier techniques are no longer applicable and more general brute force inversion methods must be resorted to for object restoration.

There are three basic linear approaches which are often used for inversion of either of the two systems, as described above. They could be referred to as:

a) continuous-continuous,
b) continuous-discrete,
c) discrete-discrete.

The first analysis method looks at the entire image restoration process in a continuous fashion (although ultimate implementation will necessarily be discrete). The second analysis method assumes the object is continuous but the image is sampled and therefore discrete. The third technique assumes completely discrete components and utilizes purely numerical analysis and linear algebraic principles for restoration. In equation form we would have the following imaging models for each of the three assumptions:

a) continuous-continuous

$$g(x, y) = \int\limits_{-\infty}^{\infty}\int f(\xi, \eta)\, h(x, y, \xi, \eta)\, d\xi\, d\eta\,, \tag{2.4}$$

b) continuous-discrete

$$g_i = \int\limits_{-\infty}^{\infty}\int h_i(\xi, \eta)\, f(\xi, \eta)\, d\xi\, d\eta \quad i = 1, \ldots, N^2\,, \tag{2.5}$$

c) discrete-discrete

$$g = [H]\, f\,. \tag{2.6}$$

The continuous-continuous model of (2.4) says that the image is simply the integration of the object and point spread function in an analog two-dimensional environment. When the model is put into the computer, the image, object, and point spread function will be sampled by N^2, N^2, N^4 points, respectively.

The continuous-discrete model of (2.5) implies that the object is continuous (as it would be in the real world) but the sensor defining the image is discrete and already sampled by N^2 points. Thus there are N^2, g_i scalar values and of course N^2, $h_i(\xi, \eta)$ different point spread functions.

Finally, the discrete-discrete system implies that the object and image are one-dimensional vectors, N^2 long, which represent the original object and image f, and g, respectively. The vectors can be raster scanned versions of two-dimensional functions or any other scanning method such that all N^2 points are obtained. The 4-dimensional function $h(\xi, \eta, x, y)$ is now reduced to a 2-dimensional array by the raster scan and thus is a matrix of size $N^2 \times N^2$.

In this chapter we will limit our discussion to the discrete-discrete notation but will modify it from time to time under additional simplifying assumption. Specifically if the degradation phenomena can be modeled as separable we can then replace (2.6) with

$$[G] = [A][F][B], \tag{2.7}$$

where $[F]$ is the sampled version of the object, $[G]$ is the sampled version of the image and $[A]$ is the column degradation of the object and $[B]$ is the row degradation of the object. Now all matrices of (2.7) are $N \times N$ greatly simplifying the computational tasks to follow.

2.2. Mathematical Representations

2.2.1. Image Representations in Orthogonal Bases

Inherent in many aspects of digital image processing is computer storage, in digital form, of a two-dimensional array of numbers representing individual brightness values taken from an original photograph, scene, or camera tube. However, a sampled and quantized image is merely a matrix of positive numbers (albeit, possibly a quite large matrix), which may then be manipulated in a digital computer by a large class of linear and nonlinear operations. Digital image processing can then be reduced to analysis of large scale matrix manipulations for many of the operations known as transformations, expansions, representations, and restorations. This section investigates some of these techniques in the context of matrix, or specifically, vector outer products.

Consider the matrix $[G]$ as an image that has been sampled and quantized, where the i-th row and j-th column correspond to the x and y spatial coordinates of a scene $g(x, y)$. Thus we see that

$$[G] = SQ\{g(x, y)\}, \tag{2.8}$$

where the nonlinear operator $SQ\{\cdots\}$ represents the sampler and quantizer. Assume that $[G]$ has the dimensions N by N. Techniques involving nonsquare image matrices have been explored in much of the literature; however, no discussion will be offered herein, as it does not significantly contribute to the philosophy of this chapter. A general separable linear transformation on an image matrix $[G]$ may be written in the form

$$[\alpha] = [U]^t [G] [V], \tag{2.9}$$

where $[\alpha]$ is termed the unitary transform domain of the image, $[U]$ and $[V]$ are unitary operators, and the superscript t denotes the matrix transpose. The unitary nature of $[U]$ and $[V]$ implies the following relations

$$[U][U]^t = I, \tag{2.10a}$$
$$[V]^t [V] = I. \tag{2.10b}$$

Hence the "inverse" transform of (2.9) may be written in the form

$$[G] = [U][\alpha][V]^t. \tag{2.11}$$

If the quantities $[U]$ and $[V]$ are written in the form

$$[U] = [\boldsymbol{u}_1 \boldsymbol{u}_2 \cdots \boldsymbol{u}_N] \tag{2.12}$$
$$[V] = [\boldsymbol{v}_1 \boldsymbol{v}_2 \cdots \boldsymbol{v}_N], \tag{2.13}$$

where the terms $\boldsymbol{u}_i$ and $\boldsymbol{v}_i$ are the vectors made up from the columns of $[U]$ and $[V]$, respectively, then

$$[G] = [\boldsymbol{u}_1 \boldsymbol{u}_2 \cdots \boldsymbol{u}_N] [\alpha] \begin{bmatrix} \boldsymbol{v}_1^t \\ \boldsymbol{v}_2^t \\ \vdots \\ \boldsymbol{v}_N^t \end{bmatrix}. \tag{2.14}$$

If the matrix α is written as a sum of the following form

$$[\alpha] = \begin{bmatrix} \alpha_{11} & 0 & 0 & \cdots & 0 \\ 0 & 0 & & & \\ \vdots & & & & \\ 0 & & & & 0 \end{bmatrix} + \begin{bmatrix} 0 & \alpha_{12} & 0 & \cdots \\ 0 & & & \\ \vdots & & & \\ 0 & \cdots & & 0 \end{bmatrix} + \cdots, \tag{2.15}$$

then it follows that

$$[G] = \sum_{i=1}^{N} \sum_{j=1}^{N} \alpha_{ij} \boldsymbol{u}_i \boldsymbol{v}_j^t. \tag{2.16}$$

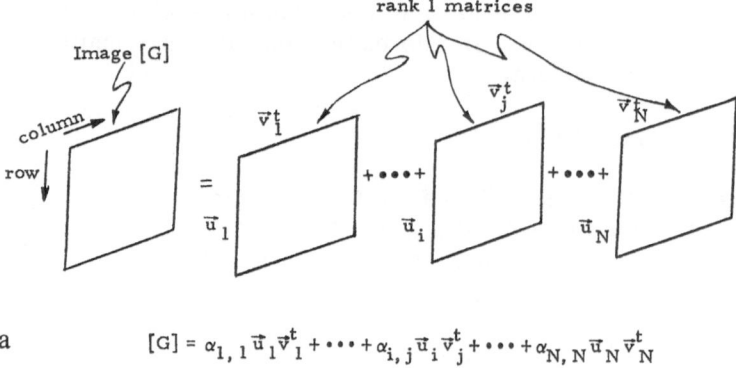

$$[G] = \alpha_{1,1} \vec{u}_1 \vec{v}_1^t + \cdots + \alpha_{i,j} \vec{u}_i \vec{v}_j^t + \cdots + \alpha_{N,N} \vec{u}_N \vec{v}_N^t$$

a

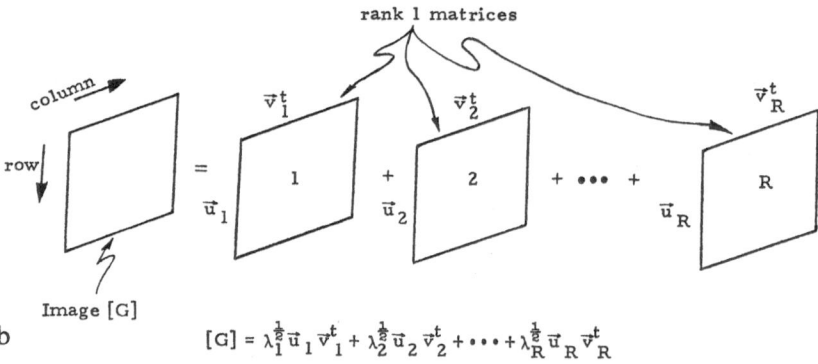

b

$$[G] = \lambda_1^{\frac{1}{2}} \vec{u}_1 \vec{v}_1^t + \lambda_2^{\frac{1}{2}} \vec{u}_2 \vec{v}_2^t + \cdots + \lambda_R^{\frac{1}{2}} \vec{u}_R \vec{v}_R^t$$

Fig. 2.2a and b. Image expansion into outer products. (a) Rank one matrix expansions, (b) singular value decomposition of $[G]$

The outer product $u_i v_j^t$ may be interpreted as an "image" so that the sum over all combinations of the outer products, appropriately weighted by the α_{ij}, regenerates the original image $[G]$. The formulation permits a particularly useful interpretation of selective coefficient retention in terms of certain transform sets.

As a graphical illustration of this process consider the example of Fig. 2.2a. Here we see that the image $[G]$ is decomposed into a sum of N^2 rank 1 matrices each weighted by the appropriate coefficient $\alpha_{i,j}$. The rank 1 matrices will represent two-dimensional basis images examples of which are trigonometric waveforms for Fourier expansions, or binary waveforms for Walsh expansions.

Selection of the transformations represented by matrix quantities $[U]$ and $[V]$ is essentially arbitrary. The quantities $[U]$ and $[V]$ may be selected from the same or different orthogonal basis functions. As an example, $[U]$ may be based on Walsh functions, and $[V]$ taken from the Fourier sines and cosines. In this particular example, the columns of the image matrix would be expanded into Walsh waves (rectangular functions) and the rows would be expanded into the complex exponentials (sines and cosines).

While it may not be readily evident from (2.16), the expansion may be interpreted as a singular value decomposition (SVD) of the image into its singular vectors under certain circumstances [2.1, 2]. If the $[\alpha]$ matrix is diagonal of rank R (i.e., if only R positive diagonal terms exist in $[\alpha]$), then

$$[G] = \sum_{i}^{R} \alpha_i \boldsymbol{u}_i \boldsymbol{v}_i^t. \tag{2.17}$$

The α_i become the square root of the singular values of $[G][G]^t$ (i.e., $\alpha_i = \lambda_i^{\frac{1}{2}}$), and $\boldsymbol{u}_i$ and $\boldsymbol{v}_i$ are the singular vectors. The more traditional approach would be the following definitions

$$[G] = [U][\Lambda]^{\frac{1}{2}}[V]^t \tag{2.18a}$$

and

$$[G][G]^t = [U][\Lambda][U]^t, \tag{2.18b}$$

$$[G]^t[G] = [V][\Lambda][V]^t, \tag{2.18c}$$

where $[\Lambda]$ is the diagonal matrix of eigenvalues of $[G][G]^t$, the columns of $[U]$ are the eigenvectors of $[G][G]^t$, and the columns of $[V]$ are the eigenvectors of $[G]^t[G]$. The symmetry and squareness of $[G][G]^t$ and $[G]^t[G]$ guarantee the realness of the λ_i and the orthogonal property of the eigenvector sets $\{\boldsymbol{u}_i\}$ and $\{\boldsymbol{v}_i\}$, respectively.

From (2.17) it is evident that the smaller R is, the fewer the degrees of freedom defining the image $[G]$; and by ordering the eigenvalues in monotonic decreasing order, we obtain the most efficient mean-square representation of the image in the fewest (truncated) sets of retained components: $\{\lambda_i^{\frac{1}{2}}, \boldsymbol{u}_i, \boldsymbol{v}_i\}$. This fact will be used in greater detail in the following section. The SVD method also provides a useful route to the pseudo-inverse of a matrix, a fact of significant interest in separable space variant point spread function restorations, a subject to be discussed later.

In brief summary, the SVD is simply the determination of the singular vectors (or eigenvectors) of $[G][G]^t$ and $[G]^t[G]$. The singular vectors and values may then be used in efficient (in the mean-square sense) representation of the image *G*. For those familiar with principal components analysis, the $[G][G]^t$ and $[G]^t[G]$ matrices can be likened to sample row and column covariance matrices, respectively, and the SVD becomes an image expansion into eigen images of such statistically defined processes. Note that we are using the image itself to define the statistics, a highly deterministic process. However, the reader is cautioned not to conclude that the SVD technique is identical to the Karhunen-Loève expansion, a fact which is not true and which will be discussed in subsequent sections.

Referring back to Fig. 2.2b we see that the SVD expansion of $[G]$ consists of only *R* terms each weighted by the appropriate square rooted singular value. These rank 1 matrices also represent orthogonal basis images but are directly tuned to the specific image at hand and therefore represent a unique and optimal basis expansion.

2.2.2. Examples of Image Representations

For the bandwidth reduction or coding objectives in digital image processing, we seek an efficient means of storage that minimizes truncation errors for large bandwidth reductions. The methods of transform

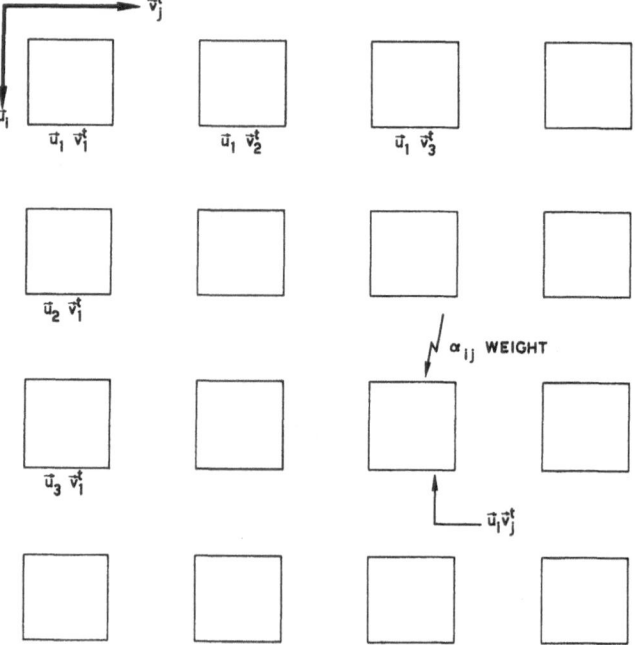

Fig. 2.3. General outer product separable kernel orthogonal expansion format

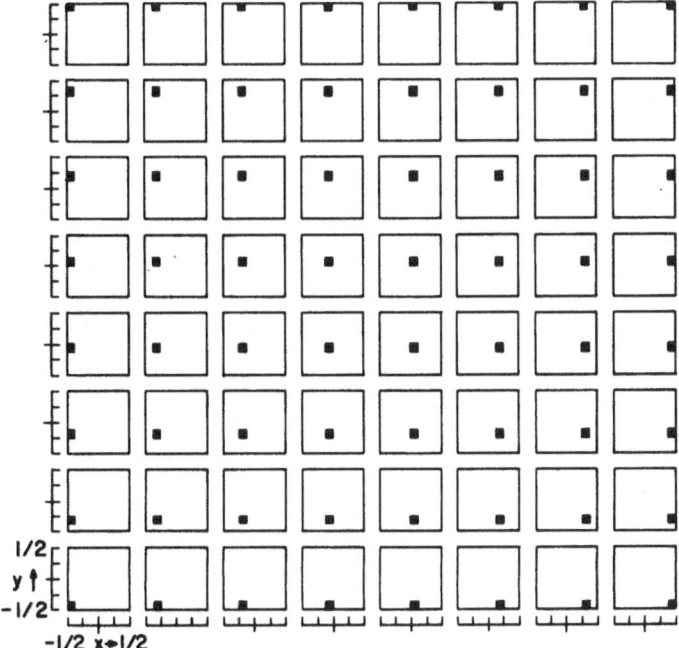

Fig. 2.4. Identity outer product expansion (black = +1, white = 0)

coding use the transform domain, the matrix $[\alpha]$ defined by (2.9), and introduce truncation by setting certain α_{ij} to zero. Thus, (2.16) has fewer than N^2 terms, and

$$[G_K] = \sum_{i,j \in \{K\}} \sum \alpha_{ij} \boldsymbol{u}_i \boldsymbol{v}_j^t, \tag{2.19}$$

where $\{K\}$ is a subset of the N^2 coefficients.

Examples of expansions into general orthogonal separable kernel transformations might best be displayed as in Fig. 2.3. The outer product images are displayed in an array showing that the image $[G]$ can be decomposed into the α_{ij} weightings of the $\boldsymbol{u}_i \boldsymbol{v}_i^t$ squares in the figure. For those orthogonal sets of basis vectors that have an all-ones vector, $\boldsymbol{1}$, the actual one-dimensional basis vector set appears in the first column and first row of the array in the figure. The possible set of orthogonal systems for decomposition of images is infinite; only a few of the simplest ones will be discussed. Figures 2.4 through 2.7 are pictorial illustrations of such outer product images, where the identity, Haar, Hadamard/Walsh, and Hadamard/other transformations are illustrated [2.3]. In all these cases, $[U] = [V]$ (i.e., the row and column

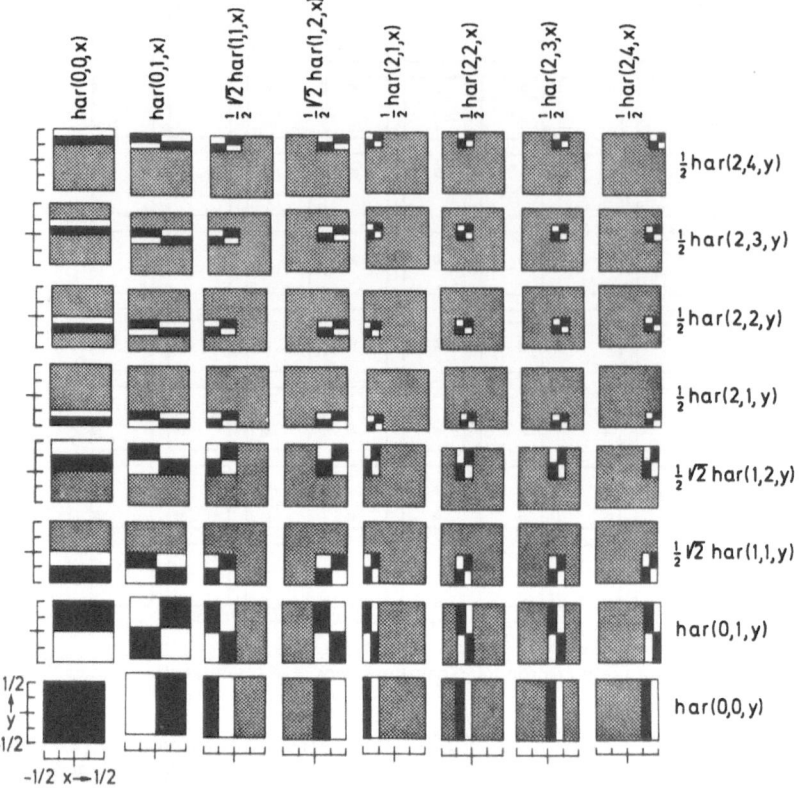

Fig. 2.5. Haar outer product expansions (black = + 1, crosshatch = 0, white = − 1)

transformations are the same), and in Figs. 2.4 through 2.7, the $[U]$ matrices [2.4] become

a) Identity

$$[U] = \begin{bmatrix} 1 & 0 & 0 & 0 & 0 & 0 & 0 & 0 \\ 0 & 1 & 0 & 0 & 0 & 0 & 0 & 0 \\ 0 & 0 & 1 & 0 & 0 & 0 & 0 & 0 \\ 0 & 0 & 0 & 1 & 0 & 0 & 0 & 0 \\ 0 & 0 & 0 & 0 & 1 & 0 & 0 & 0 \\ 0 & 0 & 0 & 0 & 0 & 1 & 0 & 0 \\ 0 & 0 & 0 & 0 & 0 & 0 & 1 & 0 \\ 0 & 0 & 0 & 0 & 0 & 0 & 0 & 1 \end{bmatrix}$$

(Note: See Fig. 2.4.)

b) Haar

$$[U] = \begin{bmatrix}
1 & -1 & 2 & 0 & 2 & 0 & 0 & 0 \\
1 & -1 & 2 & 0 & -2 & 0 & 0 & 0 \\
1 & -1 & -\sqrt{2} & 0 & 0 & 2 & 0 & 0 \\
1 & -1 & -\sqrt{2} & 0 & 0 & -2 & 0 & 0 \\
1 & 1 & 0 & 2 & 0 & 0 & 2 & 0 \\
1 & 1 & 0 & 2 & 0 & 0 & -2 & 0 \\
1 & 1 & 0 & -\sqrt{2} & 0 & 0 & 0 & 2 \\
1 & 1 & 0 & -\sqrt{2} & 0 & 0 & 0 & -2
\end{bmatrix}$$

or $[U] = [\mathscr{H}]$.

(Note: See Fig. 2.5, where the normalization factors of powers of $\sqrt{2}$ are not retained for ease in gray scale representation.)

c) Hadamard/Walsh

$$[U] = \begin{bmatrix}
1 & -1 & -1 & 1 & 1 & -1 & -1 & 1 \\
1 & -1 & -1 & 1 & -1 & 1 & 1 & -1 \\
1 & -1 & 1 & -1 & -1 & 1 & -1 & 1 \\
1 & -1 & 1 & -1 & 1 & -1 & 1 & -1 \\
1 & 1 & 1 & 1 & 1 & 1 & 1 & 1 \\
1 & 1 & 1 & 1 & -1 & -1 & -1 & -1 \\
1 & 1 & -1 & -1 & -1 & -1 & 1 & 1 \\
1 & 1 & -1 & -1 & 1 & 1 & -1 & -1
\end{bmatrix}$$

(Note: See Fig. 2.6.)

d) Hadamard/Other

$$[U] = \begin{bmatrix}
1 & 1 & 1 & -1 & -1 & 1 & 1 & 1 \\
1 & 1 & 1 & -1 & 1 & -1 & -1 & -1 \\
1 & 1 & -1 & 1 & 1 & -1 & 1 & 1 \\
1 & 1 & -1 & 1 & -1 & 1 & -1 & -1 \\
1 & -1 & 1 & 1 & 1 & 1 & -1 & 1 \\
1 & -1 & 1 & 1 & -1 & -1 & 1 & -1 \\
-1 & 1 & 1 & 1 & 1 & 1 & 1 & -1 \\
-1 & 1 & 1 & 1 & -1 & -1 & -1 & 1
\end{bmatrix}$$

(Note: See Fig. 2.7.)

e) Fourier

$$[U] = \begin{bmatrix} u \downarrow \xrightarrow{\quad x \quad} \\ e^{\frac{2\pi iux}{N}} \end{bmatrix} = \exp\left\{\frac{2\pi i}{8}\begin{bmatrix} 0 & 0 & 0 & 0 & 0 & 0 & 0 & 0 \\ 0 & 1 & 2 & 3 & 4 & 5 & 6 & 7 \\ 0 & 2 & 4 & 6 & 0 & 2 & 4 & 6 \\ 0 & 3 & 6 & 1 & 4 & 7 & 2 & 5 \\ 0 & 4 & 0 & 4 & 0 & 4 & 0 & 4 \\ 0 & 5 & 2 & 7 & 4 & 1 & 6 & 3 \\ 0 & 6 & 4 & 2 & 0 & 6 & 4 & 2 \\ 0 & 7 & 6 & 5 & 4 & 3 & 2 & 1 \end{bmatrix}\right\}$$

or $[U] = [\mathscr{F}]$.

(Note: Not shown in a figure.)

Of course, mixes between orthogonal sets are possible such that $[U] \ne [V]$; due to space limitations, they are not presented here.

As an illustration of the application of such expansions, consider larger dimensional arrays that represent true images. Figure 2.8 presents the α matrices corresponding to the expansions in the identity, Fourier, Walsh, and Haar outer products for a particular image. The transform planes, i.e., the $[\alpha]$ matrices, are presented as a componentwise logarithm of the magnitude display, allowing the small dynamic range of film for display purposes to be accommodated. Thus, the viewer sees the matrix $[\alpha']$ where $\alpha'_{ij} = \log(|\alpha_{ij}| + 1)$.

As is evident from Fig. 2.8, image energy tends to be concentrated in a few select coefficients α_{ij}; thus, only a small number of such terms need be retained for good representation of the original imagery. Because the expansions under discussion are unitary, image energy is preserved so that a few large α_{ij} coefficients represent a large percentage of the total image energy. The selection of appropriate outer product expansions and the transmission of selected coefficients are the subjects of digital transform image coding research in which storage reductions of approximately $10:1$ have been achieved [2.5].

For the SVD the off-diagonal coefficients α_{ij} are zero for $i \ne j$. Therefore, Fig. 2.9 illustrates the orthogonal system of functions utilized in the image expansion in this case. Using the SVD as a means of representation (retaining only K such outer products), we find that

$$[G_K] = \sum_{i=1}^{K} \lambda_i^{\frac{1}{2}} \boldsymbol{u}_i \boldsymbol{v}_i^t \tag{2.20}$$

and the norm between $[G]$ and $[G_K]$ is

$$\|[G] - [G_K]\| = \sum_{i=K+1}^{R} \lambda_i. \tag{2.21}$$

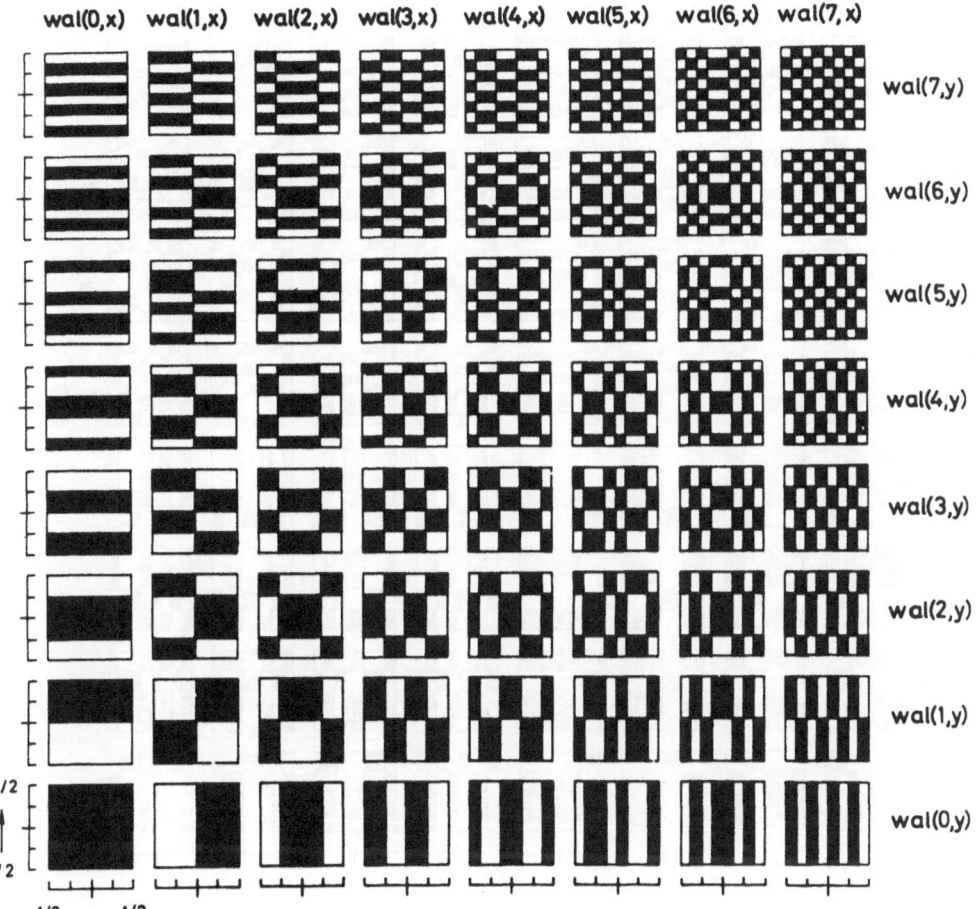

Fig. 2.6. Hadamard/Walsh outer product expansion (black = + 1, white = − 1)

Thus, a monotonic decreasing order of the eigenvalues minimizes norm or least-square truncation error.

In Figs. 2.10 through 2.13, the SVD's of images are illustrated. Selected individual singular vector outer product images are presented (in componentwise magnitude form) to illustrate the decomposition of imagery into its basic two-dimensional components. Along with individual singular vector matrices, various partial summations of these matrices are included to demonstrate the cumulative effect of the individual outer products on the reconstruction of the entire image.

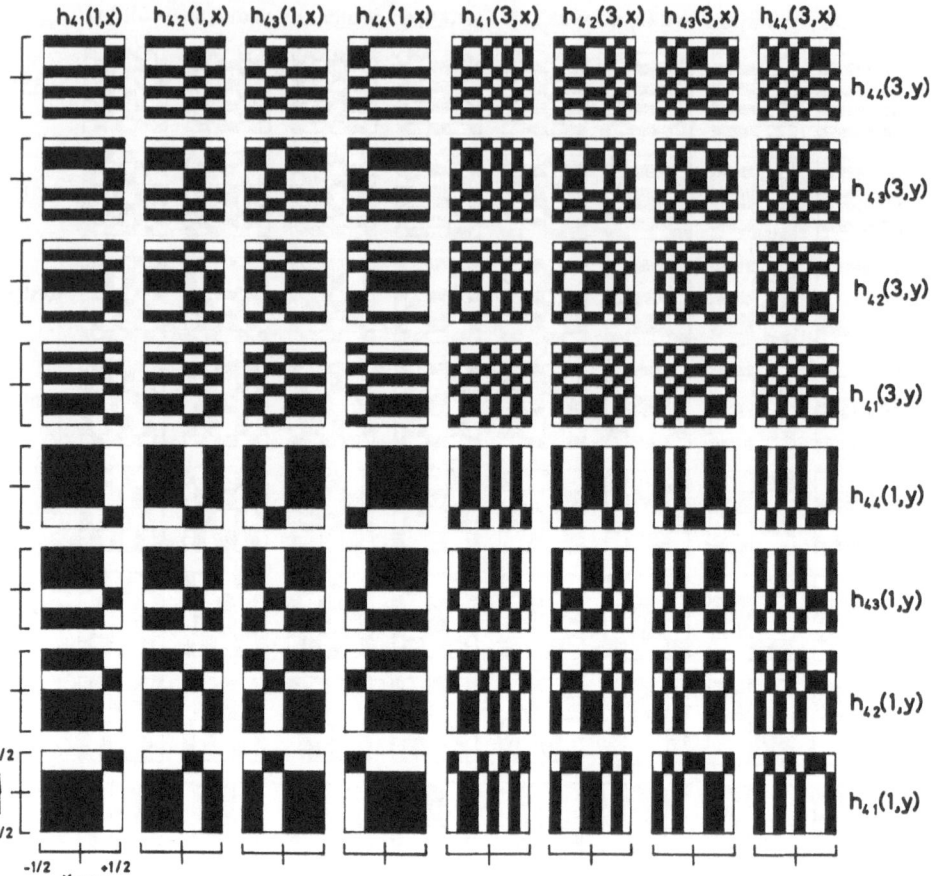

Fig. 2.7. Hadamard/Other outer product expansions (black = + 1, white = − 1)

Obviously, only a small number of terms are required for retention of significant image information.

2.2.3. Separability and Stacking Operators

The discussion presented so far has made use of an implicit assumption of rectangular separability. Thus (2.11), repeated below,

$$[G] = [U] [\alpha] [V]^t$$

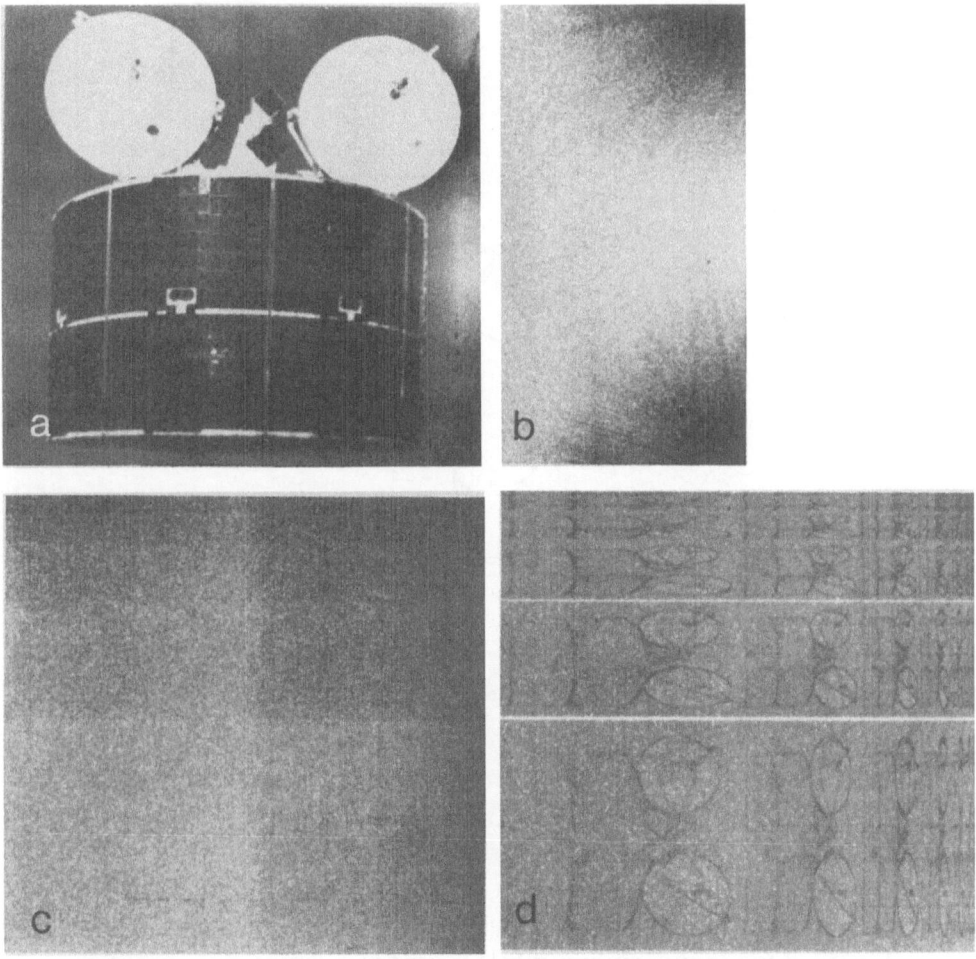

Fig. 2.8a-d. Outer product expansions of satellite. (a) Original image, (b) Fourier transform, (c) Hadamard transform, (d) Haar transform

implies that the rows of $[\alpha]$ are operated on by $[V]^t$ and the columns of $[\alpha]$ by $[U]$. Because $[U]$ and $[V]$ can be arbitrarily selected this separability assumption has some implications that should be discussed here. Specifically, the separability assumption implies a restriction on the generality of the linear relation between the transform plane $[\alpha]$ and the image $[G]$. The most general linear system relating $[\alpha]$ and $[G]$ would be given by a stacking operator [2.6] which mapped the matrices $[\alpha]$ and $[G]$ into $N^2 \times 1$ vectors $\boldsymbol{\alpha}$, $\boldsymbol{g}$ respectively. Then we

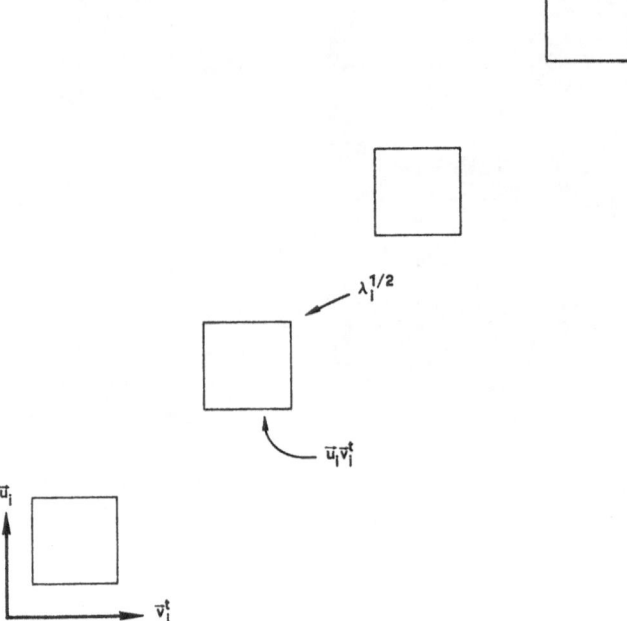

Fig. 2.9. General singular value decomposition format

would have

$$g = [H]\,\alpha.\qquad(2.22)$$

In this case $[H]$ has dimension $N^2 \times N^2$ implying N^4 entries or degrees of freedom whereas the separable case of (2.11) allows only $2N^2$ degrees of freedom. However, if $[H]$ is used to represent the operations of the separable case of (2.11) then $[H]$ becomes a Kronecker or direct product of matrices [2.7]

$$[H] = [U] \otimes [V]\qquad(2.23)$$

and takes on a block structured form. We will see that the separability assumption will allow considerable computational advantages in both coding and restoration, but will be restricted to a set of *a priori* assumptions which may not be intuitively satisfying. The important conclusion to be emphasized here is that the stacking operator allows a more general description of linear relationships between $[\alpha]$ and $[G]$ planes than allowed in the separable case.

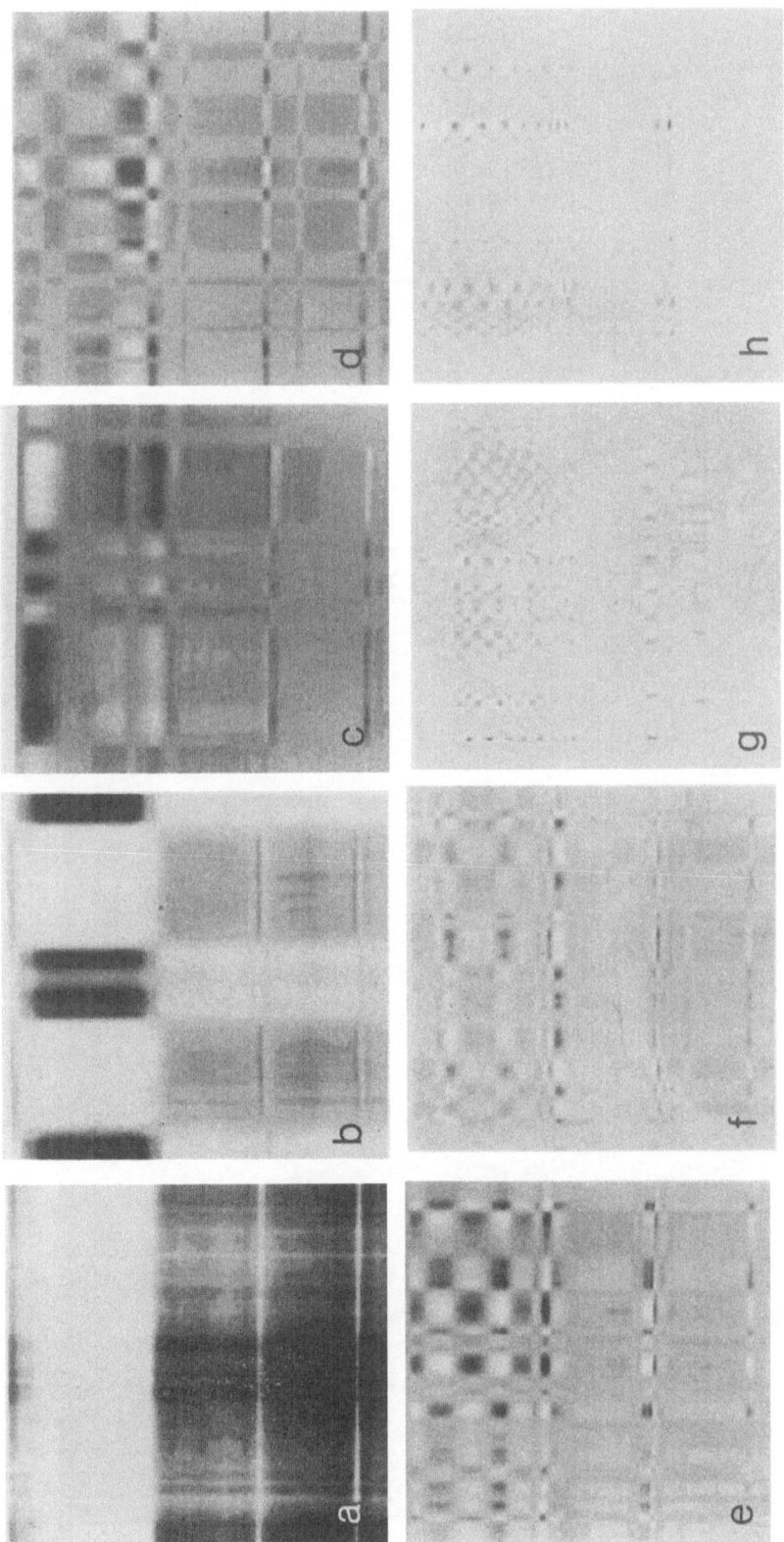

Fig. 2.10a-h. Selected singular value outer product eigenimages of satellite. (a) $u_1 v_1^t$, (b) $u_2 v_2^t$, (c) $u_4 v_4^t$, (d) $u_6 v_6^t$, (e) $u_8 v_8^t$, (f) $u_{10} v_{10}^t$, (g) $u_{20} v_{20}^t$, (h) $u_{30} v_{30}^t$

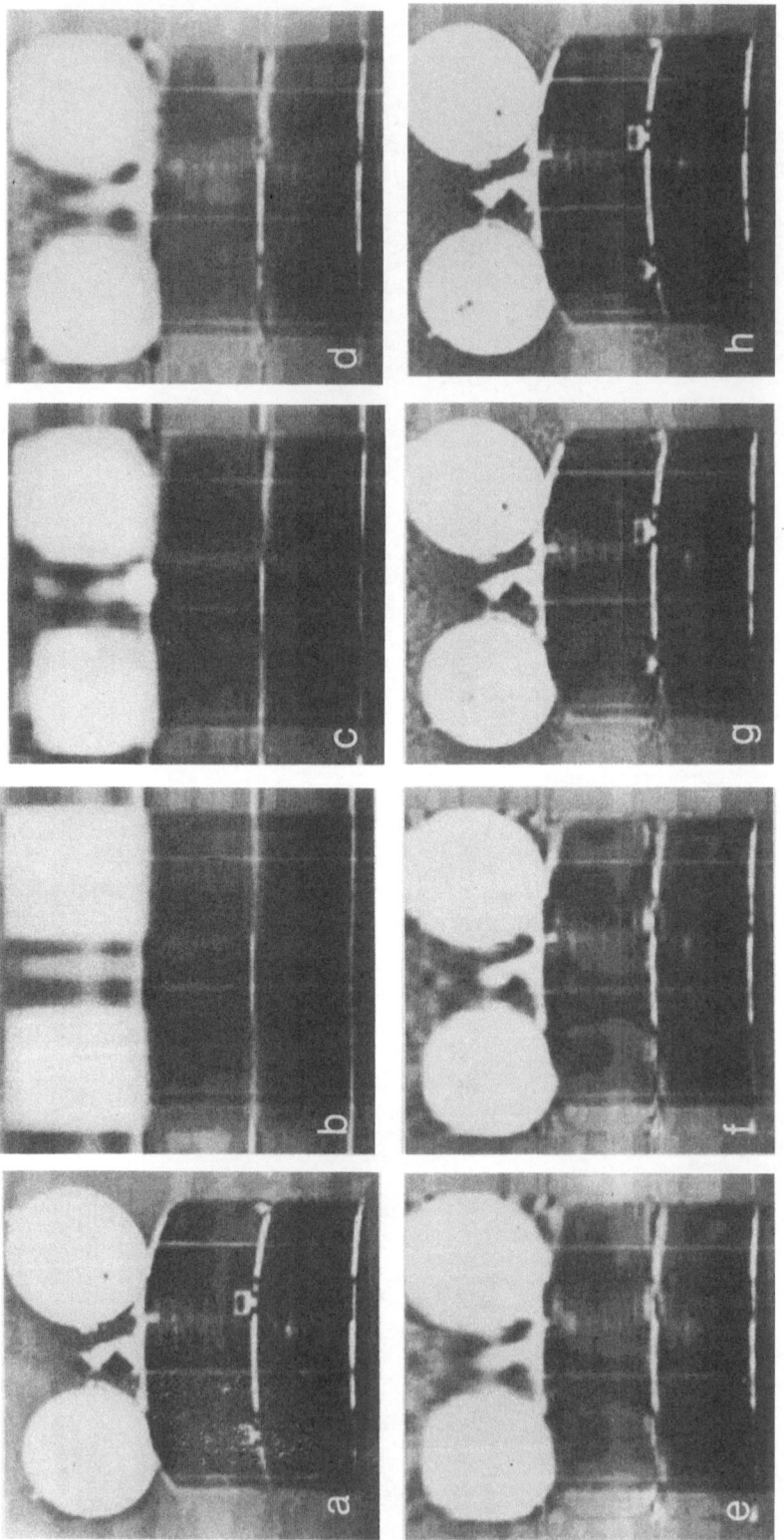

Fig. 2.11a-h. Selected partial sums of SVD expansions of satellite. (a) Original $= G_{128}$, (b) G_2, (c) G_4, (d) G_6, (e) G_8, (f) G_{10}, (g) G_{20}, (h) G_{30}

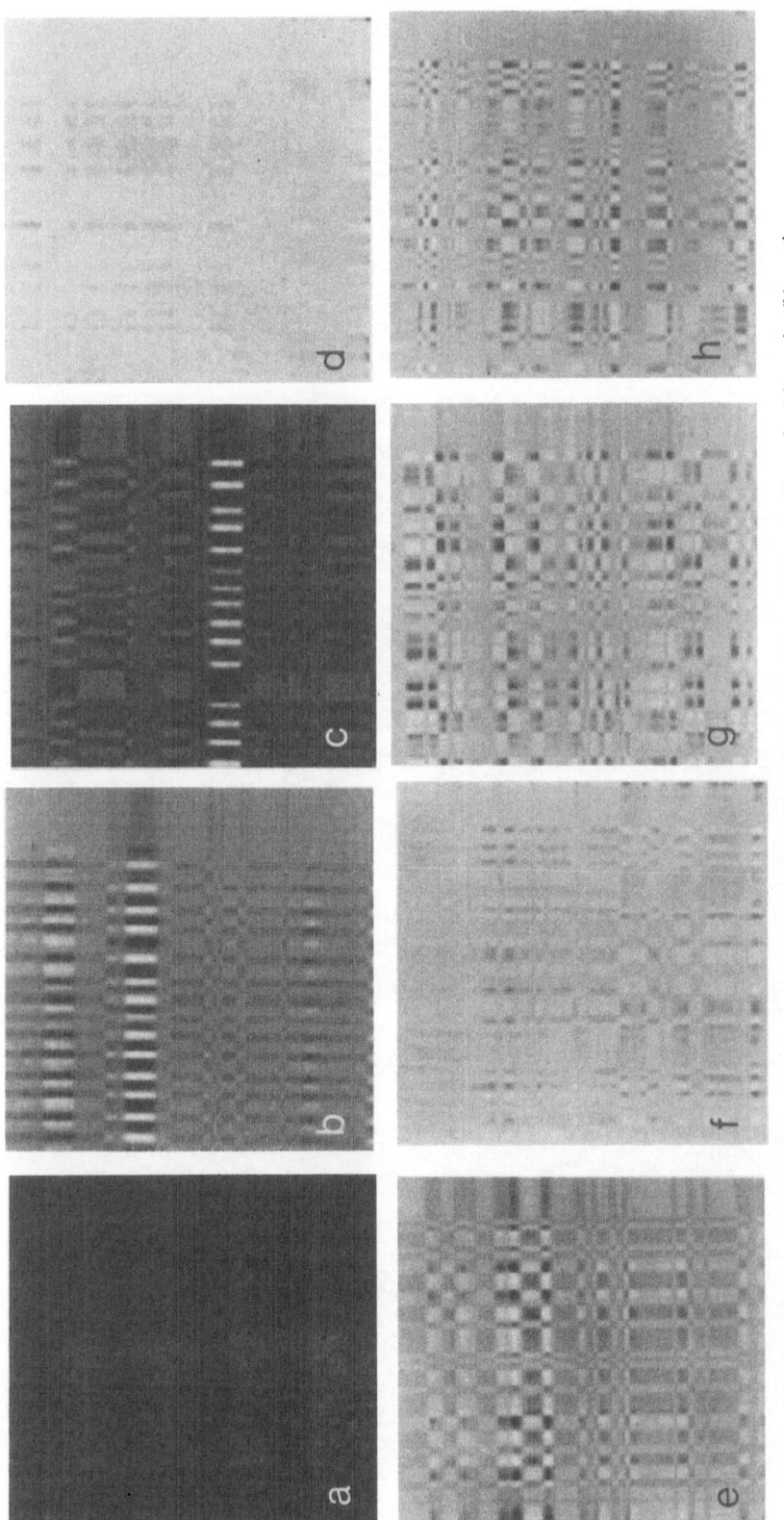

Fig. 2.12a-h. Selected singular value outer product eigenimages of text. (a) $u_1 v_1^t$, (b) $u_2 v_2^t$, (c) $u_4 v_4^t$, (d) $u_6 v_6^t$, (e) $u_8 v_8^t$, (f) $u_{10} v_{10}^t$, (g) $u_{20} v_{20}^t$, (h) $u_{30} v_{30}^t$

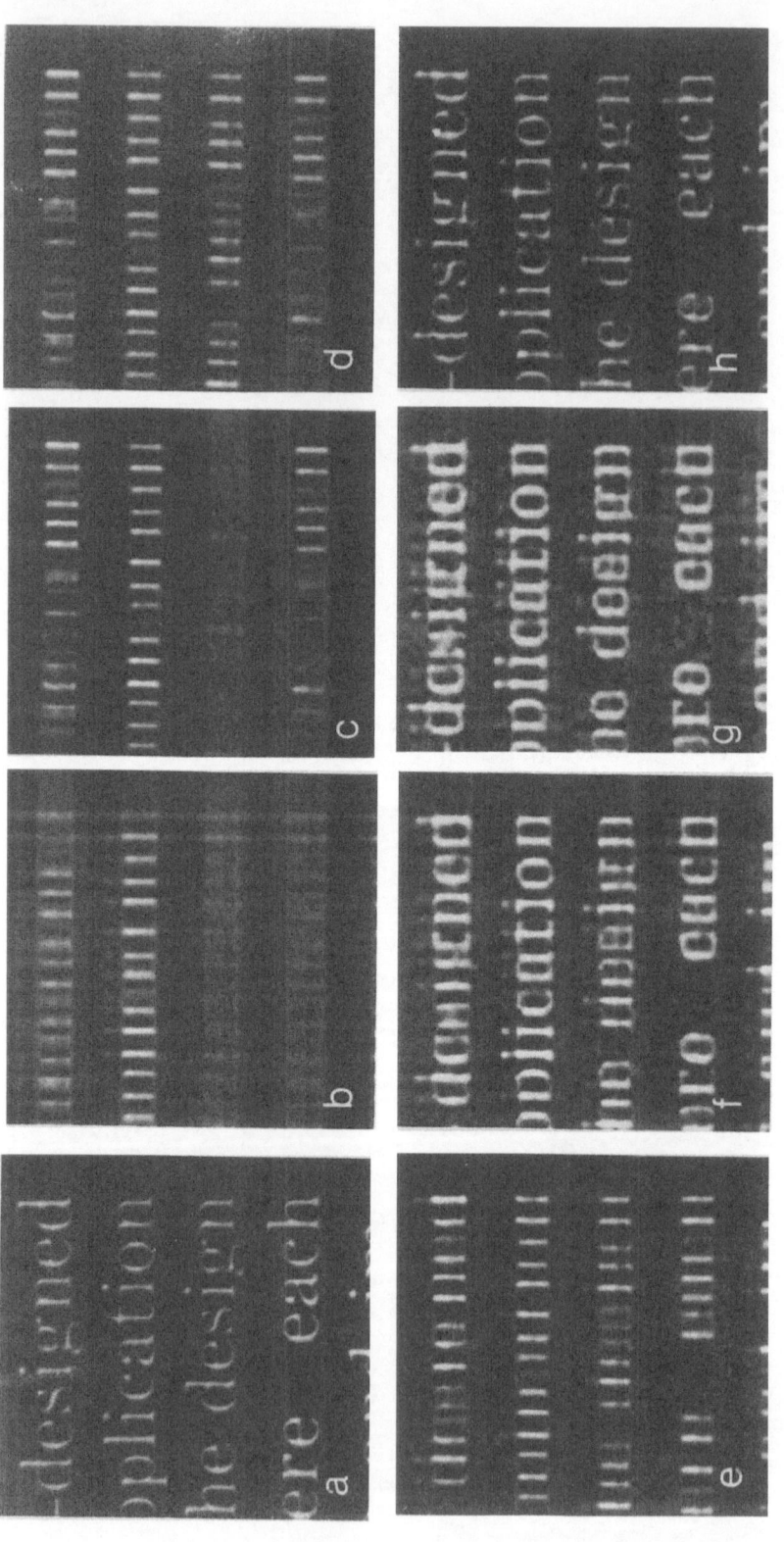

Fig. 2.13a-h. Selected partial sums of SVD expansions of text. (a) Original $= G_{128}$, (b) G_2, (c) G_4, (d) G_6, (e) G_8, (f) G_{10}, (g) G_{20}, (h) G_{30}

2.3. Image Coding

2.3.1. Spatial Image Coding

The objective of image coding is to effectively reduce the number of bits necessary to represent a picture and still maintain some fidelity quality relating the pre and post coded images. Two general approaches have been suggested, the one utilizing spatial domain techniques, the other operating in the transform domain of images. However many of the spatial techniques can be represented as two-dimensional transform methods, and this will be the approach taken here. Figure 2.14 presents the samples of an image and the general concepts of differential PCM (DPCM). Here the picture element to be sampled is predicted from the n by n previous elements and the difference between the actual and predicted value is transmitted. For an $n \times n$ predictor we have a differential at the position (i, j) given by

$$\Delta_{ij} = g_{i,j} - \sum_{k=1}^{n} p_{k,0} g_{i-k,j} - \sum_{l=1}^{n} p_{0,l} g_{i,j-l} - \sum_{k=1}^{n} \sum_{l=1}^{n} p_{k,l} g_{i-k,j-l},$$

$$(2.24)$$

where $\Delta_{i,j}$ is then coded and transmitted. For nonseparable differential PCM (Fig. 2.14a) there are $n^2 + 2n$ predictor coefficients $p_{k,l}$ and consequently $n^2 + 2n$ degrees of freedom in the predictor. (This is

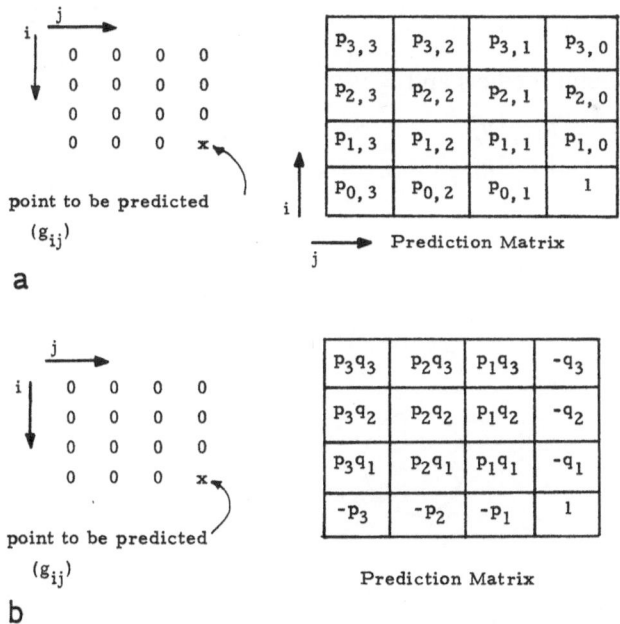

a

b

Fig. 2.14a and b. Differential PCM. (a) Nonseparable differential PCM, (b) separable differential PCM

often referred to as an $n^2 + 2n$ order predictor.) In the separable differential PCM case (Fig. 2.14b), the prediction of the $g_{i,j}$ point produces a difference

$$\Delta_{ij} = g_{ij} - \sum_{k=1}^{n} p_k g_{i-k,j} - \sum_{l=1}^{n} q_l g_{i,j-l} + \sum_{k=1}^{n} \sum_{l=1}^{n} p_k q_l g_{i-k,j-l}, \quad (2.25)$$

and we see only $2n$ unique predictor coefficients or $2n$ degrees of freedom are used (a $2n$ order predictor). For the case where we have imagery described by separable statistics [2.5], our predictor is described by a lower triangular matrix

$$[Q] = \begin{bmatrix} 1 & & & & & & & & & & \\ -q_1 & 1 & & & & & & & & & \\ -q_2 & -q_1 & 1 & & & & & & & & \\ \cdot & & & 1 & & & & & & & \\ \cdot & & & & 1 & & & & & & \\ \cdot & & & & & 1 & & & & & \\ -q_n & \cdot & \cdot & \cdot & -q_2 & -q_1 & 1 & & & & \\ \cdot & & & & & & & \cdot & & & \\ \cdot & & & & & & & & \cdot & & \\ & & \cdot & \cdot & \cdot & \cdot & \cdot & & & \cdot & \\ & & -q_n & \cdot & \cdot & \cdot & -q_2 & -q_1 & 1 & \\ & & & -q_n & \cdot & \cdot & \cdot & -q_2 & -q_1 & 1 \end{bmatrix} \quad (2.26)$$

which column predicts and

$$[P]^t = \begin{bmatrix} 1 & -p_1 & -p_2 & \cdot & \cdot & \cdot & -p_n & & & & \\ & 1 & -p_1 & & & & & \cdot & & & \\ & & 1 & & & & & & \cdot & & \\ & & & 1 & & & & & & \cdot & \\ & & & & 1 & & & & & & -p_n \\ & & & & & 1 & -p_1 & -p_2 & \cdot & \cdot & -p_n \\ & & & & & & & \cdot & & & \cdot \\ & & & & & & & & \cdot & & \cdot \\ & & & & & & & & & -p_2 & \cdot \\ & & & & & & & & 1 & -p_1 & -p_2 \\ & & & & & & & & & 1 & -p_1 \\ & & & & & & & & & & 1 \end{bmatrix} \quad (2.27)$$

which row predicts [2.8, 9]. The difference image $[\Delta]$ then becomes

$$[\Delta] = [Q] [G] [P]^t \tag{2.28}$$

and is coded and transmitted over the channel. (We have ignored boundary conditions in this argument for simplification of discussion.) The separable differential PCM predictor is adequate for imagery which is statistically separable and described by horizontal and vertical covariance matrices. Models of such systems might include n-th order Markov processes in two dimensions. The image $[\Delta]$ becomes a matrix of uncorrelated samples which can then be more efficiently coded than the original correlated samples or picture elements comprising the matrix $[G]$.

The above formulation shows that for certain processes, the separable matrix column and row operations on the image $[G]$ results in a new matrix which is more amenable to coding for bandwidth reduction. The following section also utilizes row and column matrix operations to obtain a domain where image coding can be more effectively handled. There, however, the row and column operators become unitary transforms as opposed to triangular matrices.

2.3.2. Transform Image Coding

The subject of transform image coding has been exhaustively discussed in the literature, a good survey having been presented by WINTZ [2.5] and HUANG et al. [2.10]. The major objective in transform image coding is to manipulate the image into an invertible form more amenable to coding techniques. Usually this entails providing data in a form which is more uncorrelated than the original picture elements, thereby making the bandwidth reduction process implementable with memoryless coders in the image transform domains. Often analyses are based upon second-order statistics and stationary assumptions and covariance matrices usually are used to describe such processes.

In this section we will briefly discuss a variety of transform domain coding techniques and enumerate some of the properties associated with each. Table 2.1 presents a list of the more commonly discussed transforms and will be our point of reference. In all cases the domain, (matrix or plane of data) used for coding and transmission will be described by the $[\alpha]$ matrix, i.e. the two-dimensional transform of the image matrix $[G]$. It should be emphasized that the table is not to be interpreted as being exhaustive and it should be obvious that many combinations of the transforms listed are all possible candidates for transform coding methods depending on storage, hardware, and timing configurations and constraints. Also one might intuitively assign a decreasing payoff in correlation removal as one goes down the table. While such an interpretation may be useful, one should always be wary of such generalizations, as in this case, the last two entries (hybrid and DPCM) can be made to decorrelate as well as any other transform method by increasing the order of the predictor (n).

Referring to specific entries in Table 2.1, the singular value decomposition method would only be useful if one could do fairly large computations at the source or transmitter because one must be able to compute the singular vectors which are uniquely determined by the image under question (this computation requiring on the order of $4N^3$ for total implementation). Then the actual singular (or eigen-) vectors are to be coded and transmitted, the singular values being absorbed as constants into the vectors. Because of the large computations necessary for implementation of the SVD, such techniques might only be practiced as image storage forms for large computing facilities. In terms of least squares approximation, the truncated form $[G_K]$, of (2.19) is optimum and is achieved uniquely by the truncated singular value decomposition. Therefore in terms of computer storage words (floating point form as opposed to coded form) maintaining floating point accuracy, the truncated SVD is uniquely optimum in a least squares fidelity criterion. Thus the following equation is minimized for a given K.

$$\text{Least Squares Error} = \|[G] - [G_K]\| . \tag{2.29}$$

An approach to image coding taking a statistical flavor and minimizing a mean-square truncation error for stationary imagery is given by the class of transforms described by the Karhunen-Loève, Hotelling, principal component, or factor analysis transforms. The second entry in the table refers to this class of transformation, and as described in the table, there is an additional assumption of separability in the stationary second-order statistics. The mean-square error to be minimized is given by

$$\text{Mean Square Error} = \mathscr{E}\{\|[G] - [G_K]\|^2\} , \tag{2.30}$$

where the $\mathscr{E}\{\cdots\}$ operator is a two-dimensional separable expectation operator averaging over the underlying separable statistics describing the generation of the image. The separability assumption requires the horizontal statistics be separate from the vertical statistics (an assumption, which in itself, is questionable). However the covariance matrix describing the column (vertical) statistics is given by $[\Phi_x]$ and

Table 2.1. Transform domains for coding

	Transform name	Representation	Unknown parameters (to be computed and transmitted)	Algorithmic implementation (on the order of:)	Ref.	Footnote
1	Singular value decomposition	$[G][G]^t = [U][\Lambda][U]^t$ $[G]^t[G] = [V][\Lambda][V]^t$ $[\alpha] = [\Lambda^{\frac{1}{2}}] = [U]^t[G][V]$	Basis vectors	N^3 $\left.\begin{array}{l} N^3 \\ 2N^3 \end{array}\right\}$ computed for each image	[2.1, 2]	
2	Karhunen-Loève, Hotelling, principal components, factor analysis	$[\Phi_x] = [E_x][\Lambda_x](E_x]^t$ $[\Phi_y] = [E_y][\Lambda_y][E_y]^t$ $[\alpha] = [E_x]^t[G][E_y]$	Transform coefficients	$\left.\begin{array}{l} N^3 \\ N^3 \\ 2N^3 \end{array}\right\}$ computed only once	[2.5, 10]	
3	Cosine	$[\alpha] = [\cos]^t[G][\cos]$	Transform coefficients	$2N^2\log_2 N$	[2.11]	a, b
4	Fourier	$[\alpha] = [\mathscr{F}][G][\mathscr{F}]$	Transform coefficients	$2N^2\log_2 N$ (complex)	[2.3]	a, b
5	Slant	$[\alpha] = [S]^t[G][S]$	Transform coefficients	$2N^2\log_2 N$	[2.12]	b
6	DLB	$[\alpha] = [DLB]^t[G][DLB]$	Transform coefficients	$2N^2\log_2 N$ (integer arith.)	[2.13]	a, b
7	Walsh	$[\alpha] = [W][G][W]$	Transform coefficients	$2N^2\log_2 N$ (additions)	[2.14]	b
8	Haar	$[\alpha] = [\mathscr{H}]^t[G][\mathscr{H}]$	Transform coefficients	$2(N-1)$	[2.3]	b
9	Hybrid	$[\alpha] = [Q][G][W]$	Predictor values of 1-D transform	$N^2\log N + nN^2$	[2.15]	c
10	DPCM	$[\alpha] = [Q][G][P]^t$	Predictor values	$2nN^2$	[2.9]	c

a When $N \neq 2^k$, the computations increase.

b $N = 2^k$.

c n is proportional to the order of the predictor.

the covariance matrix describing the row (horizontal) statistics is given by $[\Phi_y]$. If the stationary separable model is valid for the class of images under question, then it may be computationally expedient to compute the eigenvectors of the covariance matrices on a large computing facility (only once) remote from the image source and transmitter. Thus

$$[\Phi_x] = [E_x] [\Lambda_x] [E_x]',$$
(2.31a)

$$[\Phi_y] = [E_y] [\Lambda_y] [E_y]',$$
(2.31b)

and determination of the orthogonal $[E_x]$ and $[E_y]$ matrices require on the order of N^3 computations each. (Here $[\Lambda_x]$ and $[\Lambda_y]$ are diagonal matrices with eigenvalues of the covariance matrices on the diagonals.) Then it may be feasible to hardwire or otherwise speed up the expansion of the image $[G]$ into the decomposition coefficients determined by

$$[\alpha] = [E_x]' [G] [E_y]$$
(2.32)

at the image source or transmitter. This operation would then require $2N^3$ arithmetic computations, a savings of a factor of 2 over the SVD technique on a per image basis. It is important to realize that the two-dimensional Karhunen-Loève transform given by (2.32) results in a full matrix $[\alpha]$ whose entries are such that *on the average* the largest amount of energy of the image is stored in the fewest number of α_{ij} coefficients. Of course, the actual performance of such an algorithm in terms of the psychophysics of the human viewing response phenomena will only be as good as how well the statistical model matches the deterministic image to be transmitted.

Often the models used as valid descriptors of separable stationary processes for the generation of imagery are given by Markov statistics in which the covariance matrices $[\Phi_x]$ and $[\Phi_y]$ are Toeplitz with entries given by $\Phi_{x,ij} = \varrho_x^{|i-j|}$ and $\Phi_{y,ij} = \varrho_y^{|i-j|}$. Because such systems are diagonalized by the sines or cosines of appropriate frequency and phase shift [2.16] a discrete approximation to such a decomposition might best be given by the cosine transform, the third entry in our table. Additional advantages of this transform are a fast algorithmic implementation $(2N^2 \log_2 N$ vs $2N^3)$ as well as the fact that no eigen-decompositions are necessary. Thus the transformation is deterministically defined and could possibly lend itself to efficient hardware implementation (maybe in analog fashion). The transform will not truly diagonalize the discrete Markov process, but experimental verification indicates that it is quite close to the Karhunen-Loève statistical optimum [2.11].

If one is not satisfied with the Markov model as a description of the image generation statistic, but if the stationarity assumption is to be maintained, then $[\Phi_x]$ and $[\Phi_y]$ are still of the Toeplitz form. For large sample arrays (large N) and for fairly rapid decorrelation between not too distant picture samples, the Toeplitz covariance matrices can be closely approximated by circulant matrices, the differences being in the end points of the matrices

$$[\Phi_x] \approx [C_x], \tag{2.33a}$$

$$[\Phi_y] \approx [C_y]. \tag{2.33b}$$

Such "circularized" statistics are diagonalized by the discrete Fourier transform, an example of which was presented earlier. Thus

$$[C_x] = [\mathscr{F}][D_x][\mathscr{F}], \tag{2.34a}$$

$$[C_y] = [\mathscr{F}][D_y][\mathscr{F}], \tag{2.34b}$$

where

$$[\mathscr{F}] = \overset{x\downarrow}{\left[\exp\left(\frac{2\pi i u x}{N}\right)\right]}^{\overset{u}{\longrightarrow}} \tag{2.35}$$

and $[D_x]$ and $[D_y]$ are diagonal matrices whose non-zero entries (eigenvalues) are given by the Fourier series expansion of the first row of the circulant [2.7]. Thus the two-dimensional Fourier transform might be an appealing domain in which to do image coding as stationary Toeplitz processes are "almost" uncorrelated in the Fourier domain. The larger the value of N and the closer $[\Phi]$ is to a diagonal determines how close the Fourier transform comes to decorrelating the data. The obvious advantage of using the Fourier transform is the availability of a fast $(2N^2 \log_2 N$ vs $2N^3)$ deterministic algorithm although complex arithmetic is implied.

The Slant, DLB, Walsh, and Haar transforms, entries 5, 6, 7, 8, respectively, in Table 2.1, all have their own statistical characteristics whose separable covariance matrices are given by

$$[\Phi_x^U] = [U][\Phi_x][U]^t, \tag{2.36a}$$

$$[\Phi_y^U] = [U][\Phi_y][U]^t, \tag{2.36b}$$

where $[U] = [S]$ or $[DLB]$ or $[W]$ or $[\mathscr{H}]$, respectively. If the sum of the absolute values of the off-diagonal elements are a measure of how poorly the transform decorrelates the process, then the slant transform is probably best while the Haar transform is poorest for the group of four transforms under discussion. Computationally speaking the slant, DLB, and Walsh transforms all require on the order of $2N^2 \log_2 N$ computations where computations refers to floating point multiplications, integer multiplications, and only additions, respectively. The Haar requires only $2(N-1)$ computations but does not do a very good job of decorrelating the data. Both the Walsh and Haar transforms, examples of which were presented earlier, are implementable in only additions or subtractions, and as such are extremely attractive for efficient hardware implementation. However, facetiously speaking, the identity transform is quite attractive in hardware implementation (i.e. it requires no operations) but obviously does not decorrelate the data at all. Therefore the tradeoff of correlation versus implementation is one of subjective evaluation[1].

The transform coding scheme next listed in the table (entry number 9) is often referred to as a "hybrid" system because it utilizes both spatial and transform techniques. Specifically a one-dimensional unitary transform (cosine, Fourier, Walsh, etc.) is taken along the rows of an image and these one-dimensional transform coefficients are then encoded in a DPCM technique across the transformed rows. The motivation for the hybrid technique is usually speed of operation as well as equipment and memory reduction. Consequently realtime television bandwidth reduction systems become feasible utilizing such techniques and only nN storage locations are necessary (here n is the order of the predictor). Of course, the number of computations is reduced, $N^2 \log N + nN^2$ vs $2N^2 \log N$ for a fast two-dimensional transform.

The completely spatial technique of two-dimensional DPCM is the final entry in the table and refers to the technique discussed in the previous section. For the two-dimensional DPCM system only n^2 storage locations are needed and $2nN^2$ computations are necessary. It is clear that when $n = \log_2 N$, we have an equal number of computations for both DPCM and regular fast transform methods. When $n = N$, we then have the case of full predictor matrices $(2N^3)$ and are doing the same number of computations as a full blown Karhunen-Loève expansion. Therefore the power in the DPCM technique is in having a small order predictor (small n) which effectively decorrelates the data and thereby saves computations and storage requirements.

---

[1] A rate distortion method for transform evaluation has also been proposed by PEARL et al. [2.17]. However, for practical analytic solutions only Gaussian sources seem tractable. This assumption is suspect due to the requirement of infinite tails in the density function.

The above discussion has been directed toward the use of the various transform domains as more uncorrelated sources for coding as a means of achieving a bandwidth reduction in image transmission. The actual coding algorithms have not been discussed as they do not fit into the context of this chapter and the interested reader is referred to [2.5, 8–17]. Additional concerns must also be dispelled in the task of image coding. One would like to observe the coder under various channel conditions to investigate the performance in the presence of noise. Also, hardware and timing constraints may impose additional criteria on the performances of the system. Finally, the ultimate receiver, i.e. the viewer and his parameters, must be considered in designing the image coding system.

2.4. Object Restoration

In the introduction to this chapter the model (Fig. 2.1) of general linear imaging systems was briefly presented as an aid in developing notational familiarity with the material which followed. Referring to the discrete-discrete model of (2.6) and using the stacking operator [2.6, 18] to "vectorize" our object f and image g, we obtain (2.6)

$$g = [H]f$$

which in the continuous-continuous model, was described as (2.1)

$$g(x, y) = \iint\limits_{-\infty}^{\infty} f(\xi, \eta)\, h(x, y, \xi, \eta)\, d\xi\, d\eta\,.$$

The subject of object restoration will be the attempted inversion of the degradation introduced in the imaging system upon the object $f(\xi, \eta)$. The solution of the system described in the above equation has been investigated in one dimension by HANSON [2.19] and VARAH [2.20] utilizing the SVD as an aid in obtaining a pseudo-inversion of the equation. Their work was done in a discrete space using a quadrature integration rule to obtain a vector space equivalent representation. SONDHI [2.21] suggested the use of the SVD as a tool for image processing restoration of spatially invariant degradations. TREITEL and SHANKS [2.22] have provided planar filters using the SVD approach similarly described here. Four specific assumptions concerning the impulse response $h(x, y, \xi, \eta)$ will be made and the implications of each of these assumptions will then be evaluated in the context of two-dimensional transformations and the inversion of the associated degradation phenomenon. The four specific assumptions, in order of increasing complexity are:

a) Separable space invariant point spread function (SSIPSF)

$$h(x, y, \xi, \eta) = a(x - \xi)b(y - \eta).$$ (2.37)

b) Separable space variant point spread function (SSVPSF)

$$h(x, y, \xi, \eta) = a(x, \xi)b(y, \eta).$$ (2.38)

c) Nonseparable space invariant point spread function (NSIPSF)

$$h(x, y, \xi, \eta) = h(x - \xi, y - \eta).$$ (2.39)

d) Nonseparable space variant point spread function (NSVPSF)

$$h(x, y, \xi, \eta) = h(x, y, \xi, \eta).$$ (2.40)

The following sections use the algebra of matrices and outer products as a tool in restoration processes where the point spread function matrix $[H]$ of (2.6) is not of specific concern. We assume we have analytic knowledge of $[H]$ and desire recovery of f. From a modeling viewpoint, we would like the imaging system to conserve energy, and, because the scalar elements of f and g are themselves energy measurements, we have

$$\sum_{i=1}^{N^2} f_i = \sum_{i=1}^{N^2} g_i.$$ (2.41)

Similarly, an impulse or Dirac-Delta function (point source of light) anywhere in f should result in the same amount of energy in g independent of its position in f. Therefore, we describe a unity gain system

$$\sum_{i=1}^{N^2} h_{ij} = 1; \quad j = 1, \cdots N^2.$$ (2.42)

Because of the nature of energy sensing devices, (i.e. only non-negative values are obtainable), we also have

$$f_i \geq 0$$
$$g_i \geq 0$$ (2.43)
$$h_{ij} \geq 0.$$

Because the $[H]$ matrix is nonnegative, we know [2.23] the following is componentwise nonnegative

$$[H_1] = \lambda_1^{\frac{1}{2}} \boldsymbol{u}_1 \boldsymbol{v}_1^t \tag{2.44}$$

where λ_1, $\boldsymbol{u}_1$, $\boldsymbol{v}_1^t$ are taken from the set $\{\lambda_i, \boldsymbol{u}_i, \boldsymbol{v}_i\}$ defined by the SVD of $[H]$.

2.4.1. Separable Space Invariant Point Spread Functions

This imaging situation is probably a good first-order approximation of well corrected linear systems. The space invariant point spread function refers to the fact that the functional form of the impulse response does not depend on where the point source of light (impulse response) occurs in the original object plane. The separable assumption refers to the rectangular coordinate separability of the impulse response resulting in

$$h(x, y, \xi, \eta) = a(x - \xi)\, b(y - \eta). \tag{2.37}$$

In the vector notation of (2.6)

$$[H] = [A] \otimes [B]^t, \tag{2.45}$$

where both $[A]$ and $[B]$ are Toeplitz and $\otimes$ is the direct or Kronecker product of matrices. Because of the rectangular separability of the impulse response we can use unstacked matrix notation

$$[G] = [A][F][B], \tag{2.46}$$

where $[A]$ blurs the columns of the object $[F]$, and $[B]$ blurs the rows of $[F]$. Because $[F]$ may not be square and also because the blur matrices may be singular and therefore irretrievably lose some aspect of the object, the inversion of the degradation will necessarily have to be a pseudo-inverse. However if we "circularize" the problem such that the blur matrices become circulants, equivalent to implying that $[F]$ and $[G]$ are periodically repeated throughout the plane, then the Fourier matrix of (2.35) diagonalizes the blur matrices. This implies circular convolution and

$$[A] = [\mathscr{F}][\Lambda_a^{\frac{1}{2}}][\mathscr{F}], \tag{2.47}$$

$$[B] = [\mathscr{F}][\Lambda_b^{\frac{1}{2}}][\mathscr{F}]. \tag{2.48}$$

Inserting into (2.46) we obtain

$$[G] = [\mathscr{F}] [\Lambda_a^{\frac{1}{2}}] [\mathscr{F}] [F] [\mathscr{F}] [\Lambda_b^{\frac{1}{2}}] [\mathscr{F}]. \tag{2.49}$$

Because $[\mathscr{F}]$ is unitary and the $[\Lambda]$ matrices are diagonal, we can solve for the object as

$$[F] = [\mathscr{F}]^* [\Lambda_a^{-\frac{1}{2}}] [\mathscr{F}]^* [G] [\mathscr{F}]^* [\Lambda_b^{-\frac{1}{2}}] [\mathscr{F}]^*, \tag{2.50a}$$

where * implies complex conjugation. However, the center portion of this equation is simply the two-dimensional Fourier transform of the image $[G]$ weighted by the inverse Fourier coefficients of the blur matrix. Thus

$$[F] = [\mathscr{F}]^* [\alpha] [\mathscr{F}]^*, \tag{2.50b}$$

where

$$\alpha_{ij} = \lambda_{ai}^{-\frac{1}{2}} \lambda_{bj}^{-\frac{1}{2}} u_i^t [G] u_j, \tag{2.51}$$

and u_i is the i-th column of the Fourier matrix making $u_i^t [G] u_j$ equal to the ij-th Fourier coefficient of the image. The matrix $[\alpha]$ can then be interpreted as the inverse filtered Fourier transform of the image. The object then becomes the two-dimensional transform of $[\alpha]$, as illustrated in (2.50b). This analysis then results in traditional inverse Fourier filtering for removal of separable space invariant blur.

2.4.2. Separable Space Variant Point Spread Functions

In this situation the point spread function changes its shape as it explores the object plane although still being separable in rectangular coordinates. Examples of such imaging systems include side looking radars and rectangular anode radiographic imaging devices.

The SVPSF can be represented in separable form

$$h(x, y, \xi, \eta) = a(x, \xi) b(\eta, y) \tag{2.38}$$

and we see that the discrete matrix representation again becomes

$$[H] = [A] \otimes [B]^t, \tag{2.45}$$

where $[H]$ is $N^2 \times N^2$, $[A]$ and $[B]$ are both $N \times N$ but not necessarily Toeplitz as in the SIPSF case. The beauty of being able to represent the SVPSF as a separable phenomenon is one of computational simplicity. Now in matrix notation it is not necessary to use the lexicographic or stack notation. Rather the image $[G]$ becomes

$$[G] = [A][F][B] \tag{2.46}$$

as before where the object $[F]$ is operated upon by the column operator $[A]$ and the row operator $[B]$ in a separable fashion. The model of the above equation has assumed a noiseless environment and the rank of the image $[G]$ is determined by the minimum of the ranks of $[A]$, $[F]$, and $[B]$. If one were desirous of inverting the space variant blur introduced by $[A]$ and $[B]$, one would obtain $[F]$ as

$$[F] = [A]^{-1}[G][B]^{-1} \tag{2.51a}$$

provided the inverse matrices existed. This proviso is in fact highly unlikely as most imaging systems irretrievably remove some portion of the object which of course cannot be recovered. In this case it might be desirable to obtain a minimum norm estimate of the object $[\hat{F}]$ by pseudo-inverting the blur matrices to recover up to the ranks of the individual blurs. Notationally the pseudoinverse estimate can be represented as

$$[\hat{F}] = [A]^{+}[G][B]^{+}, \tag{2.51b}$$

where $[A]^{+}$ and $[B]^{+}$ are the pseudo-inverses of $[A]$ and $[B]$, respectively. Possibly the easiest way to discuss the pseudo-inverse of a matrix is in its singular value decomposition (SVD) form where

$$[A] = [U_a][\Lambda_a]^{\frac{1}{2}}[V_a]^{t} \tag{2.51c}$$

and

$$[B] = [U_b][\Lambda_b]^{\frac{1}{2}}[V_b]^{t}. \tag{2.51d}$$

Then

$$[A]^{+} = [V_a][\Lambda_a]^{-\frac{1}{2}}[U_a]^{t} \tag{2.51e}$$

and

$$[B]^{+} = [V_b][\Lambda_b]^{-\frac{1}{2}}[U_b]^{t}. \tag{2.51f}$$

Here, the $[\Lambda_a]^{-\frac{1}{2}}$ and $[\Lambda_b]^{-\frac{1}{2}}$ matrices are diagonal with nonzero entries being reciprocals of the $[\Lambda_a]^{\frac{1}{2}}$ and $[\Lambda_b]^{\frac{1}{2}}$ nonzero entries.

In vector notation

$$[A] = \sum_{i=1}^{R_a} \lambda_{ai}^{\frac{1}{2}} u_{ai} v_{ai}^t, \tag{2.51g}$$

$$[B] = \sum_{i=1}^{R_b} \lambda_{bi}^{\frac{1}{2}} u_{bi} v_{bi}^t, \tag{2.51h}$$

and

$$[A]^+ = \sum_{i=1}^{R_a} \lambda_{ai}^{-\frac{1}{2}} v_{ai} u_{ai}^t, \tag{2.52a}$$

$$[B]^+ = \sum_{i=1}^{R_b} \lambda_{bi}^{-\frac{1}{2}} v_{bi} u_{bi}^t, \tag{2.52b}$$

and R_a, R_b are the respective ranks of the blur matrices $[A]$ and $[B]$. Using the SVD expansions and pseudo-inverses described here, we can now get our estimate of the object as

$$[\hat{F}] = [V_a] [A_a]^{-\frac{1}{2}} [U_a]^t [G] [V_b] [A_b]^{-\frac{1}{2}} [U_b]^t \tag{2.53a}$$

or

$$= [V_a] [\alpha] [U_b]^t \tag{2.53b}$$

or

$$= \sum_i^{R_a} \sum_j^{R_b} \alpha_{ij} v_{ai} u_{bj}^t, \tag{2.53c}$$

where the matrix $[\alpha]$ is given by

$$[\alpha] = [A_a]^{-\frac{1}{2}} [U_a]^t [G] [V_b] [A_b]^{-\frac{1}{2}}. \tag{2.54}$$

It is instructive to investigate the ij-th entry of $[\alpha]$ which becomes

$$\alpha_{ij} = \lambda_{ai}^{-\frac{1}{2}} \lambda_{bj}^{-\frac{1}{2}} u_{ai}^t [G] v_{bj}. \tag{2.55}$$

Here we note the similarity to the Fourier case for SIPSF's in the previous sections. However, the u_i and v_j vectors are no longer columns of the Fourier matrix but are vectors from the basis system defined by the blur matrices themselves.

Computationally it is numerically quite difficult to determine R_a and R_b due to computer and round-off noise. Thus a truncated estimate of the object becomes appealing where

$$[\hat{F}_K] = [V_a]\,[\alpha_K]\,[U_b]^t \tag{2.56a}$$

or

$$= \sum_{i=1}^{K} \sum_{j=1}^{K} \alpha_{ij} v_{ai} u_{bj}^t. \tag{2.56b}$$

Note that these representations indicate that the expansion of the truncated estimate of the object $[\hat{F}_K]$ is expanded into a two-dimensional basis system defined by the column and row orthogonal eigenvectors of the respective blur matrices. Because we are obtaining a pseudo-inverse which is an optimal minimum norm least-square estimate of the original object [2], the expansion defined by the $[V_a]$ and $[U_b]^t$ spaces are the most efficient for such inversion. To shed a little insight into this statement, consider the separable space invariant point spread function (SIPSF) example discussed earlier where $[A]$ and $[B]$ are circulant matrices. In this case $[U_a] = [U_b] = [V_a] = [V_b]$ $= [\text{discrete Fourier matrix}] = [\mathscr{F}]$ with entries $\exp(2\pi iux/N)$. Thus $[\hat{F}_K]$ is expanded into the Fourier representation where traditional inverse filtering occurs.

The above analysis has been described for the noise-free case in which the generalized or pseudo-inverse results in the optimal mean square estimate filter. Generalization to the additive noise case and separable (SVPSF) will result in a generalized pseudo-inverse Weiner filter where the noise power spectrum would be represented in the $[V_a]$, $[U_b]^t$ space rather than the two-dimensional Fourier space.

The above theory has been put to the simulation test on matrices all of the size $N = 128$. In order that the reader may obtain a better feel for the phenomenon of a separable space variant point spread function imaging system, an object with 16 point sources of light is used as an input (see Figs. 2.15 and 2.16). One particular separable space variant blur might be the one, as illustrated in the figures, in which the lower left corner is in greater focus than the upper right corner. Because of the obvious space variant nature of the blur, Fourier techniques are not applicable. In the simulation the horizontal and vertical blur were made equal. The blurred image $[G]$ is shown in each of Figs. 2.15 and 2.16 for two different degrees of distortion. The condition numbers (ratio of maximum to minimum singular value) associated with the distortions are

Severe: $C([A]) = 10^{16}$

Moderate: $C([A]) = 10^{14}$

Ideal: $C([I]) = 1$,

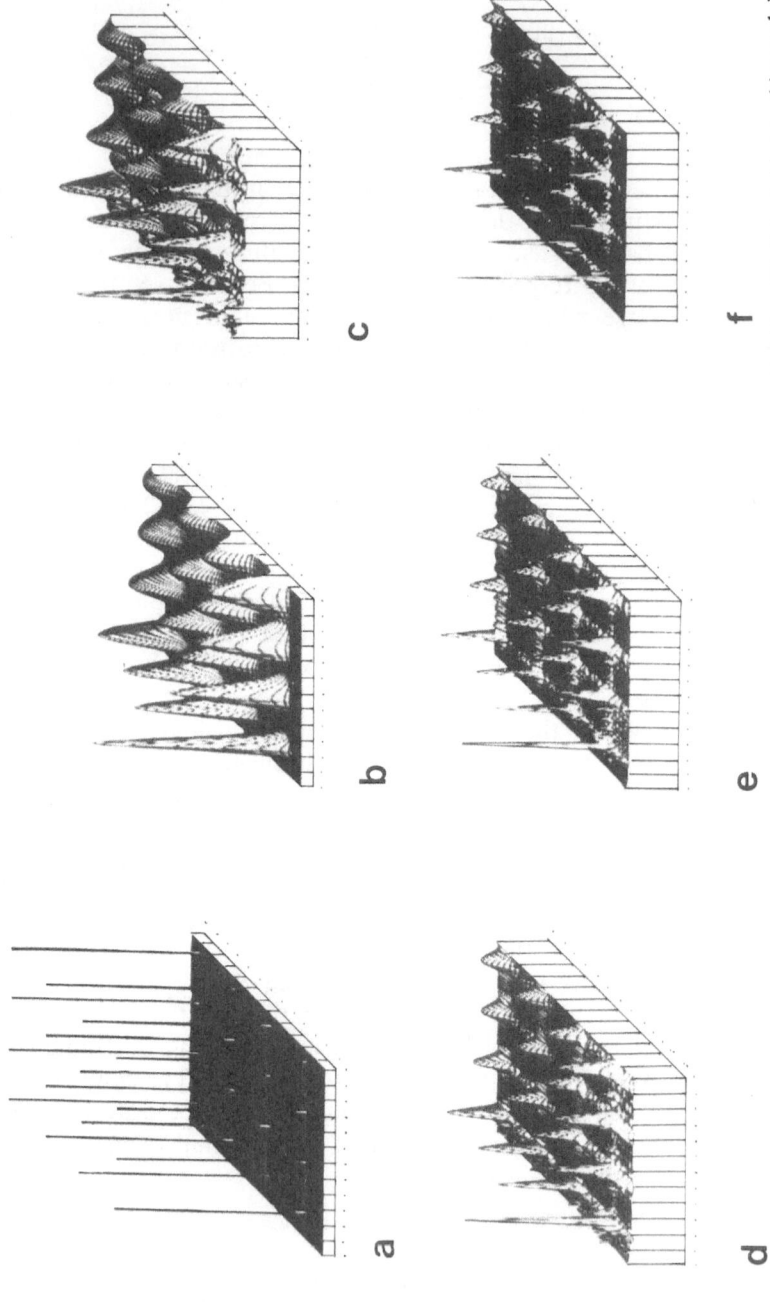

Fig. 2.15a-f. Pseudoinversion of a separable space variant point spread function: severe distortion. (a) Object, (b) distorted image, (c) pseudoinverse: first 21 eigenvalues, (d) pseudoinverse: first 31 eigenvalues, (e) pseudoinverse: first 41 eigenvalues, (f) pseudoinverse: first 51 eigenvalues

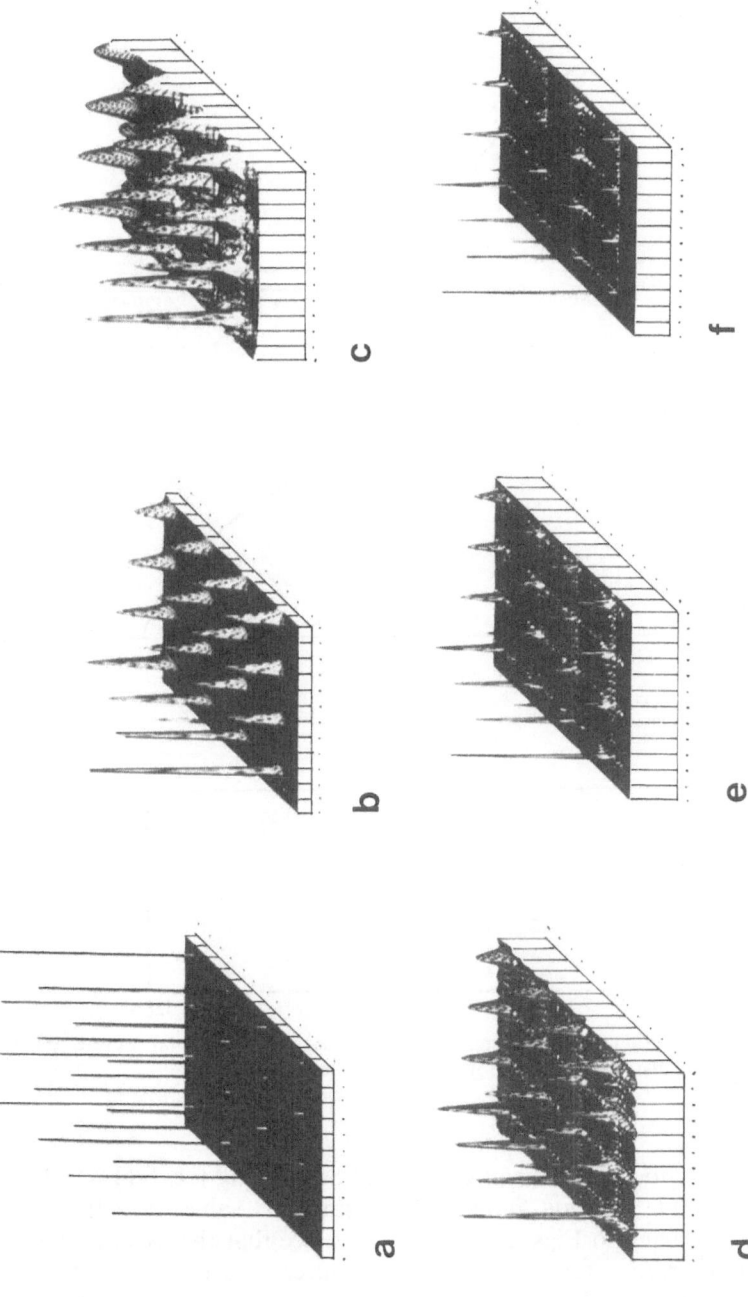

Fig. 2.16a-f. Pseudoinversion of a separable space variant point spread function: moderate distortion. (a) Object, (b) distorted image, (c) pseudoinverse: first 21 eigenvalues, (d) pseudoinverse: first 41 eigenvalues, (e) pseudoinverse: first 61 eigenvalues, (f) pseudoinverse: first 81 eigenvalues

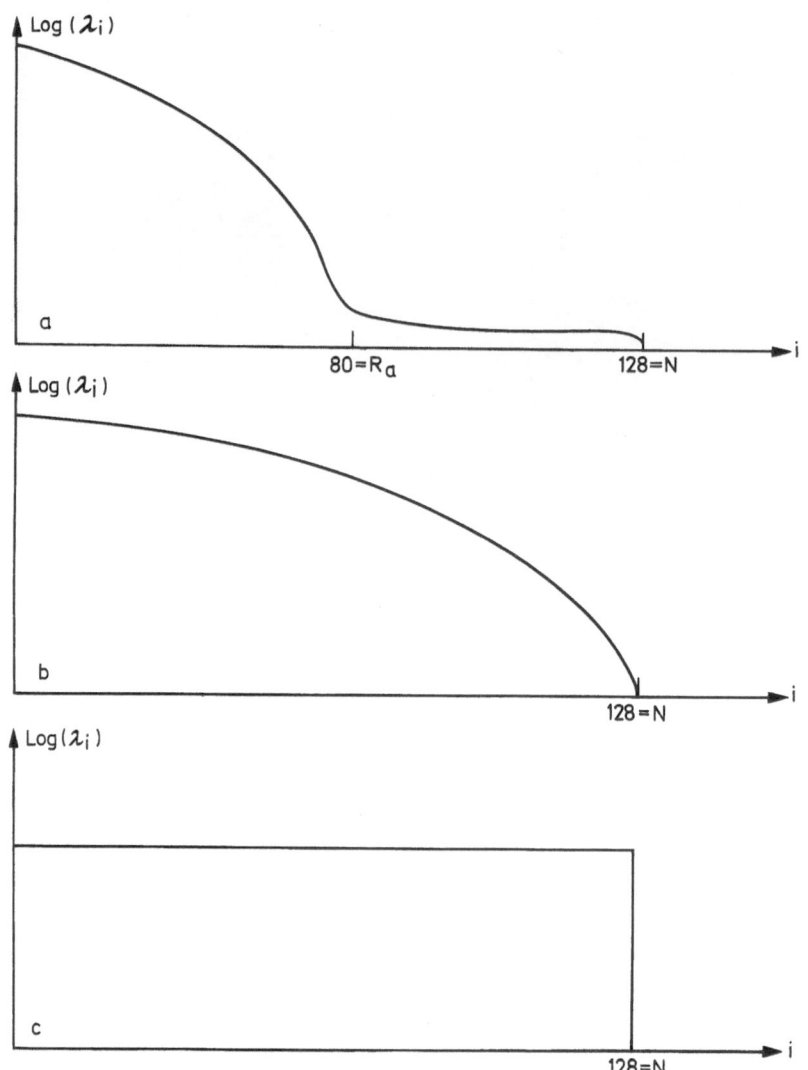

Fig. 2.17a-c. Singular value maps for various distortions. (a) Severe distortion, (b) moderate distortion, (c) Ideal imaging (no distortion)

where by "ideal" we mean distortionless imaging (i.e. $[A] = [B] = [I]$). Figure 2.17 presents plots of the singular values for these cases. Referring back to Figs. 2.15 and 2.16 we see that the pseudo-inversions of the blurred array of object point sources are restored with greater and greater precision at each increased value of K. In the severe

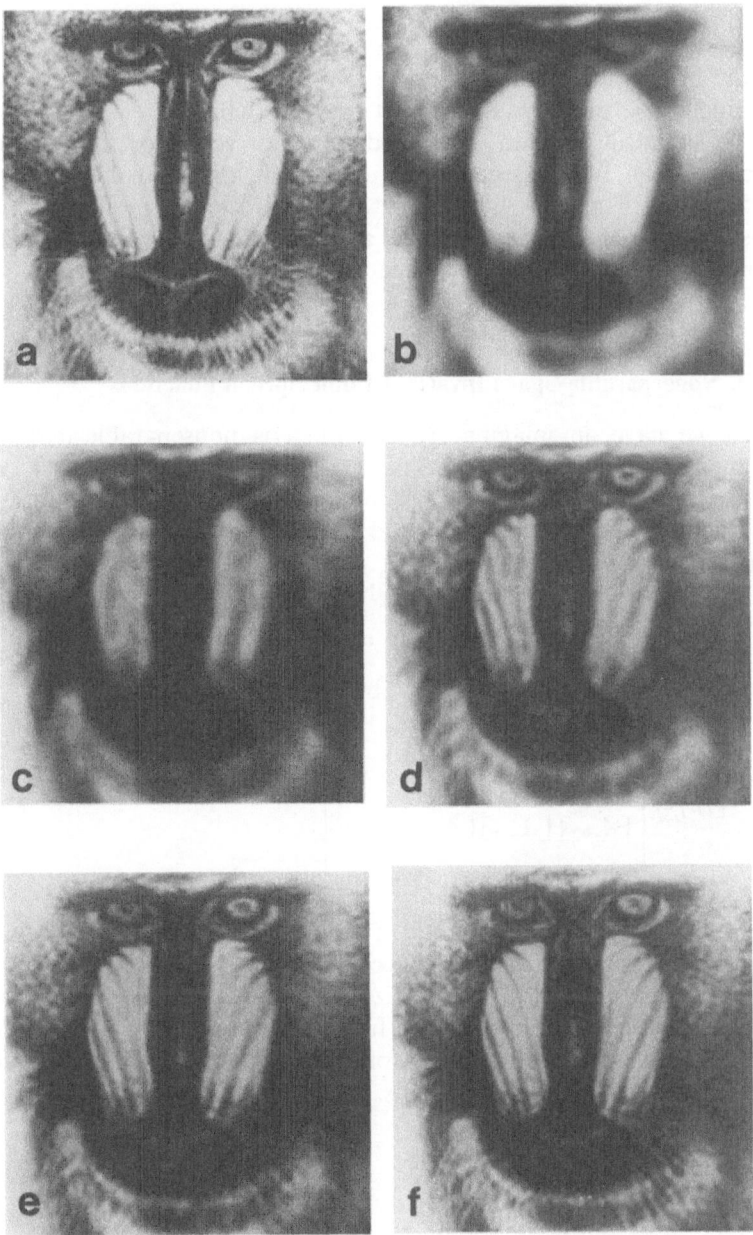

Fig. 2.18a-f. Pseudoinversion for a continuous tone image subject to a (moderate) space variant distortion. (a) Object, (b) distorted image, (c) pseudoinverse: first 21 eigenvalues, (d) pseudoinverse: first 41 eigenvalues, (e) pseudoinverse: first 61 eigenvalues, (f) pseudoinverse: first 81 eigenvalues

distoration case the pseudo-inversion visually blows up at about $K = 56$ where $C_{56}([A]) = 10^6$ and computer noise seems to dominate. In the moderate distortion case $C_{56}([A]) = 20$ and the inversion remains stable out beyond the 81st estimate where $C_{81}[A] = 10^3$.

A "Mandrill Baboon" was used as a test scene for the moderate space variant blur distortion as discussed above. The results are presented in Fig. 2.18. The moderate distoration is seen to remove considerable detail from the object and as higher-order pseudo-inverse estimates are achieved more and more facial detail is recovered.

2.4.3. Nonseparable Space Invariant Point Spread Functions

When we have an imaging system which is nonseparable (either in rectangular or any other coordinate system) our computational tasks increase considerably. In the nonseparable SIPSF case we must use the stacking operator notation and the point spread function matrix $[H]$ associated with (2.39)

$$h(x, y, \xi, \eta) = h(x - \xi, y - \eta) \tag{2.39}$$

becomes block Toeplitz of size N^2 by N^2.

$$[H] = \begin{bmatrix} [A_1] [A_2] \cdots \\ [A_{-1}] [A_1] [A_2] \cdots \\ [A_{-2}] [A_{-1}] [A_1] [A_2] \cdots \\ \vdots \end{bmatrix}, \tag{2.57}$$

where the $[A_i]$ are individually Toeplitz. Unfortunately we cannot equate $[H]$ with a single Kronecker product as separability does not hold. However, if we "circularize" our individual Toeplitz matrices to implement circular (rather than full) convolution, then $[H]$ becomes block circulant and additional simplifications result [2.24, 25]. Because $[H]$ is a block circulant matrix, it is diagonalized by a matrix $[\mathscr{F}_{N^2}]$ where

$$[H] = [\mathscr{F}_{N^2}] [D] [\mathscr{F}_{N^2}]. \tag{2.58}$$

$[D]$ is diagonal and $[\mathscr{F}_{N^2}]$ is block Fourier (i.e. made up of the Kronecker of two $N \times N$ Fourier matrices). Thus

$$[\mathscr{F}_{N^2}] = [\mathscr{F}] \otimes [\mathscr{F}]. \tag{2.59}$$

As a paranthetical comment the columns or rows of $[\mathscr{F}_{N^2}]$ are generalized Walsh functions [2.26] and the reader should be aware that $[\mathscr{F}_{N^2}]$ is not a Fourier matrix of size $N^2 \times N^2$ with N^2 roots of unity as entries but is a matrix $N^2 \times N^2$ with N roots of unity as entries. Thus $[\mathscr{F}_{N^2}]$ implies separability (from the above equation) as far as implementation is concerned on a two-variable image. Equivalently $[\mathscr{F}_{N^2}]$ is the stacked version of a two-dimensional Fourier transform matrix. It does not imply separability of the point spread function, however. Returning to our restoration problem

$$g = [H]f, \tag{2.60a}$$

$$g = [\mathscr{F}_{N^2}] [D] [\mathscr{F}_{N^2}]f \tag{2.60b}$$

but because $[\mathscr{F}_{N^2}]$ is unitary (the Kronecker product of unitary matrices is unitary), we see

$$\hat{f} = [\mathscr{F}_{N^2}]^* [D]^+ [\mathscr{F}_{N^2}]^* g, \tag{2.61}$$

where $[D]^+$ implies a diagonal matrix of entries equal to the reciprocals of the entries in $[D]$ but we do not divide by zero and therefore replace zero entries with zero. Upon reflection it is seen that the above equation can be implemented on a computer by performing a separable two-dimensional transform on g (i.e. $[\mathscr{F}_{N^2}]^* g =$ stacked $[\mathscr{F}]^* [G] [\mathscr{F}]^*$) a point by point multiplication by the N^2 entries of $[D]^+$ and a separable two-dimensional transform on the resulting vector $[D]^+ [\mathscr{F}_{N^2}] g$. This, of course, is the typical filtering operation utilized in the vast majority of all Fourier domain image processing filtering algorithms. Thus two-dimensional Fourier transforming, point by point multiplication in the Fourier domain and inverse transforming is vectorially equivalent to (2.61). Consequently for nonseparable SIPSF's implementation of the restoration algorithm can be done separably up to the point by point filter multiplication, but because this multiplication implies only scalar operations, (i.e. no vector sums), we do not bother to go to the stacked notation in the computer.

2.4.4. Nonseparable Space Variant Point Spread Functions

For this particular type of imaging system there is very little we can say about the point spread function matrix $h(x, y, \xi, \eta)$ without further analytic knowledge.

One interesting extremal condition for the SVD representation of $[H]$ occurs in the perfect imaging situation. In this case,

$$[H] = [I],$$

and the norm between $[H]$ and $[H_K]$ becomes

$$\|[H] - [H_K]\| = \sum_{i=K+1}^{N^2} \lambda_i,$$

but all the eigenvalues of $[H]$ are unity, and consequently,

$$\|[H] - [H_K]\| = N^2 - K.$$

Note that, due to the multiplicity of eigenvalues, there is no unique eigenvector outer product expansion. This unfortunately, is a very poor approximation as a function of K, and indeed, in the perfect imaging limit, SVD does not appear attractive. At the other extreme, where $[H]$ is defined by the all-one's matrix to within a scale factor of $(N^2)^{-1}$, we have

$$h_{ij} = 1 \forall ij$$

and

$$\|[H] - [H_1]\| = 0,$$

because

$$[H] = \mathbf{1}\mathbf{1}^t.$$

In other words, $[H]$ has rank $R = 1$ and is perfectly represented by one singular value and its associated outer product. Of course, this is equivalent to leaving the lens totally defocused (i.e. no imaging). It is believed that a type of continuum exists in transversing between these two extreme conditions on $[H]$.

In our restoration objective we wish to invert the $N^2 \times N^2$ matrix $[H]$ to achieve a better estimate of the object f. Thus,

$$\hat{f} = [H]^+ g \tag{2.62}$$

and because

$$[H]^+ = \sum_{i=1}^{R} \lambda_i^{-\frac{1}{2}} v_i u_i^t, \tag{2.63}$$

then a form for the object f is

$$\hat{f} = [H]^{+} g = \sum_{i=1}^{R} \lambda_i^{-\frac{1}{2}} (u_i^t g) \, v_i. \tag{2.65}$$

The inner product of $(u_i^t g)$ is a scalar weighting (along with $\lambda_i^{-\frac{1}{2}}$) on the singular vector v_i, and the k-th estimate of f becomes

$$\hat{f}_k = \hat{f}_{k-1} + \lambda_k^{-\frac{1}{2}} (u_k^t g) v_k,$$

where a partial summation formulation was used to suggest a convergence to the true object if $[H]$ is truly non-singular. In addition, the partial computation of $\hat{f}$ implies that simultaneous computer storage of the entire set of images and point spread function matrices is not necessary.

Without further knowledge of $[H]$ it is difficult to invert this most general case of imaging. However, because of the large degree of computational savings obtained by the separable assumptions on the imaging system in previous sections, it may be expedient to postulate an approximation to the most general $[H]$ by a series of separable functions. Thus

$$h(x, y, \xi, \eta) \cong \sum_{i=1}^{I} a_i(x, \xi) \, b_i(y, \eta) \tag{2.66a}$$

or

$$[H] \cong \sum_{i=1}^{I} [A_i] \otimes [B_i]^t \tag{2.66b}$$

implying

$$[G] \cong \sum_{i=1}^{I} [A_i] [F] [B_i]. \tag{2.67}$$

However, the inverse of $[H]$ is the inverse of a sum of Kronecker's rather than a sum of inverses, leaving doubt as to the usefulness of the assumption of (2.66a). A polynomial power series approximation

$$h(x, y, \xi, \eta) \cong \sum_{i=1}^{I} c_i [A]^i \otimes [B]^{it} \tag{2.68a}$$

or

$$[H] \cong \sum_{i=1}^{I} c_i [A]^i \otimes [B]^{it} \tag{2.68b}$$

implying

$$[G] \cong \sum_{i=1}^{I} c_i [A]^i [F] [B]^i \tag{2.68c}$$

also suffers from the inversion of a sum series approximation. Thus the most general imaging system will be computationally difficult to invert.

2.5. Conclusions

This chapter has addressed certain aspects of two-dimensional signal processing as applied to digital computer image processing. The central theme for the chapter is two-dimensional transforms and their uses and implementations. The common thread throughout the chapter has been a quadratic form matrix notation where the image $[G]$ was operated upon by two matrices $[U]$ and $[V]$ to form a two-dimensional transform domain

$$[\alpha] = [U] [G] [V]^t .$$

Such separable image decompositions were investigated in an image representation context under mathematical considerations and then again under image coding objectives. Both spatial and transform domain image coding processes were analyzed using the two-dimensional transform notation, the former implying predictor matrices, the latter implying unitary matrices.

The central theme of two-dimensional transforms was then carried over to object restoration problems where the quadratic form became

$$[G] = [A] [F] [B] ,$$

where the object $[F]$ was blurred by $[A]$ and $[B]$ to form the image $[G]$. Inversion of the blur phenomena was discussed under four general imaging circumstances and when separable assumptions did not hold, the Kronecker operator was introduced to show the reader how the separable assumption fit into the most general imaging system formulation.

In conclusion, it is hoped that the reader has found the generality of the matrix formulation of two-dimensional transforms both informative and revealing. Simply stated, the matrix algebra approach coupled with linear imaging system models makes two-dimensional transform a very powerful tool for many forms of two variable signal processing. Space limitations do not permit elaboration on the technique to feature extraction procedures, but similar approaches apply equally well in that aspect of image processing.

Acknowledgements

This chapter could not have been written without the invaluable support of many contributing individuals and organizations. In general, the author would like to acknowledge support from the Image Processing Institute at the University of Southern California (under the direction of Professor WILLIAM K. PRATT), the Optical Systems Division at The Aerospace Corporation (under the direction of Mr. ALAN BOARDMAN), and the Image Processing Group at the Los Alamos Scientific Laboratories (under the direction of Drs. D. JANNEY and B. HUNT).

Portions of this work were supported by The Aerospace Corporation internal R&D fee-sponsored research funds and by the Advanced Research Projects Agency of the Department of Defense Contract No. F 08606–72–C–0008 and monitored by the Air Force Eastern Test Range, Patrick AFB, Florida.

Specific acknowledgement must be made to Professor HABIBI of USC whose transform domain formulation of both hybrid and two dimensional DPCM systems the author so liberally plagarized. In addition, the singular value decomposition routines applied to image processing were made available from Mr. C. L. PATTERSON of The Aerospace Corporation. The interested reader is referred to a paper by PATTERSON and BUECHLER [2.27] for further discussion on the SVD imaging material. Finally, the images in this chapter were all generated at the Aerospace Corporation.

References

2.1. G. H. GOLUB, C. REINSCH: Numer. Math. **14**, 403 (1970).
2.2. A. ALBERT: *Regression and the Moore-Penrose Pseudoinverse* (Academic Press, New York, 1972).
2.3. H. C. ANDREWS: *Computer Techniques in Image Processing* (Academic Press, New York, 1970), Chap. 5.
2.4. H. F. HARMUTH: *Transmission of Information by Orthogonal Functions*, 2nd ed. (Springer, Berlin-Heidelberg-New York, 1972).
2.5. P. A. WINTZ: Proc. IEEE **60**, 809 (1972).
2.6. M. P. EKSTROM: IEEE Trans. Computers C-**22**, 322 (1973).

2.7. R. BELLMAN: *Introduction to Matrix Analysis* (McGraw-Hill Book Company, Inc., New York, 1960).
2.8. A. HABIBI, R. S. HERSHEL: IEEE Trans. Communication COM-**22**, 692 (1974).
2.9. A. HABIBI: Nat. Telecomm. Conf. Record **1**, 12 D-1 (November, 1973).
2.10. T. S. HUANG, W. F. SCHRIEBER, O. J. TRETIAK: Proc. IEEE **59**, 1586 (1971).
2.11. N. AHMED, T. NATARAYAN, K. R. RAO: IEEE Trans. Computers C-**23**, 90 (1974).
2.12. W. CHEN: USCEE Rept. 441 (May, 1973).
2.13. R. M. HARALICK, K. SHANMUGAN: IEEE Trans. Systems, Man and Cybernetics SMC-**4**, 16 (1974).
2.14. W. K. PRATT, J. KANE, H. C. ANDREWS: Proc. IEEE **57**, 58 (1969).
2.15. A. HABIBI: IEEE Trans. Communication COM-**22**, 614 (1974).
2.16. W. D. RAY, E. M. DRIVER: IEEE Trans. Inform. Theory IT-**16**, 663 (1970).
2.17. J. PEARL, H. C. ANDREWS, W. K. PRATT: IEEE Trans. Communication COM-**20**, 411 (1972).
2.18. D. NISSEN: Econometrics **36**, 603 (1968).
2.19. R. J. HANSON: SIAM J. Numer. Anal. **8**, 616 (1971).
2.20. J. M. VARAH: SIAM J. Numer. Anal. **10**, 257 (1973).
2.21. M. M. SONDHI: Proc. IEEE **60**, 842 (1972).
2.22. S. TREITEL, J. L. SHANKS: IEEE Trans. Geoscience Electronics GE-**9**, 10 (1971).
2.23. J. TODD: *Survey of Numerical Analysis* (McGraw-Hill Book Co., Inc., New York, 1958), Chapt. 8.
2.24. M. P. EKSTROM: IEEE Trans. Computers C-**23**, 320 (1974).
2.25. B. R. HUNT: IEEE Trans. Computers C-**22**, 805 (1973).
2.26. H. E. CHRESTENSON: Pacific J. Mathematics **5**, 17 (1955).
2.27. C. L. PATTERSON, G. B. BUECHLER: Computer **7**, 46 (1974).

Further Reference with Title

H. C. ANDREWS, C. L. PATTERSON: Outer products and their uses in digital image processing. Am. Math. Monthly **82**, No. 1, 1–15 (1975).

3. Two-Dimensional Nonrecursive Filters

J. G. FIASCONARO

With 17 Figures

This chapter deals primarily with four techniques for designing two-dimensional nonrecursive digital filters. These methods include: the use of window functions, frequency sampling, the straightforward application of linear programming, and a new algorithm that was developed by the author. The theory required to understand these four algorithms is presented in Section 3.1. That section discusses general two-dimensional discrete systems, some of the aspects of the theory of linear approximation, and linear programming as it applies to the filter design problem. Section 3.2 contains a detailed description of the four algorithms and gives some examples of filters designed with two of the techniques. A brief summary and some conclusions are presented in Section 3.3.

The first three techniques (i.e. the use of window functions, frequency sampling, and the straightforward use of linear programming) are direct extensions of techniques for designing one-dimensional nonrecursive digital filters [3.1–5]. Of these three, only the third one can be used to design filters that are optimal in the sense that the maximum absolute value (Tchebycheff norm) of the error has been minimized. Other techniques for designing optimal one-dimensional nonrecursive digital filters include: the second algorithm of REMES [3.6], and methods developed by HERRMANN [3.7], HOFSTETTER et al. [3.8], and PARKS and McCLELLAN [3.9]. Unfortunately, none of these techniques can be readily extended to the two-dimensional case.

The straightforward linear programming approach does not work as well in the two-dimensional case as it does in the one-dimensional case. This results from the fact that many more constraint points are required in the former case than in the latter case for a given number of unknown unit sample response coefficients. The new algorithm for designing optimal two-dimensional filters was developed in an effort to solve this problem. The new algorithm is iterative in nature and it uses linear programming as an optimization technique; however, only a small number of constraint points are required at each iteration.

The new algorithm is superior to the straightforward application of linear programming because a) "good" approximations can generally be obtained rapidly and these approximations can be made better simply by performing additional iterations, b) best approximations can be obtained on extremely dense sets of points, and c) the new algorithm provides a way to break up one large linear programming problem into a series of smaller problems. Unfortunately, the new algorithm is not as computationally efficient as was hoped and with the computation facilities at hand (IBM 360/67), it was limited to filters with roughly no more than 50 independent unit sample response coefficients.

3.1. Theory

3.1.1. Two-Dimensional Discrete Systems

A two-dimensional discrete system is a mapping $\mathscr{L}$ which maps an input array $f(n, m)$ into an output array $g(n, m) = \mathscr{L}\{f(n, m)\}$ where n and m range over the positive and negative integers. If $g_1 = \mathscr{L}\{f_1\}$ and $g_2 = \mathscr{L}\{f_2\}$ and

$$cg_1 + g_2 = \mathscr{L}\{cf_1 + f_2\}$$

for an arbitrary constant c, then the system is linear. Furthermore if $g = \mathscr{L}\{f\}$ and

$$g(n - n_0, m - m_0) = \mathscr{L}\{f(n - n_0, m - m_0)\}$$

for arbitrary integers n_0 and m_0, then the system is shift-invariant.

The unit sample $u_0(n, m)$ is defined by

$$u_0(n, m) = \begin{cases} 1 & \text{for} \quad n = 0, m = 0 \\ 0 & \text{otherwise} . \end{cases}$$

Clearly, any array $f(n, m)$ can be written as the summation of constants times the shifted unit sample as follows

$$f(n, m) = \sum_{k=-\infty}^{\infty} \sum_{l=-\infty}^{\infty} f_{kl} u_0(n - k, m - l),$$

where the constant $f_{kl} = f(k, l)$. So if $g(n, m) = \mathscr{L}\{f(n, m)\}$ then

$$g(n, m) = \sum_{k=-\infty}^{\infty} \sum_{l=-\infty}^{\infty} f_{kl} \mathscr{L}\{u_0(n - k, m - l)\} .$$

The unit sample response, $h(n, m)$, of the system is defined by

$$h(n, m) = \mathcal{L}\{u_0(n, m)\} .$$

Therefore,

$$g(n, m) = \sum_{k=-\infty}^{\infty} \sum_{l=-\infty}^{\infty} f(k, l)\, h(n-k, m-l) . \tag{3.1}$$

Thus the output of the system is the input to the system convolved with the unit sample response of the system. It can also be shown that

$$g(n, m) = \sum_{k=-\infty}^{\infty} \sum_{l=-\infty}^{\infty} f(n-k, m-l)\, h(k, l) . \tag{3.2}$$

If $h(n, m)$ is zero outside of a finite area, then the filter is called non-recursive. In this case the infinite limits on the summations in (3.1) and (3.2) can be replaced by finite limits. Thus each element of the output array of a nonrecursive filter is a finite weighted sum of elements of the input array.

A digital filter is said to be stable if the output remains bounded for all bounded inputs. A digital filter is stable if and only if

$$\sum_{n=-\infty}^{\infty} \sum_{m=-\infty}^{\infty} |h(n, m)| < \infty .$$

In actual practice this test can be quite difficult to apply for filters that are not nonrecursive [3.10–11]. This problem does not exist for non-recursive filters. In this case the sum must always be bounded because $h(n, m)$ is zero outside of a finite area.

If the input to the system is given by

$$f(n, m) = e^{j(\mu n + v m)} ,$$

where $j = \sqrt{-1}$, then it can easily be shown

$$g(n, m) = e^{j(\mu n + v m)} \sum_{k=-\infty}^{\infty} \sum_{l=-\infty}^{\infty} h(k, l)\, e^{-j(\mu k + v l)} . \tag{3.3}$$

Thus the complex exponential is an eigenfunction of the system. The double summation in (3.3) is interpreted as the system function $H(\mu, v)$ which is continuous and periodic in both μ and v with a period of 2π. It is also possible to interpret the double summation in (3.3) as a two-dimensional Fourier series expansion for $H(\mu, v)$. Consequently, $h(k, l)$ can be expressed in terms of $H(\mu, v)$ as

$$h(k, l) = \frac{1}{4\pi^2} \int\limits_{-\pi}^{\pi} \int\limits_{-\pi}^{\pi} H(\mu, v) \, e^{J(\mu k + vl)} \, d\mu \, dv \, .$$

The two-dimensional z-transform is a generalization of the system function. The z-transform of any array $f(n, m)$ is defined by

$$F(z_1, z_2) = \sum_{n=-\infty}^{\infty} \sum_{m=-\infty}^{\infty} f(n, m) \, z_1^{-n} \, z_2^{-m} \, ,$$

where z_1 and z_2 are complex variables. The system function is a special case of the z-transform with $z_1 = \exp(j\mu)$ and $z_2 = \exp(jv)$. The z-transform of the output of a digital filter is the product of the z-tansform of the input times the z-transform of the unit sample response, that is

$$G(z_1, z_2) = F(z_1, z_2) \, H(z_1, z_2) \, .$$

The inverse z-transform $g(n, m)$ is obtained by successive application of the (one-dimensional) inverse z_1-transform and inverse z_2-transform to $G(z_1, z_2)$ [3.12].

All of the theory of two-dimensional discrete systems presented thus far has been a direct extension of the theory of one-dimensional discrete systems [3.13]. However, there are some areas where this extension is impossible. For one-dimensional systems, if it is possible to express the z-transform as a rational function, then the numerator and denominator can be factored to obtain the poles and zeros of the z-transform. Once the pole locations are known, the inverse z-transform can be computed for each possible region of convergence of the z-transform.

For two-dimensional systems there are two problems. First, it is in general impossible to factor a polynomial in two (or more) variables and such a polynomial is generally zero for a continuum of values of the two variables rather than at a specific number of discrete points. Second, $z_1 z_2$-space is four dimensional and so it is impossible to visualize regions of convergence which can be easily visualized in the z-plane for functions of one variable. These two problems have prevented the theory of multi-dimensional z-transforms from becoming as richly developed as the theory of one-dimensional z-transforms.

When working with a digital computer, one would like to consider arrays that are zero outside of some finite area. In addition, one would like the transform of an array to be another array and not a continuous function. This problem is solved by introducing the discrete Fourier transform. If $f(n, m)$ is zero for $n < 0$ and $n \geq N$ and for $m < 0$ and $m \geq N$ then its discrete Fourier transform (DFT) is given by

$$F(k, l) = \sum_{n=0}^{N-1} \sum_{m=0}^{N-1} f(n, m)\, W^{-(nk + ml)},$$

where $W = \exp[j(2\pi/N)]$. It should be noted that the transform produces only N^2 distinct values of $F(k, l)$ and that these values are samples of the z-transform of $f(n, m)$ evaluated at equally spaced intervals for $|z_1| = |z_2| = 1$. The inverse DFT is given by

$$f(n, m) = \frac{1}{N^2} \sum_{k=0}^{N-1} \sum_{l=0}^{N-1} F(k, l)\, W^{(nk + ml)}.$$

Furthermore, given that the DFT of $f(n, m)$ is $F(k, l)$ and that the DFT of $h(n, m)$ is $H(k, l)$, then the inverse DFT of $H(k, l)\, F(k, l)$ is the periodic convolution of $f(n, m)$ and $h(n, m)$, analogous to the one-dimensional case. This can be made to correspond to a linear convolution by suitably augmenting the arrays $f(n, m)$ and $h(n, m)$ with zeros before taking their DFT's.

The two dimensional DFT is generally computed using the one-dimensional fast Fourier transform (FFT) algorithm. This is possible because the DFT can be written as

$$F(k, l) = \sum_{m=0}^{N-1} \left(\sum_{n=0}^{N-1} f(n, m)\, W^{-nk} \right) W^{-ml}.$$

The sum in parentheses is the one-dimensional DFT of the rows of the array $f(n, m)$ and it transforms this array into a new array $f'(k, m)$ whose columns are then transformed by another one-dimensional DFT to produce $F(k, l)$. In general, $2N$ one-dimensional DFT's must be computed. However, it is frequently possible to reduce this number substantially by taking advantage of certain properties of the array being transformed.

3.1.2. Approximation Theory

The frequency response of a digital filter is given by the system function which is defined in the preceding section as

$$H(\mu, v) = \sum_{n=-\infty}^{N} \sum_{m=-\infty}^{N} h(n, m)\, e^{-j(\mu n + vm)}$$

and which is equal to the z-transform of $h(n, m)$ evaluated for $z_1 = \exp(j\mu)$ and $z_2 = \exp(jv)$. If the filter is nonrecursive then the frequency response can be written as

$$H(\mu, v) = \sum_{n=-N}^{N} \sum_{m=-N}^{N} h(n, m)\, e^{-j(\mu n + vm)}, \qquad (3.4)$$

where it is assumed that $h(n, m)$ is zero outside of a square region with $2N+1$ samples on each side and centered at the origin. The filter design problem for nonrecursive filters consists of determining the unit sample response coefficients $h(n, m)$ which produce a desired frequency response $H(\mu, v)$.

If the unit sample response coefficients are restricted to be pure real, then it follows that

$$H(\mu, v) = H^*(-\mu, -v)$$

where the asterisk implies complex conjugation. If the frequency response is also assumed to be pure real (so that the filter has zero phase), then

$$h(n, m) = h(-n, -m)$$

and (3.4) can be rewritten as

$$H(\mu, v) = \sum_{n=-N}^{\infty} \sum_{m=-N}^{\infty} h(n, m) \cos(n\mu + mv).$$

However, since $h(n, m) = h(-n, -m)$ this equation can be rewritten in a number of different ways. One possibility is the following

$$H(\mu, v) = h(0, 0) + 2 \sum_{n=1}^{N} h(n, 0) \cos n\mu$$

$$+ 2 \sum_{m=1}^{N} \sum_{n=-N}^{N} h(n, m) \cos(n\mu + mv). \qquad (3.5)$$

This situation is shown in Fig. 3.1a for the case of a 7×7 unit sample response. It is at times desirable to place further restrictions on the symmetry of the unit sample response and the frequency response. If it is assumed that

$$h(n, m) = h(|n|, |m|)$$

then

$$H(\mu, v) = H(|\mu|, |v|)$$

and the frequency response can be expressed as (see Fig. 3.1 b)

$$H(\mu, v) = h(0, 0) + 2 \sum_{n=1}^{N} h(n, 0) \cos n\mu + 2 \sum_{m=1}^{N} h(0, m) \cos mv$$

$$+ 2 \sum_{m=1}^{N} \sum_{n=1}^{N} h(n, m) [\cos(n\mu + mv) + \cos(n\mu - mv)] \,.$$

(3.6)

And finally, if it is assumed that

$$h(n, m) = h(m, n) \quad \text{and} \quad h(n, m) = h(|n|, |m|)$$

then

$$H(\mu, v) = H(v, \mu) \quad \text{and} \quad H(\mu, v) = H(|\mu|, |v|)$$

and the frequency response can be expressed as (see Fig. 3.1 c)

$$H(\mu, v) = h(0, 0) + 2 \sum_{n=1}^{N} h(n, 0) [\cos n\mu + \cos nv] + 2 \sum_{n=1}^{N} h(n, n)$$

$$\cdot [\cos(n\mu + nv) + \cos(n\mu - nv)] + 2 \sum_{m=1}^{N} \sum_{n=m+1}^{N} h(n, m) \quad (3.7)$$

$$\cdot [\cos(n\mu + mv) + \cos(n\mu - mv) + \cos(m\mu + nv)$$

$$+ \cos(m\mu - nv)] \,.$$

Can (3.4) through (3.7) be used to realize "ideal" filters exactly? For common types of "ideal" filters the answer to this question is no. For example, $H(\mu, v)$ for an ideal circular low-pass filter is given by

$$H(\mu, v) = \begin{cases} 1 & \text{for } \sqrt{\mu^2 + v^2} \leq R \\ 0 & \text{otherwise} \end{cases}$$

over one period, for example for $-\pi \leq \mu \leq \pi$ and $-\pi \leq v \leq \pi$. The corresponding unit sample response is given by

$$h(n, m) = \frac{R J_1(R\sqrt{n^2 + m^2})}{2\pi\sqrt{n^2 + m^2}} \,,$$

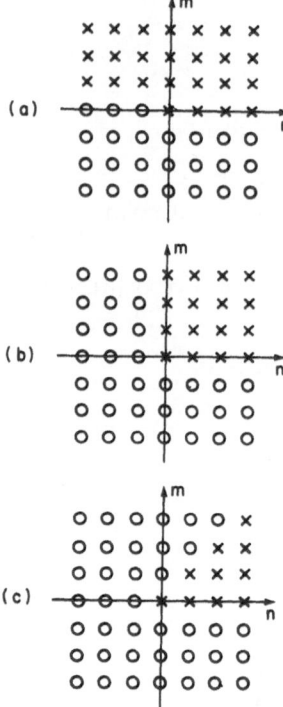

× INDEPENDENT UNIT SAMPLE RESPONSE COEFFICIENT

O COEFFICIENT DETERMINED BY SYMMETRY

Fig. 3.1a–c. Symmetry constraints

where J_1 is the Bessel function of the first kind, order one. Similarly, the frequency response of an ideal circular band-pass filter is given by

$$H(\mu, v) = \begin{cases} 1 & \text{for } R_1 \leq \sqrt{\mu^2 + v^2} \leq R_2 \\ 0 & \text{otherwise} \end{cases}$$

and the unit sample response is

$$h(n, m) = \frac{R_2 J_1(R_2 \sqrt{m^2 + m^2})}{2\pi \sqrt{n^2 + m^2}} - \frac{R_1 J_1(R_1 \sqrt{n^2 + m^2})}{2\pi \sqrt{n^2 + m^2}}.$$

An ideal circular high-pass filter has a frequency response given by

$$H(\mu, v) = \begin{cases} 0 & \text{for } \mu^2 + v^2 \leq R^2 \\ 1 & \text{otherwise} \end{cases}$$

and so its unit sample response is given by

$$h(n, m) = \begin{cases} 1 & \text{for } n = 0 \text{ and } m = 0 \\ -\dfrac{R J_1(R\sqrt{n^2 + m^2})}{2\pi\sqrt{n^2 + m^2}} & \text{otherwise}. \end{cases}$$

One type of ideal two-dimensional differentiator has a frequency response given by

$$H(\mu, v) = \mu^2 + v^2.$$

The corresponding unit sample response is

$$h(n, m) = \begin{cases} \dfrac{2\pi^2}{3} & \text{for } n = 0 \text{ and } m = 0 \\ \dfrac{2}{n^2} \cos n\pi & \text{for } n \neq 0 \text{ and } m = 0 \\ \dfrac{2}{m^2} \cos m\pi & \text{for } n = 0 \text{ and } m \neq 0 \\ 0 & \text{otherwise}. \end{cases}$$

In all of these cases the unit sample response is of infinite extent, i.e., there is no finite area outside of which the unit sample response is zero. Consequently, these filters are not nonrecursive and so cannot be realized with (3.4) through (3.7).

Because the frequency response of such filters cannot be realized exactly it must be approximated. As a result, some measure of the closeness of an approximate frequency response to the desired frequency response, and some criterion for deciding on a closest or best approximation is needed. A natural choice for both of these is the Tchebycheff norm [3.6] which, for a function F, is defined as a maximum absolute value of F and written as $\|F\|$. If $H(\mu, v)$ is an approximation to an ideal frequency response $F(\mu, v)$ then the measure of the closeness of $H(\mu, v)$ to $F(\mu, v)$ is the Tchebycheff norm of the error function, that is

$$\|\text{Error}\| = \max |F(\mu, v) - H(\mu, v)|.$$

The best approximation to $F(\mu, v)$ is that $H(\mu, v)$ which minimizes the Tchebycheff norm of the error function.

Some of the basic results from the theory of linear Tchebycheff approximation apply to the filter design problem as it has been stated thus far. In particular, three theorems are important: the existence theorem, the characterization theorem, and the unicity theorem. These theorems will be quoted here without proofs since these are available in the literature [3.6]. All of these theorems are phrased in terms of the two-dimensional filter design problem. This is done strictly for notational convenience. All of the results which apply to the two-dimensional case apply with equal force to the 3-, 4-, or n-dimensional case.

Existence Theorem. A finite dimensional linear subspace of a normed linear space contains at least one point of minimum distance from a fixed point.

In order to see how this theorem applies to the filter design problem it is necessary to observe that the set c of all functions that are continuous on a closed subset of the region $-\pi \leq \mu \leq \pi$, $-\pi \leq v \leq \pi$ constitutes a normed linear space if the Tchebycheff norm is used and addition and scalar multiplication are defined by

$$(f + g)(\mu, v) = f(\mu, v) + g(\mu, v)$$
$$(\lambda f)(\mu, v) = \lambda f(\mu, v)$$

for any real constant λ, and for f and g elements of c. Furthermore, (3.5) through (3.7) define finite dimensional linear subsets of this normed linear space. So if the "ideal" filter is a member of the normed linear space there is at least one approximation which is closest (in the sense of the Tchebycheff norm) to the "ideal" filter.

Equations (3.5) through (3.7) can be expressed as generalized polynomials of the form

$$H(\mu, v) = \sum_{i=1}^{n} c_i g_i(\mu, v),$$

where the c_i are the unit sample response coefficients, and the $g_i(\mu, v)$ are the various cosine functions. With this notation in mind and assuming that the ideal frequency response is $f(\mu, v)$ it is possible to state the characterization theorem.

Characterization Theorem. In order that the coefficients $c_1, ..., c_n$ shall render the Tchebycheff norm of $r(\mu, v) = \Sigma c_i g_i(\mu, v)^1 - f(\mu, v)$ a minimum, it is necessary and sufficient that the origin of n space shall lie in the convex hull of the point set

$$\{r(\mu, v) [g_1(\mu, v), ..., g_n(\mu, v)] : |r(\mu, v)| = \|r(\mu, v)\|\}.$$

---

[1] When the limits on the summation are not specified, the limits of 1 and n are implied.

The convex hull of a set A is defined as the set of points p which are expressible as finite sums of the form $p = \Sigma \theta_i f_i$ with $f_i \in A$, $\theta_i \geq 0$, and $\Sigma \theta_i = 1$. Expressed another way, the theorem states the $\Sigma c_i g_i$ is a best approximation if and only if there are, say, m points for which $|r(\mu_i, v_i)| = \|r(\mu, v)\|$ and

$$\begin{bmatrix} g_1(\mu_1, v_1) & g_1(\mu_2, v_2) & \cdots & g_1(\mu_m, v_m) \\ g_2(\mu_1, v_1) & g_2(\mu_2, v_2) & \cdots & g_2(\mu_m, v_m) \\ \vdots & \vdots & \vdots \\ g_n(\mu_1, v_1) & g_n(\mu_2, v_2) & \cdots & g_n(\mu_m, v_m) \end{bmatrix} \begin{bmatrix} \theta_1 r(\mu_1, v_1) \\ \theta_2 r(\mu_2, v_2) \\ \vdots \\ \theta_m r(\mu_m, v_m) \end{bmatrix} = \begin{bmatrix} 0 \\ 0 \\ \vdots \\ 0 \end{bmatrix},$$

where it is clear that the restriction that $\Sigma \theta_i = 1$ can be replaced by $\Sigma \theta_i > 0$. This theorem provides a way of recognizing whether or not a given approximation is a best approximation. However, no information about actually calculating a best approximation is provided. This problem is discussed later in this section and in the next section.

Unicity Theorem (Haar). The best approximation to a continuous function f by a generalized polynomial $\Sigma c_i g_i$ is unique for all choices of f if and only if $\{g_i, ..., g_n\}$ satisfies the Haar condition.

A set of continuous functions $g_1, ..., g_n$ satisfies the Haar condition if each determinant

$$D = \begin{vmatrix} g_1(\mu_1, v_1) & \cdots & g_n(\mu_1, v_1) \\ \vdots & & \vdots \\ g_1(\mu_n, v_n) & \cdots & g_n(\mu_n, v_n) \end{vmatrix}$$

made up from n distinct points is nonzero. Expressed another way, a system of functions $\{g_1, ..., g_n\}$ satisfies the Haar condition if and only if no generalized polynomial $\Sigma c_1 g_1$ has more than $n - 1$ roots. A system of functions which satisfies the Haar condition is sometimes called a Tchebycheff system.

The question immediately arises. Do the cosine functions of (3.5) through (3.7) satisfy the Haar condition on the region $-\pi \leq \mu \leq \pi$, $-\pi \leq v \leq \pi$? The answer is no because of the following argument [3.14]. Assume that n points have been chosen such that the determinant D is nonzero. Now move two points, say (μ_1, v_1) and (μ_2, v_2), in a continuous manner such that they always remain in the region $-\pi \leq \mu \leq \pi$, $-\pi \leq v \leq \pi$, there are always n distinct points, and (μ_1, v_1) ends up where (μ_2, v_2) started and (μ_2, v_2) ends up where (μ_1, v_1) started. It is clear that this interchanges two rows of the determinant D and so it changes the sign of the determinant. Since the determinant was nonzero at the start, varied in a continuous manner, and changed sign, it must have gone through zero. Therefore, the functions in question do not satisfy the Haar condition. A formal proof of this is available in the literature [3.15] and it is shown that in general there are no nontrivial Tchebycheff systems for functions of more than one variable.

If the set of functions $\{g_1, ..., g_n\}$ satisfies the Haar condition, not only is the best approximation unique but it is also possible to prove a very powerful characterization theorem known as the alternation theorem [3.6]. This theorem applies only to the one-dimensional approximation problem.

Alternation Theorem. Let $g_1, ..., g_n$ be a system of functions continuous on $[a, b]$ and satisfying the Haar condition, and let X be any closed subset of $[a, b]$. In order that a certain generalized polynomial $P = \Sigma c_i g_i$ shall be a best approximation on X to a given function f that is continuous on X it is necessary and sufficient that the error function $r = f - P$ exhibit on X at least $n + 1$ "alternations" thus: $r(x_i) = -r(x_{i-1}) = \pm \|r\|$, with $x_0 < \cdots < x_n$ and $x_i \in X$.

Two sets of functions that are fundamental to the one-dimensional filter design problem are $\{1, x, x^2, ..., x^n\}$ and $\{1, \cos x, \sin x, \cos 2x, \sin 2x, ..., \cos nx, \sin nx\}$. The first satisfies the Haar condition on any interval and the second satisfies the Haar condition on the interval $[0, 2\pi]$, see [3.16]. As a result of this, the one-dimensional filter design problem is considerably easier to handle than the two-dimensional filter design problem.

The two-dimensional filter design problem can be formulated in a slightly different way [3.17] by starting with a (possibly infinite) system of incompatible equations of the form

$$F(\mu, v) = \sum_{i=1}^{n} c_i g_i(\mu, v),$$

where $F(\mu, v)$ is an ideal (unrealizable) frequency response. The goal is to find a set $\{c_1, ..., c_n\}$ which minimizes the Tchebycheff norm of the function Δ defined by

$$\Delta = F(\mu, v) - \sum_{i=1}^{n} c_i g_i(\mu, v).$$

A finite subsystem

$$\Delta_m = F(\mu_m, v_m) - \sum_{i=1}^{n} c_i g_i(\mu_m, v_m) \qquad m = 1, ..., M$$

of the original set of equations is called limiting if all of the Δ_m are nonzero and the absolute values of all of the Δ_m cannot be simultaneously reduced for any choice of the set $\{c_1, ..., c_n\}$. To investigate how the absolute values of the Δ_m vary with changes in the set $\{c_1, ..., c_n\}$, it is helpful to rewrite the subsystem of equations in the form

$$\operatorname{sign}\Delta_m \cdot \Delta_m = \operatorname{sign}\Delta_m \cdot F(\mu_m, v_m) - \operatorname{sign}\Delta_m \sum_{i=1}^{n} (c_i + d_i)\, g_i(\mu_m, v_m)\,,$$

where the d_i represent the changes in the c_i. In this form, the left-hand side of each equation is equal to the absolute value of Δ_m. It will certainly be impossible to simultaneously reduce the absolute value of all of the Δ_m if the set of inequalities

$$\operatorname{sign}\Delta_m \cdot \sum_{i=1}^{n} d_i g_i(\mu_m, v_m) > 0 \qquad m = 1, ..., M$$

is incompatible, i.e., if all of these inequalities cannot be simultaneously satisfied for any choice of the set $\{d_1, ..., d_n\}$. It can be shown (using the Theorem on Linear Inequalities [3.6] that this set of inequalities is incompatible if and only if the origin of n-space lies in the convex hull of the set $\{\operatorname{sign}\Delta_m[g_1(\mu_m, v_m), ..., g_n(\mu_m, v_m)]\}$. With this background it is possible to state and prove the following theorem which is a version of the generalized de La Vallee Poussin theorem.

Theorem. If P is a generalized polynomial, i.e., $P(\mu, v) = \Sigma c_i g_i(\mu, v)$, and X is a finite set of points for which the system of equations

$$\Delta_m = F(\mu_m, v_m) - \sum_{i=1}^{n} c_i g_i(\mu_m, v_m) \qquad m = 1, ..., M$$

is limiting, then

$$\min_X |P - F| \leq \max_X |P^* - F| \leq \max_X |P - F|\,,$$

where P^* is a best Tchebycheff approximation to F on the set X.

Proof. The right-hand inequality is trivial since P is a possible best approximation to F on X. To prove the left-hand inequality assume that

$$\max_X |P^* - F| < \min_X |P - F|\,.$$

This implies that all of the values of $|P - F|$ have been simultaneously reduced in going from P to P^*. However, this violates the condition that the system

$$\varDelta_m = F(\mu_m, v_m) - \sum_{i=1}^{n} c_i g_i(\mu_m, v_m) \qquad m = 1, \ldots, M$$

is limiting. Therefore, the left-hand inequality is proved.

The generalized de La Vallee Poussin theorem is the basis for an algorithm which can be used to calculate a best Tchebycheff approximation $P^*(\mu, v)$ to an ideal frequency response $F(\mu, v)$ on a (possibly infinite) set of points X. The algorithm can be described as follows. Let P_k be a best approximation to F on $X_k \subseteq X$ with

$$P_k(\mu, v) = \sum_{i=1}^{n} c_i^k g_i(\mu, v)$$
$$\|E_k\| = \max_{X} |P_k - F|$$
$$\delta_k = \max_{X_k} |P_k - F|.$$

Determine a new set of points $X_{k+1} \subseteq X$ such that

$$\max_{X_{k+1}} |P_k - F| = \|E_k\|$$
$$\min_{X_{k+1}} |P_k - F| \geq \delta_k$$

and such that the system of equations

$$\varDelta = F(\mu, v) - \sum_{i=1}^{n} c_i g_i(\mu, v) \quad \text{for all } (\mu, v) \in X_{k+1}$$

is limiting. By the theorem it follows that

$$\min_{X_{k+1}} |P_k - F| \leq \max_{X_{k+1}} |P_{k+1} - F| \leq \max_{X_{k+1}} |P_k - F|$$

or

$$\delta_k \leq \delta_{k+1} \leq \|E_k\|, \tag{3.8}$$

where P_{k+1} is a best approximation to F on X_{k+1} and

$$\delta_{k+1} = \max_{X_{k+1}} |P_{k+1} - F|.$$

What this means is that the value of δ must increase (or remain the same—a condition never observed in practice) from iteration k to iteration $k+1$. The algorithm must converge because a best approximation, P^*, exist and

$$\max_X |P^* - F| = \delta^* < \infty .$$

The exact method for determining the set X_{k+1} from P_k and X_k will become clear after the discussion of linear programming.

Algorithms such as this one can be grouped into three broad categories [3.18]: minimizing methods, maximizing methods, and mini-max methods. A minimizing method is one which guarantees a decrease in the upper bound on the error from one iteration to the next. A maximizing method guarantees an increase in the lower bound to the error and a mini-max method guarantees both an increase in the lower bound and a decrease in the upper bound. It is clear from (3.8) that the algorithm just described is a maximizing method because δ_k is a lower bound on the error δ^* and δ_k is guaranteed to increase from iteration k to iteration $k+1$; however, $\|E_k\|$, an upper bound on the error, is not guaranteed to decrease.

3.1.3. Linear Programming

An optimization method known as linear programming [3.19–20] can be used to solve the filter design problem that is described in the preceding section. The most general linear programming problem can be formulated as follows:

$$\text{minimize } z = C_1^T X_1 + C_2^T X_2 + d \tag{3.9a}$$

subject to

$$p_1: \ A_{11} X_1 + A_{12} X_2 \geqq B_1 , \tag{3.9b}$$

$$p_2: \ A_{21} X_1 + A_{22} X_2 = B_2 , \tag{3.9c}$$

$$n_1: \ X_1 \geqq 0 , \tag{3.9d}$$

$$n_2: \ X_2 \text{ free,} \tag{3.9e}$$

where p_1 is the number of inequality constraints, p_2 is the number of equality constraints, n_1 is the number of normal variables, and n_2 is the number of free (unrestricted) variables. The objective function z is made up of a constant cost term d and a linear function of the variables.

The vectors C_1 and C_2 (C_1^T and C_2^T are the transpose of C_1 and C_2, respectively) are referred to collectively as the cost vector, the vectors B_1 and B_2 as the constraint vector, the vectors X_1 and X_2 as the program vector, and the matrices A_{11}, A_{12}, A_{21}, and A_{22} as the coefficient matrix [3.20].

This linear programming problem is called the primal problem Associated with each primal problem is another linear programming problem called the dual problem which can be formulated as follows:

$$\text{minimize } z' = -B_1^T U_1 - B_2^T U_2 - d \tag{3.10a}$$

subject to

$$p_1: \ U_1 \geqq 0, \tag{3.10b}$$

$$p_2: \ U_2 \text{ free,} \tag{3.10c}$$

$$n_1: \ -A_{11}^T U_1 - A_{21}^T U_2 \geqq -C_1, \tag{3.10d}$$

$$n_2: \ -A_{12}^T U_1 - A_{22}^T U_2 = -C_2. \tag{3.10e}$$

There is one normal variable in the dual problem for each inequality constraint in the primal problem, one free variable for each equality constraint, one inequality constraint for each normal variable and one equality constraint for each free variable [3.20].

There are many relationships between the primal problem and the dual problem [3.20]. Two of these relationships are important here. First, it can easily be shown that the dual of the dual problem is the primal problem. Second, it can be shown that if either the primal or the dual problem has an optimal solution then so does the other and

$$z_{opt} + z'_{opt} = 0.$$

Thus the duality relationship provides two ways of formulating any linear optimization problem. It is frequently more efficient to solve one rather than the other, as will be seen shortly.

The algorithm used to solve linear programming problems is called the simplex algorithm [3.19, 20]. Most of the details of this algorithm are unimportant here. However, it should be noted that this algorithm produces one of three mutually exclusive results: first, the optimal solution (if the problem has a finite optimal solution); second, an indication that the problem is feasible but the optimal solution is unbounded; or third, an indication that the problem is infeasible. The problem is feasible if all of the constraints can be simultaneously satisfied for some (not necessarily optimal) choice of the variables. The problem is infeasible if all of the constraints cannot be simultaneously satisfied for any choice of the variables. In the process of arriving at the result, additional variables called slack variables are added to the problem. In the case of the primal problem $p_1 + p_2$ variables are added while in the case of the dual problem $n_1 + n_2$ variables are added so that in both cases the total number of variables is $n_1 + n_2 + p_1 + p_2$. Details concerning these variables are not important. However, if the problem has a finite optimal solution, the number of nonzero variables in the optimal program vector is $p_1 + p_2$ for the primal problem and $n_1 + n_2$ for the dual problem. These nonzero variables are called basis variables and they are an important part of an algorithm (described in Subsection 3.2.4) for designing nonrecursive digital filters.

The goal in solving the filter design problem is to choose unit sample response coefficients for (3.5) through (3.7) so that δ is minimized in the equations

$$H(\mu, v) \leq F(\mu, v) + \delta$$
$$H(\mu, v) \geq F(\mu, v) - \delta,$$

where $F(\mu, v)$ is the desired frequency response, and μ and v range over some subset of the region $-\pi \leq \mu \leq \pi$, $-\pi \leq v \leq \pi$. In some cases there are infinitely many points (μ, v) over which the best approximation is desired. In order to apply linear programming to such problems it is necessary to select some finite subset of the infinite set. This process of choosing a finite subset is called discretization and the errors introduced by this are called discretization errors. It can be shown [3.6] that these errors diminish as the finite subset becomes more and more dense in the infinite set. The effect of discretization on the filter design problem is discussed further in Subsection 3.2.5.

Assuming that some finite set of p_1 points has been picked and that $H(\mu, v)$ is given by (3.5) then the filter design problem can be formulated as a primal linear programming as follows, see (3.9),

minimize $z = X_1$

subject to

$2p_1:\ A_{11} X_1 + A_{12} X_2 \geq B_1$

$n_1:\ X_1 \geq 0$

$n_2:\ X_2$ free,

where

$$X_1 = \delta$$
$$X_2^T = [h(0, 0), h(1, 0), ..., h(N, 0), h(-N, 1), ..., h(N, 1), ...,$$
$$\cdot h(-N, N), ..., h(N, N)]$$
$$B_1^T = [F(\mu_1, v_1), -F(\mu_1, v_1), ..., F(\mu_{p_1}, v_{p_1}), -F(\mu_{p_1}, v_{p_1})].$$

A_{11} is a $2p_1 \times 1$ matrix all of whose elements are unity, and the matrix A_{12} is given in Fig. 3.2. These are no equality constraints, so $p_2 = 0$. In this case, $n_1 = 1$ and n_2 equals the number of independent unit sample response coefficients. The formulation for (3.6) and (3.7) is exactly analogous.

In the above formulation of the filter design problem as a primal linear programming problem there are $2N^2 + 2N + 2$ variables and $2p_1$ constraints. In most filter design problems there are more constraints than variables. However, the solution to a linear programming problem is usually achieved more efficiently if the number of variables is greater than the number of constraints. This can be achieved by formulating the problem as the dual of the above (primal) problem, see (3.10),

$$\text{minimize } z' = -B_1^T U_1$$

subject to

$$2p_1: \ U_1 \geqq 0$$
$$n_1: \ -A_{11}^T U_1 \geqq -1$$
$$n_2: \ -A_{12}^T U_1 = 0.$$

As indicated previously, the optimal program vector for a dual linear programming problem contains $n_1 + n_2$ nonzero elements. Consequently, all but these $n_1 + n_2$ columns of the constraint matrix for the dual problem could be thrown out without affecting the solution. Furthermore, it turns out that the constraints in the primal problem that correspond to these $n_1 + n_2$ columns are satisfied with equality [i.e., $|F(\mu, v) - H(\mu, v)| = \delta$]. Because all of the $n_1 + n_2$ nonzero elements of the program vector are positive and because the absolute value of the error at each of the constraint points equals δ, the equality constraints for the dual problem (with $U_1 = U_1$ opt) constitute an explicit statement of the fact that zero is in the convex hull of the set $\{r(\mu, v) [g_1(\mu, v), ..., g_n(\mu, v)]: |r(\mu, v)| = \|r(\mu, v)\|\}$ as expressed in the characterization theorem (Subsection 3.1.2).

$$
\begin{bmatrix}
1 & 2\cos\mu_1\cdots & 2\cos N\mu_1 & 2\cos(-N\mu_1+\nu_1)\cdots & 2\cos(N\mu_1+\nu_1)\cdots & \cdots & 2\cos(-N\mu_1+N\nu_1)\cdots & 2\cos(N\mu_1+N\nu_1) \\
-1 & -2\cos\mu_1\cdots & -2\cos N\mu_1 & -2\cos(-N\mu_1+\nu_1)\cdots & -2\cos(N\mu_1+\nu_1)\cdots & \cdots & -2\cos(-N\mu_1+N\nu_1)\cdots & -2\cos(N\mu_1+N\nu_1) \\
\cdot & \cdot & \cdot & \cdot & \cdot & & \cdot & \cdot \\
\cdot & \cdot & \cdot & \cdot & \cdot & & \cdot & \cdot \\
\cdot & \cdot & \cdot & \cdot & \cdot & & \cdot & \cdot \\
1 & 2\cos\mu_{p_1}\cdots & 2\cos N\mu_{p_1} & 2\cos(-N\mu_{p_1}+\nu_{p_1})\cdots & 2\cos(N\mu_{p_1}+\nu_{p_1})\cdots & \cdots & 2\cos(-N\mu_{p_1}+N\nu_{p_1})\cdots & 2\cos(N\mu_{p_1}+N\nu_{p_1}) \\
-1 & -2\cos\mu_{p_1}\cdots & -2\cos N\mu_{p_1} & -2\cos(-N\mu_{p_1}+\nu_{p_1})\cdots & -2\cos(N\mu_{p_1}+\nu_{p_1})\cdots & \cdots & -2\cos(-N\mu_{p_1}+N\nu_{p_1})\cdots & -2\cos(N\mu_{p_1}+N\nu_{p_1})
\end{bmatrix}
$$

Fig. 3.2. Constraint submatrix A_{12}

The linear programming software package used by the author was an IBM programming system called Mathematical Programming System/ 360 (program number 360 A-CO-14X) which was run on an IBM 360/67 computer. This system utilizes the revised simplex algorithm, allows the user to add his own Fortran programs to the system, and incorporates features which capitalize on the structure of the System/360 computer to produce fast and accurate solutions to linear programming problems.

3.2. Algorithms

3.2.1. Window Functions

As indicated in Subsection 3.1.2, the unit sample responses for many types of ideal two-dimensional filters are of infinite extent. As a result, it is impossible to realize these filters exactly using nonrecursive techniques. One way to produce a nonrecursive approximation to the desired frequency response is to truncate the unit sample response to the desired size. This can be viewed as multiplying the unit sample response by a window function that is unity over some finite area and zero elsewhere. In general, however, a better approximation can be obtained if the window function is not restricted to be unity inside the nonzero area.

The problem of determining good two-dimensional window functions has been studied by HUANG [3.21] who proved the following result. If a symmetrical one-dimensional window $w(x)$ (where x is a continuous variable) and a two-dimensional window $w_2(x, y)$ are related by

$$w_2(x, y) = w(\sqrt{x^2 + y^2}),$$

then their Fourier transforms $W(\mu)$ and $W_2(\mu, v)$ satisfy

$$\frac{1}{2\pi} W_2(\mu, v) \otimes H_2(\mu, v) = W(\mu) \otimes H(\mu),$$

where

$$H(\mu) = \begin{cases} 1 & \text{for } \mu \geq 0 \\ 0 & \text{for } \mu < 0 \end{cases}$$

$$H_2(\mu, v) = \begin{cases} 1 & \text{for } \mu \geq 0 \text{ and all } v \\ 0 & \text{for } \mu < 0 \text{ and all } v \end{cases}$$

and $\otimes$ denotes convolution.

Since this result deals with functions of a continuous variable instead of two discrete variables, it cannot be applied without qualification to the design of two-dimensional digital filters. The first thing that must be considered is the effect of forming a continuous two-dimensional window function from a continuous one-dimensional window function by using the relation $w_2(x, y) = w(\sqrt{x^2 + y^2})$. This obviously rotates the one-dimensional function about the origin in the xy-plane. However, it does not simply rotate the frequency response of the one-dimensional window about the origin in the μv-plane. The two-dimensional frequency response, which is circularly symmetrical, is given by

$$W(\mu, v) = W(\sqrt{\mu^2 + v^2}) = W(\varrho) = 2\pi \int\limits_0^\infty w(r)\, r J_0(\varrho r)\, dr ,$$

where $r = \sqrt{x^2 + y^2}$, and J_0 is the Bessel function of the first kind, order zero. This special case of the Fourier transform for circularly symmetrical functions is called the Hankel transform [3.22]. As an example, consider the one-dimensional rectangular window

$$w(t) = \begin{cases} 1 & \text{for } |t| \leq a \\ 0 & \text{otherwise} \end{cases}$$

whose frequency response is given by

$$W(\mu) = \frac{2 \sin a\mu}{\mu} .$$

Rotating this function about the origin in the xy-plane produces the function

$$W_2(x, y) = \begin{cases} 1 & \text{for } \sqrt{x^2 + y^2} \leq a \\ 0 & \text{otherwise} \end{cases}$$

whose frequency response is given by

$$W_2(\mu, v) = \frac{2\pi a J_1(a\sqrt{\mu^2 + v^2})}{\sqrt{\mu^2 + v^2}} .$$

This is similar to but certainly not equal to the one-dimensional frequency response rotated about the origin in the μv-plane.

The next thing that must be considered is the effect of sampling the two-dimensional window function. This can be viewed as multiplying the window by an array of impulses which is equivalent to convolving the frequency response of the window with another array of impulses. Since the frequency response of the window function cannot be band limited, a certain amount of aliasing will occur. Because of this effect, the window function must be sampled on a reasonably dense grid. As a result, this technique cannot be used to obtain digital window functions containing a small number of samples.

The last thing that must be considered is the effect of convolving the frequency response of the window function with the actual frequency response of the ideal filter instead of the step function $H_2(\mu, v)$ defined above. A good window function has a frequency response that is an approximation to an impulse, i.e., the frequency response has a narrow central lobe and side lobes whose amplitudes are considerably less than the amplitude of the central lobe. If, just to consider a specific example, this technique is to be applied to the design of a low-pass filter with an ideal frequency response given by

$$H(\mu, v) = \begin{cases} 1 & \text{for } \sqrt{\mu^2 + v^2} \leq R \\ 0 & \text{otherwise}, \end{cases}$$

then the width of the central lobe of the frequency response of the window function must be much smaller than R. If this is so then the periodic convolution of the frequency response of the window function with $H(\mu, v)$ will be approximately equal to the linear convolution stated in HUANG's result. With these qualifications if $w(x)$ is a good one-dimensional window function then $w_2(x, y) = w(\sqrt{x^2 + y^2})$ is a good circularly symmetrical two-dimensional window function.

In order to use this technique to design filters it is necessary to have the unit sample response for both the ideal filter and the window function. In most cases, ideal filters are specified by their frequency response rather than their unit sample response. Thus, the unit sample response must be determined from the equation

$$h(n, m) = \frac{1}{4\pi^2} \int_{-\pi}^{\pi} \int_{-\pi}^{\pi} H(\mu, v) \, e^{j(\mu n + vm)} \, d\mu . dv, \tag{3.11}$$

where $H(\mu, v)$ is the ideal frequency response. The unit sample responses for several ideal filters (low-pass, band-pass, and high-pass filters, and a differentiator) are given in Subsection 3.1.2.

The simplest two-dimensional window function is the rectangular window whose unit sample response is given by

$$w(n, m) = \begin{cases} 1 & \text{for } -N \leq n \leq N \text{ and } -M \leq m \leq M \\ 0 & \text{otherwise}. \end{cases}$$

The frequency response corresponding to this window function is

$$W(\mu, v) = \frac{\sin(2N + 1)\dfrac{\mu}{2}}{\sin \dfrac{\mu}{2}} \cdot \frac{\sin(2M + 1)\dfrac{v}{2}}{\sin \dfrac{v}{2}}.$$

The similarity to the one-dimensional case is apparent. As in the one-dimensional case, this window is not very good. Better two-dimensional windows can be obtained from known one-dimensional continuous window functions. One of the simplest of these one-dimensional windows is the Hamming window which is given by

$$w(t) = \begin{cases} 0.54 + 0.46 \cos(\pi t/\tau) & \text{for } t < \tau \\ 0 & \text{otherwise}. \end{cases}$$

KAISER'S windows [3.3] constitute a more complicated family of window functions given by

$$w(t) = \begin{cases} \dfrac{I_0[\alpha\sqrt{(\tau^2 - t^2)}]}{I_0(\alpha t)} & \text{for } |t| < \tau \\ 0 & \text{otherwise}, \end{cases}$$

where I_0 is the modified Bessel function of the first kind, order zero, and α is a constant. The width of the central lobe of the frequency response and the amplitude of the side lobes (relative to the central lobe) can be controlled by varying the product $\alpha\tau$.

It is easy to show that multiplying the unit sample response of an ideal filter by a window function corresponds to the periodic convolution of their respective frequency responses. If $w(n, m)$ and $f(n, m)$ are the unit sample responses of the window and ideal filter, respectively, then the unit sample response of the nonrecursive filter is given by

$$g(n, m) = w(n, m) \cdot f(n, m).$$

Therefore, the frequency response of the nonrecursive filter is

$$G(\mu, v) = \sum_{m=-\infty}^{\infty} \sum_{n=-\infty}^{\infty} \left[\frac{1}{4\pi^2} \int_{-\pi}^{\pi} \int_{-\pi}^{\pi} W(\alpha, \beta) \, e^{j(\alpha n + \beta m)} \, d\alpha \, d\beta \right]$$
$$\cdot f(n, m) \, e^{-j(\mu n + vm)} ,$$

where $w(n, m)$ has been replaced by the term in brackets. Interchanging the order of summation and integration and combining the two exponential terms leads to the desired result

$$G(\mu, v) = \frac{1}{4\pi^2} \int_{-\pi}^{\pi} \int_{-\pi}^{\pi} W(\alpha, \beta) \, F(\mu - \alpha, v - \beta) \, d\alpha \, d\beta .$$

In addition to the fact that the window function technique does not produce optimal filters, the technique has several other shortcomings. To begin with, it may be difficult if not impossible to determine the unit sample response of the ideal filter in closed form using (3.11). The ideal filters whose unit sample responses are given in Subsection 3.1.2 are easy to characterize analytically. The frequency response is nonzero inside of an area that is easily defined and the response is easily characterized inside that area. It is very easy to define ideal filters for which the integration in (3.11) would be extremely difficult. On the other hand, since most two-dimensional window functions are derived from one-dimensional windows, this unit sample response is usually easily expressed in closed form.

Even after the unit sample response has been expressed in closed form, the difficulties are not necessarily over. The examples given here indicate that generating samples of the window function may involve calculating complicated functions such as Bessel functions and modified Bessel functions. This is certainly not impossible but it is time consuming for large windows.

Another problem with the window function technique is that the locations of the band edges of the resulting filter are hard to control. As mentioned, the frequency response of the nonrecursive filter is the (periodic) convolution of the frequency response of the ideal filter with the frequency response of the window function. The band edges of the ideal filters are usually very sharp but this convolution has the effect of smearing out this sharp edge into a gradual transition band. Thus, the choice of the band edge locations in the ideal filter has to be made with this smearing effect in mind. In addition, it is hard to design narrow-band or wideband nonrecursive filters this way because of the smearing effect of the convolution.

There is also the problem mentioned earlier. Namely, the actual frequency response of the digital window function is an aliased version of the frequency response of the corresponding continuous two-dimensional window function. This generally means that large window functions must be used.

RABINER [3.23] has used this technique to design nonrecursive filters. The ideal filter was a circularly symmetrical low-pass filter. Two window functions were used. The first window was a circularly symmetrical zero-one window and the second was a Kaiser window. As expected, there was much less ripple in the frequency response of the filter designed with the Kaiser window than there was in the frequency response of the other nonrecursive filter.

3.2.2. Frequency Sampling

A nonrecursive digital filter is completely characterized either by its unit sample response or by an equivalent number of uniformly spaced samples of its frequency response. These samples of the frequency response are given by the discrete Fourier transform. The frequency sampling technique for designing nonrecursive filters utilizes the fact the the (continuous) frequency response of such a filter can be written as a linear function of these frequency samples. Therefore a linear optimization technique such as linear programming can be used to optimize some aspect of the frequency response. The unit sample response can then be determined by the inverse DFT relationship. Only a few of the frequency samples are usually made variables and the rest are set to the value that the ideal filter has at that frequency. More will be said about this restriction shortly.

In order to show how the frequency response of a nonrecursive filter can be written as a linear function of its frequency samples, it is helpful to introduce a version of the DFT that is slightly different from the one given in Subsection 3.1.1. It will be assumed that the unit sample response of the filter is zero outside of the area $-N \leq n \leq N$ and $-N \leq m \leq N$. Therefore, the z-transform of the unit sample response is given by

$$H(z_1, z_2) = \sum_{n=-N}^{N} \sum_{m=-N}^{N} h(n, m) z_1^{-n} z_2^{-m}. \tag{3.12}$$

The DFT, which corresponds to uniformly spaced samples of the z-transform, is given by

$$H(k, l) = \sum_{n=-N}^{N} \sum_{m=-N}^{N} h(n, m) \exp\left[-j \frac{2\pi}{Q} (nk + ml)\right],$$

where $Q = 2N + 1$, $-N \leq k \leq N$, and $-N \leq l \leq N$. It is easy to verify that in this case the inverse DFT is given by

$$h(n, m) = \frac{1}{Q^2} \sum_{k=-N}^{N} \sum_{l=-N}^{N} H(k, l) \exp\left[j \frac{2\pi}{Q}(nk + ml)\right]. \tag{3.13}$$

By combining (3.13) with (3.12) one arrives at the result

$$H(z_1, z_2) = \frac{z_1^N z_2^N (1 - z_1^{-Q})(1 - z_2^{-Q})}{Q^2} \sum_{k=-N}^{N} \sum_{l=-N}^{N}$$

$$\cdot \frac{H(k, l) \exp\left[j \frac{2\pi}{Q} N(k + l)\right]}{\left[1 - z_1^{-1} \exp\left(-j \frac{2\pi}{Q} k\right)\right]\left[1 - z_2^{-1} \exp\left(-j \frac{2\pi}{Q} l\right)\right]}.$$

The frequency response of the filter, which is obtained by letting $z_1 = \exp(j\mu)$ and $z_2 = \exp(jv)$, is given by

$$H(\mu, v) = \frac{e^{jN(\mu + v)}(1 - e^{-j\mu Q})(1 - e^{-jvQ})}{Q^2} \sum_{k=-N}^{N} \sum_{l=-N}^{N}$$

$$\cdot \frac{H(k, l) \exp\left[j \frac{2\pi N}{Q}(k + l)\right]}{\left\{1 - \exp\left[-j\left(\mu + \frac{2\pi}{Q} k\right)\right]\right\}\left\{1 - \exp\left[-j\left(v + \frac{2\pi}{Q} l\right)\right]\right\}}. \tag{3.14}$$

In order to put (3.14) into a form so that linear programming can be used, it is necessary to impose two constraints. The unit sample response and the frequency responses must be constrained to be pure real. With these two restrictions, the DFT samples satisfy the relationship

$$H(k, l) = H(-k, -l).$$

Using this relationship between the frequency samples it is possible to write (3.14) as

$$H(\mu, v) = \frac{1}{Q^2} \left\{ H(0,0)\phi_1(\mu)\phi_1(v) \right.$$

$$+ \sum_{k=1}^{N} H(k,0)\phi_1(v)[\phi_2(\mu,k) + \phi_2(\mu, -k)] \tag{3.15}$$

$$\left. + \sum_{k=-N}^{N} \sum_{l=1}^{N} H(k,l)[\phi_2(\mu,k)\phi_2(v,l) + \phi_2(\mu, -k)\phi_2(v, -l)] \right\},$$

where

$$\phi_1(x) = \frac{\sin Q \dfrac{x}{2}}{\sin \dfrac{x}{2}}$$

and

$$\phi_2(x, y) = \frac{\sin \left[Q\left(\dfrac{x}{2} + \dfrac{\pi y}{Q}\right)\right]}{\sin \left(\dfrac{x}{2} + \dfrac{\pi y}{Q}\right)}.$$

At first glance, this does not appear to be an improvement over (3.14). However, it is important to note that now all of the coefficients of the frequency samples are real and so linear programming can be applied.

It is possible to reduce the number of independent frequency samples by imposing symmetry constraints on the frequency response of the filter. In particular, if the desired frequency response is circularly symmetrical, then the frequency samples can be constrained so that

$$H(k, l) = H(|k|, |l|)$$

and

$$H(k, l) = H(l, k).$$

This situation is depicted in Fig. 3.3 for a 9×9 array of frequency samples. When using the frequency sampling technique, one generally restricts the number of variable frequency samples even further by assigning fixed values to most of them. In the case of a low-pass filter, the frequency samples in the pass band are set to unity, those in the stop band are set to zero, and only those in the transition band are allowed to be variables. An example of a 9×9 low-pass filter is shown in Fig. 3.4. If there are M frequency samples in the transition band, then (3.15) can be rewritten as

$$H(\mu, v) = H_F(\mu, v) + \sum_{k=1}^{M} T_k H_k(\mu, v), \qquad (3.16)$$

where $H_F(\mu, v)$ represents the contribution of the frequency samples that are set to fixed values, the T_k are the samples in the transition band, and

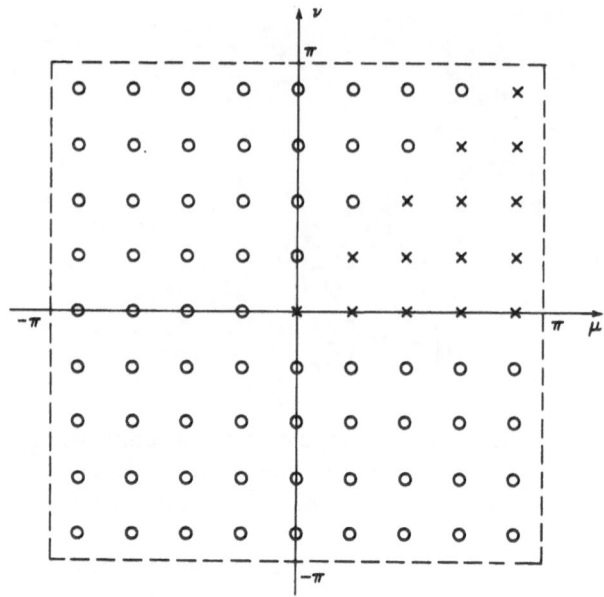

× INDEPENDENT FREQUENCY SAMPLE

O SAMPLE DETERMINED BY SYMMETRY

Fig. 3.3. Frequency domain symmetry constraints

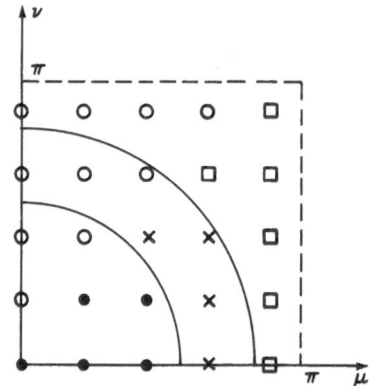

× VARIABLE FREQUENCY SAMPLES

● SAMPLES SET TO UNITY

□ SAMPLES SET TO ZERO

O SAMPLES DETERMINED BY SYMMETRY

Fig. 3.4. Frequency sampling low-pass filter specification

the summation represents the contribution to the frequency response made by the samples in the transition band. In general, the filters obtained with this method will not be best approximations to the desired frequency response because only the transition-band samples are allowed to vary.

Linear programming can be used to obtain optimal values for the frequency samples in the transition band and to minimize the ripple in the pass band and stop band if the constraint equations are of the form

$$H(\mu, v) \leqq F(\mu, v) + \delta$$
$$H(\mu, v) \geqq F(\mu, v) - \delta \, .$$

where $H(\mu, v)$ is given by (3.16), and $F(\mu, v)$ is the desired frequency response. A pair of constraint equations must be written for each constraint point in both the pass band and the stop band. No constraint equations are written for points in the transition band. The set of constraint points in the frequency plane must be dense enough to insure that the resulting continuous frequency response will be close to the frequency response at the constraint points but not so dense as to make the problem computationally unsolvable. Hu [3.24, 25] has found that a set of $8M \times 8M$ constraint points is sufficient if the size of the array of frequency samples is $M \times M$. Only one-eight of these constraint points need to be used because of the symmetry constraints discussed earlier.

After optimal values of the transition-band frequency samples have been determined, the unit sample response of the filter can be determined from (3.13), the inverse DFT relationship.

Hu [3.25] has used this method to design a variety of circularly symmetrical low-pass filters. The size of the array of frequency samples for each filter was 25×25 but because of the assumed symmetry only 91 of these were potentially independent variables. Most of these 91 frequency samples were set to either unity or zero with only from 2 to 16 variable frequency samples in the transition band. The width of the transition band was varied from $0.08\,\pi$ to $0.24\,\pi$ and the edge of the pass band was varied from zero to $0.88\,\pi$. The stop-band ripple was minimized with the pass-band ripple unconstrained. The attenuation in the stop band varied from 10 dB for narrow transition bands to about 70 dB for wide transition bands. The number of constraint points varied from about 1500 to about 5000 depending on the location and width of the transition band. Computation time ranged from 20 to 30 min. The work was done on a GE 635 computer and the linear programming package that was used was the IBM scientific subroutine APMM [3.26].

3.2.3. Straightforward Application of Linear Programming

As indicated previously, if the unit sample response and the frequency response of a nonrecursive digital filter are constrained to be real then the (continuous) frequency response can be written as a linear function of either the unit sample response coefficients or the DFT frequency samples. In terms of the unit sample response coefficients, the frequency response is given by (3.5). The relationship between the frequency response and the DFT frequency samples is given by (3.15). Using either of these relationships, it is straightforward to write constraint equations of the form

$$H(\mu, v) \leqq F(\mu, v) + \delta$$
$$H(\mu, v) \geqq F(\mu, v) - \delta \,,$$

where $F(\mu, v)$ is the desired frequency response. Linear programming can be used to minimize the ripple δ subject to the constraint matrix determined by writing a pair of constraint equations for each constraint point. The set of constraint points is made sufficiently dense to insure that the frequency response between constraint points will be close to the values at the constraint points but not so dense as to render the problem computationally unsolvable.

It is also possible to write the constraint equations in the more general form

$$H(\mu, v) \leqq F(\mu, v) + W(\mu, v)\, \delta$$
$$H(\mu, v) \geqq F(\mu, v) - W(\mu, v)\, \delta \,,$$

where $W(\mu, v)$ is a real-valued weighting function. This weighting function can be used to affect the error function of the resulting approximation in some desired way. In the case of filters (such as low-pass, high-pass, and band-pass filters) with distinct pass bands and stop bands, this weighting function can be used to obtain different amounts of ripple in each of the bands. The weighting function used to achieve this is given by

$$W(\mu, v) = \begin{cases} 1 & \text{for constraint points in the pass band} \\ \alpha & \text{for constraint points in the stop band.} \end{cases}$$

The resulting ripple in the stop band will be (at most) α times larger than the resulting ripple in the pass band.

If the unit sample response of the filter is zero outside of the region $-N \leqq n \leqq N$, $-N \leqq m \leqq N$ and the unit sample response and the frequency response are constrained to be real then the number of independent unit sample response coefficients or DFT frequency samples is equal to $2N^2 + 2N + 1$ and constraint points must fill the region $-\pi \leq \mu \leq \pi$, $0 \leq v \leq \pi$. In general, one must use all of these variables in determining a best approximation to a desired frequency response. However, as N gets large the number of variables grows rapidly and the time required to obtain the solution using linear programming grows even faster. Consequently, it is useful to impose the symmetry relations discussed earlier in order to reduce both the number of variables and the required number of constraint points. If it is assumed that $h(n, m) = h(|n|, |m|)$ then the number of variables is reduced to $N^2 + 2N + 1$ and the area which must be filled with constraint points is halved. If it is further assumed that $h(n, m) = h(m, n)$ then the number of variables is reduced to $(N^2 + 3N + 2)/2$ and the area which must be filled with constraint points is halved again. The saving in computation time resulting from imposing these symmetry relations is dramatic. More will be said about this later.

If it is only assumed that the unit sample response and the frequency response of the filter are real then the dense set of constraint points must fill the area $-\pi \leq \mu \leq \pi$, $0 \leq v \leq \pi$. In actual practice a 256×256 point FFT is used to evaluate the "continuous" frequency response of the filter and so the area actually used is $-\pi \leq \mu \leq (127/128)\,\pi$, $0 \leq v \leq (255/256)\,\pi$. The constraint points are placed on a rectangular grid with the spacing between points equal to π/KN where the unit sample responses goes from $-N$ to N in both directions, and K is an integer constant (generally chosen to be between 5 and 15). If, just to consider a specific example, a low-pass filter with circular band edges is to be designed with a pass-band edge at R_1 and a stop-band edge at R_2 then only those grid points satisfying both

$$\sqrt{\mu^2 + v^2} < R_1 - \frac{\pi}{KN} \tag{3.17a}$$

and

$$\sqrt{\mu^2 + v^2} > R_2 + \frac{\pi}{KN} \tag{3.17b}$$

are actually used as constraint points. In addition, however, constraint points are placed directly on each band edge. The total number of band-edge constraint points is the integer part of

$$Q_1 = R_1(KN + 1) + 1$$

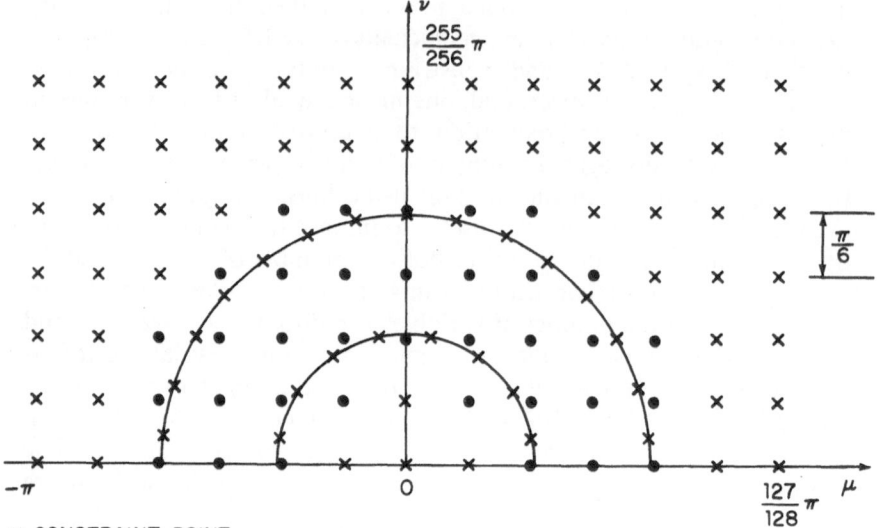

Fig. 3.5. Constraint point locations (Case 1)

for the pass-band edge and

$$Q_2 = R_2(KN + 1) + 1$$

for the stop-band edge. These points are uniformly spaced around the band edge with an angular spacing of

$$\Delta\theta_1 = \frac{\pi}{Q_1} \quad \text{and} \quad \Delta\theta_2 = \frac{\pi}{Q_2}$$

starting at an angle

$$\theta_1 = 0.5\frac{\pi}{Q_1} \quad \text{and} \quad \theta_2 = 0.5\frac{\pi}{Q_2}.$$

An example of the placement of constraint points for $N = 2$, $K = 3$, $R_1 = 1.1$ and $R_2 = 2.0$ is shown in Fig. 3.5.

If it is assumed that the unit sample response and the frequency response are real and in addition that $h(n, m) = h(|n|, |m|)$ and $h(n, m) = h(m, n)$ then the set of constraint points must fill half of the area

$0 \leq \mu \leq \pi$, $0 \leq v \leq \pi$. In actual practice, a 256×256 point FFT is used to evaluate the frequency response over the area $0 \leq \mu \leq (255/256)\,\pi$, $0 \leq v \leq (255/256)\,\pi$. As in the last case, constraint points are placed on a rectangular grid with the spacing between points equal to π/KN. In addition, considering a low-pass filter, only those grid points satisfying both (3.17a) and (3.17b) are actually used as constraint points. As before, though, constraint points are placed directly on the band edges. In this case the total number of band-edge constraint points is the integer part of

$$Q_1 = \frac{KNR_1}{4} + 1$$

for the pass-band edge and

$$Q_2 = \frac{KNR_2}{4} + 1$$

for the stop-band edge. If the value of either Q_1 or Q_2, as determined by these formulas, is less than two, it is set equal to two. These points are uniformly spaced (starting at zero) along the band edges with an angular spacing of

$$\varDelta\theta_1 = \frac{\pi}{4(Q_1 - 1)} \quad \text{and} \quad \varDelta\theta_2 = \frac{\pi}{4(Q_2 - 1)} .$$

An example of the placement of constraint points for this case and once again for $N = 2$, $K = 3$, $R_1 = 1.1$, and $R_2 = 2.0$ is shown in Fig. 3.6.

In all of the filter designs done this way by the author, the frequency response was expressed as a linear function of the unit sample response coefficients. Hu [3.25], however, has done some examples with the frequency response expressed as a linear function of the DFT frequency samples, see (3.15). The dense set of constraint points was the same one that was used for the designs using the frequency sampling technique. If the size of the array of frequency samples is $M \times M$ then the constraint points are chosen from a grid of $8M \times 8M$ points. Only those points in half of one quadrant are used as constraint points, and the grid points in the transition band are ignored.

After the number of constraint points and their locations have been chosen, the actual optimization can be carried out. The details of this, however, depend strongly on the particular linear programming package being used and so these details will not be described here. The linear

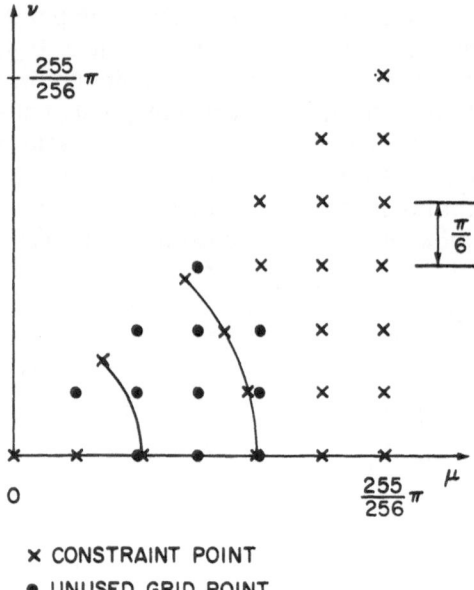

x CONSTRAINT POINT
• UNUSED GRID POINT

Fig. 3.6. Constraint point locations (Case 2)

programming software package used by the author was an IBM pro-programming system called Mathematical Programming System/360 (program number 360 A-CO-14X) which was run on an IBM 360/67 computer.

3.2.4. The New Algorithm

There are several methods for designing optimal one-dimensional non-recursive digital filters which are iterative in nature and which use only a very small set of points at each iteration to determine the best approximation to the desired frequency response on the interval $[0, 2\pi]$. The second algorithm of REMES [3.6], HOFSTETTER'S algorithm [3.8], and the method of PARKS and McCLELLAN [3.9] all have this property. Unfortunately, none of these methods can be extended directly to the two-dimensional case. All three methods either explicitly or implicitly re-quire that the functions used in the approximation satisfy the Haar condi-tion which is not the case for the two-dimensional filter design problem (see Subsection 3.1.2). In addition, both HOFSTETTER'S algo-rithm and the method of PARKS and McCLELLAN rely on the Lagrange interpolation formula which cannot be extended to the general two-dimensional case [3.27–29]. However, if a method similar to these could be found for the two-dimensional approximation problem then the number of constraint equations required in a solution utilizing linear programming could be greatly reduced at the expense of having to iterate through several solutions to arrive at an actual best approximation to the desired frequency response.

The search for such an iterative method for the multi-dimensional case was motivated both by the existence of the one-dimensional techniques and by the following statement. If a generalized polynomial

$$P(x) = \sum_{i=1}^{n} c_i g_i(x)$$

(see Subsection 3.1.2 for an explanation of this notation) is a best approximation to a function F on a set X, then there is a finite subset X_0 containing at most $n + 1$ points such that P is a best approximation to F on X_0 [3.6]. A nearly complete description of an iterative algorithm for the multi-dimensional case is presented at the end of Subsection 3.1.2. The remaining details are presented here.

At each iteration of the algorithm, a best approximation to the desired frequency response is found (using linear programming and expressing the frequency response as a function of the unit sample response coefficients) on a "thin" set of constraint points. It is generally the case that the maximum absolute value of the error on the thin set of constraint points is strictly less than the maximum absolute value of the error on the entire area of interest in the frequency domain and that the approximation is not a best approximation to the desired frequency response. The goal is to choose another thin set of constraint points so that when the optimal solution is found on this new set of points the maximum absolute value of the error on the new thin set will be greater than the maximum absolute value of the error on the old thin set.

In order to start this process, an initial solution on a thin set of constraint points must be determined. The locations of these initial constraint points are not crucial. The method used to determine their locations is exactly the one that was described in Subsection 3.2.3 for determining a dense set of constraint points. The constraint points are placed on a rectangular grid with the spacing between points equal to π/KN where the unit sample response goes from $-N$ to N in both directions and K is an integer constant. In this case, however, the constant K is chosen to be two rather than the larger values used when an approximation is obtained by solving only one linear programming problem. Constraint points are placed along the band edges exactly as described previously and as illustrated in Figs. 3.5 and 3.6. As before, the grid points in the transition band are not used as constraint points. Except for this initial choice of constraint points, each iteration is identical and a typical iteration will now be described.

Assuming that a solution has just been obtained on some set of constraint points, the major activities performed during each iteration can be described as follows:

1) Save the solution to the linear programming problem. The parts of the solution that are of interest are the unit sample response co-efficients, the maximum absolute value of the error, and the variables making up the basis for the dual problem (see Subsection 3.1.3). The basis variables are used in determining the next set of constraint points.

2) Evaluate the "continuous" frequency response in the appropriate area in the frequency domain. The frequency response is actually evaluated only on a 256×256 point grid using an FFT.

3) Evaluate the frequency response along the band edges (assuming as in the low-pass case that there are band edges). This calculation is also done on a dense set of points but unfortunately the FFT cannot be used.

4) Find the local maxima and minima in the error function in the area of interest in the frequency domain including the band edges. This is essential to determining the next set of constraint points. It is also undoubtedly the least straightforward part of the entire algorithm.

5) Generate various output plots. These plots allow one to follow the algorithm from one iteration to the next.

6) Generate data for and solve the next linear programming problem.

The first step in the algorithm is to save the relevant parts of the solution to the last linear programming problem. The unit sample response coefficients, the maximum absolute value of the error, and the variables making up the basis for the dual problem (see Subsection 3.1.3) constitute the information that must be saved. The solution provided by the linear programming package MPS/360 contains both the optimal program vector for the linear programming problem that was solved (i.e., for either the primal problem or the dual problem) and the optimal program vector for the dual of the linear programming problem that was solved. If the linear programming problem was formulated in primal form then the unit sample response coefficients and the maximum ab-solute value of the error are the elements of the optimal program vector. The maximum absolute value of the error is also the optimal value of the objective function. The basis variables for the dual problem are the nonzero elements of the optimal program vector for the dual of the primal problem. If the linear programming problem was formulated in dual form, then the basis variables for the dual problem are the nonzero elements of the optimal program vector. The unit sample response coefficients and the maximum absolute value of the error are the negatives of the elements of the optimal program vector for the dual of the dual problem. The maximum absolute value of the error is also the negative of the optimal value of the objective function.

The unit sample response coefficients must be saved so that the frequency response can be evaluated. The maximum absolute value of the error is saved at each iteration so that it can be compared with the value obtained on the previous iteration. If this value decreases, the algorithm is terminated. According to the theory, this value must increase from one iteration to the next; and it nearly always does. However, in three out of approximately 250 observed iterations this value decreased slightly. In all three instances, though, the algorithm was very near convergence when the absolute value of the error decreased and so this problem has been attributed to accumulated numerical inaccuracies. The basis variables for the dual problem are saved so that the corresponding constraint points can be included in the set of constraint points for the next iteration. One of the theoretical requirements for the new set of constraint points is that the system of equations

$$\Delta = F(\mu, v) - \sum_{i=1}^{n} c_i g_i(\mu, v)$$

must be limiting on this new set of points (see Subsection 3.1.2). This system of equations is limiting if all of the values of Δ are nonzero and if the absolute values of all of the Δ's cannot be simultaneously reduced for any choice of the set $\{c_1, ..., c_n\}$. It is certainly the case that all of the values of Δ are nonzero because the absolute value of each Δ equals the maximum absolute value of the error. In addition, all of the values of Δ cannot be simultaneously reduced in absolute value because the solution to the preceding linear programming problem is optimal. Since any system of equations that contains a limiting system is itself limiting, it is clear that the theoretical requirement for the set of constraint points for the next iteration will be met as long as the constraint points corresponding to the basis variables are included in the new set. At each iteration, the locations of the constraint points corresponding to the basis variables are compared to the locations of the constraint points corresponding to the basis variables for the preceding iteration. If all of these locations are unchanged, then the algorithm has converged and it is terminated. This is the usual termination procedure.

The second step in the algorithm is the evaluation of the "continuous" frequency response of the filter. A 256×256 array of samples of the frequency response is calculated using the FFT. The decision to use an array size of 256 × 256 was a compromise. On the one hand, the array must be dense enough so that the local maxima and minima in the error function can be located accurately enough [step 4) of the algorithm]. On the other hand, the array must fit conveniently into the available computer memory. This two-dimensional FFT is calculated using the one-dimensional FFT, as described in Subsection 3.1.1. The FFT algorithm that is used is a simplified version of one written by BRENNER [3.30]. The algorithm was modified to do a 256-point, one-dimensional, double-precision, forward transform. All of the FFT calculations are done in double-precision arithmetic until the final result is obtained. The final result is stored in a single-precision 256 × 256 array. This is done to minimize the effect of computation noise on the selection of local minima and maxima in the error function [step 4)].

If it is only assumed that the unit sample response and the frequency response are real then the samples of the frequency response are uniformly spaced over the area $-\pi \leq \mu \leq (127/128)\pi$, $0 \leq \nu \leq (255/256)\pi$. In this case the unit sample response coefficients satisfy the symmetry relation $h(n, m) = h(-n, -m)$. Because of this symmetry relation, it is only necessary to calculate $N + 1$ (one-dimensional) column transforms. The other N column transforms can be determined by symmetry. The 256 samples of the frequency response in the range $0 \leq \nu \leq (255/256)\pi$ can be determined as the first half of a 512-point transform. This 512-point transform can be calculated as the transform of two 256-point sequences by applying the fundamental notion behind the decimation-in-time algorithm for the FFT. However, since both of these sequences are real they can be combined, their transforms can be computed simultaneously, and the results can be separated after the calculation by using even-odd separation. After the column transforms have been calculated, it is necessary to do the row transforms. Each row is (in general) complex. However, since it is known that the result of each row transform is real, pairs of rows can be combined, transformed together, and the results can be separated after the calculation. Thus, the total number of one-dimensional, 256-point transforms that must be computed is $129 + N$.

If it is assumed that the unit sample response and the frequency response are real and in addition that $h(n, m) = h(|n|, |m|)$ and $h(n, m) = h(m, n)$ then the samples of the frequency response are uniformly spaced in the area $0 \leq \mu \leq (255/256)\pi$, $0 \leq \nu \leq (255/256)\pi$. In this case the column transforms are computed exactly as in the previous case. A small savings could have been realized at the expense of program complexity by taking advantage of the fact that each column is now not only real but also even and so the transform of each column is also real and even. In order to calculate the row transforms efficiently, however, it is necessary to take advantage of the fact that each row is also real and even. Once again, the desired samples of the frequency response correspond to the first half of a 512-point transform. However, there is a method [3.31] for computing the transform of two real, even, 512-point sequences with one 256-point transform. Therefore, the total number of one-dimensional 256-point transforms that must be computed is again equal to $129 + N$.

The third step in the algorithm is the evaluation of the frequency response of the filter along the band edges. Many filters (such as low-pass, high-pass and band-pass filters) have obvious band edges. However, there are filters, such as the differentiator mentioned earlier, which have no band edges and in these cases this step of the algorithm is bypassed. In general, the shape of a band edge can be arbitrary but all of the (low-pass) filters designed with this algorithm have circular band edges. Because of this, the FFT algorithm cannot be used to perform this calculation. Instead, the expression for the frequency response in terms of the unit sample response coefficients must be evaluated directly. This does not cause any great loss of efficiency because only a small number of samples of the frequency response are evaluated.

If it is only assumed that the unit sample response and the frequency response are real then the frequency response must be evaluated around a circle arc of 180°. If the pass-band edge of the filter is at a radius of R_1 and the stop-band edge is at a radius of R_2 then the total number of samples of the frequency response that are calculated is

$$Q_1 = 2 \text{ (Integer Part of } [64 \, R_1]) + 1$$

for the pass-band edge and

$$Q_2 = 2 \text{ (Integer Part of } [64 \, R_2]) + 1$$

for the stop-band edge. These samples are uniformly spaced around the band edges with an angular spacing of

$$\Delta\theta_1 = \frac{\pi}{(Q_1 - 1)} \quad \text{and} \quad \Delta\theta_2 = \frac{\pi}{(Q_2 - 1)}.$$

Since Q_1 and Q_2 are guaranteed to be odd, a sample is always calculated at 0°, 90°, and 180°.

If it is assumed that the unit sample response and the frequency response are real and in addition that $h(n, m) = h(|n|, |m|)$ and $h(n, m) = h(m, n)$ then the frequency response must be evaluated around a circle arc of only 45°. In this case the total number of samples of the frequency response that are calculated is

$$Q_1 = 2 \text{ (Integer Part of } [16 R_1]) + 1$$

for the pass-band edge and

$$Q_2 = 2 \text{ (Integer Part of } [16 R_2]) + 1$$

for the stop-band edge. The angular spacing between samples is

$$\Delta\theta_1 = \frac{\pi}{4(Q_1 - 1)} \quad \text{and} \quad \Delta\theta_2 = \frac{\pi}{4(Q_2 - 1)}.$$

The first sample is always at $0°$ and the last is always at $45°$.

The fourth step in the algorithm is the determination of the local maxima and minima in the error function. A point at which the current error function attains a local maximum or minimum is included in the set of constraint points for the next iteration if the absolute value of the error at this point is greater than or equal to the maximum absolute value of the error on the current set of constraint points. This guarantees that the two conditions

$$\max_{X_{k+1}} |P_k - F| = \|E_k\|$$

and

$$\min_{X_{k+1}} |P_k - F| \geq \delta_k$$

stated in the theoretical development of the algorithm (Subsection 3.1.2) will be satisfied. The locations of these local maxima and minima and the constraint points (mentioned earlier) corresponding to the basis variables for the dual problem make up the complete set of constraint points for the next iteration. At first glance, it would appear that finding these local maxima and minima should be a fairly easy job. As it turns out, however, this is the least straightforward part of the entire algorithm.

To begin with, it was found that if the absolute value of the error at each local maximum or minimum was compared directly with δ_k (i.e., with the maximum absolute value of the error on the current set of constraint points), then it was possible to incorrectly exclude some points for which the computed absolute value of the error was just slightly less than δ_k. To remedy this, the absolute value of the error at each point is compared with $\delta_T = 0.999 \, \delta_k$ rather than with δ_k itself. This has been found to work quite satisfactorily.

Secondly, it was found that a fairly complex rule was needed to select maxima and minima. In the simplest case, a point along an edge can be considered to be a local maximum or minimum if the product of the differences between this point and its two adjacent neighbors is positive. In addition, a point somewhere in the middle of the frequency plane can be considered to be a local maximum or minimum if the differences between this point and its eight adjacent neighbors all have the same sign. Unfortunately, it was found that this simple rule did not work well because it is very sensitive to noise generated in the calculation of the frequency response. Many modifications to this rule were tried. The one that will be described now was found to produce the best results.

In the case where the frequency response and the unit sample response are assumed only to be real, the search for local maxima and minima goes as follows. First, the corners of the frequency plane, $(\mu, v) = (-\pi, 0)$ and $(\mu, v) = [-\pi, (255/256)\pi)$, are checked. If the absolute value of the error at either of these points is greater than or equal to δ_T, the point is considered to be a local maximum or minimum. Next, the edge $\mu = -\pi$, $0 < v < (255/256)\pi$ is checked. The differences between each point along the edge and its two adjacent neighbors are computed. A difference is set to zero if the absolute value of the difference is less than $\delta_k \times 10^{-5}$. If the product of the differences is not positive, the point is not a local maximum and minimum. If this product is positive then a) the point is accepted as a local maximum if the error at the point is positive and both differences are negative, or b) it is accepted as a local minimum if the error is negative and both differences are positive. If the error and both differences all have the same sign, the point is rejected. If a string of points is such that all of the inter-point differences are zero, then this same rule is applied to the nearest nonzero differences. However, now the midpoint of the string of points is chosen as the location of the maximum or minimum if a point is to be chosen at all. After checking this edge, the edge $-\pi \leq \mu \leq (127/128)\pi$, $v = 0$ and the edge $-\pi \leq \mu \leq (127/128)\pi$, $v = (255/256)\pi$ are checked using this same rule. The only difference here is that the points $(\mu, v) = [(127/128)\pi, 0]$ and $(\mu, v) = [(127/128)\pi, (255/256)\pi]$ are not considered to be corner points because the actual corner points are available on the edge $\mu = \pi$, $0 \leq v \leq (255/256)\pi$ which is identical to the edge $\mu = -\pi$, $0 \leq v \leq (255/256)\pi$. Points along the edge at $\mu = \pi$ need not be included in the new set of constraint points, however, because these constraints would be identical to the constraints for points along the edge at $\mu = -\pi$. After the edges and corners have been checked, it is necessary to check the interior area $-\pi < \mu \leq (127/128)\pi$, $0 < v < (255/256)\pi$. The rule used here is the following. If the nonzero differences between a point and its eight adjacent neighbors do not all have the same sign, then the point is rejected. If all of the signs are positive and the error at the point is negative, then the point is considered to be a local minimum. If all of the signs are negative and the error at the point is positive, then the point is considered to be a local maximum. The points along the line $\mu = (127/128)\,\pi$, $0 < \nu < (255/256)\,\pi$ are included in this calculation and are not considered to be an edge since the true edge at $\mu = \pi$ is available. The rule for selecting local maxima and minima along the band edges is the same as the rule for the edge $\mu = -\pi$, $0 \leq \nu \leq (255/256)\,\pi$. In this case, however, no points are treated like corner points since pairs of neighbors are available either explicitly or implicitly for each points.

In the case where the frequency response and the unit sample response are assumed to be real and where, in addition, it is assumed that $h(n, m) = h(|n|, |m|)$ and $h(n, m) = h(m, n)$, the rules for finding local maxima and minima at corners, along edges, and in the interior area of the frequency response are exactly the same as those just described. The 256×256 array of samples of the frequency response fills the area $0 \leq \mu \leq (255/256)\,\pi$, $0 \leq \nu \leq (255/256)\,\pi$. The three points $(\mu, \nu) = (0, 0)$, $(\mu, \nu) = [(255/256)\,\pi, 0]$, and $(\mu, \nu) = [(255/256)\,\pi, (255/256)\,\pi]$ are considered to be corners. The edges are $0 < \mu < (255/256)\,\pi$, $\nu = 0$ and $\mu = (255/256)\,\pi$, $0 \leq \nu \leq (255/256)\,\pi$ and the pass-band and stop-band edges, if present. The interior points occupy the triangular shaped area $0 < \mu < (255/256)\,\pi$, $0 < \nu \leq \mu$.

If local minima and maxima occurred only at isolated points in the frequency domain and if the variation of the error function along an edge could be guaranteed to be as large as the maximum absolute value of the error on the current set of constraint points, then the problem of finding local maxima and minima would be much easier. These two conditions are always satisfied in the one-dimensional case. This is another reason why the two-dimensional filter design problem is inherently harder than the one-dimensional filter design problem. In the two-dimensional case, local minima and maxima can occur not only at isolated points but also along ridges. A ridge exists in a function if the amplitude of the function is constant along a line. If a ridge exists in the error function for an iteration, then too many points along the ridge get picked as local maxima (or minima) and therefore as constraint points. The consequences of having a nearly flat edge are the same. If an edge is nearly flat, the signal-to-noise ratio is just too small to allow accurate selection of minima and maxima. Both of these conditions are observed quite frequently in the design of low-pass filters but they have not occurred in any of the differentiators that have been designed. From a theoretical point of view, these extra constraint points present no problem. However, from a practical point of view, these extra constraint points are troublesome because the time required to set up and solve a linear programming problem is proportional to the number of constraints.

The fifth step in the algorithm is the generation of output plots. The plots that have been found useful can be placed into the following four categories:

1) A diagram showing separately the locations of the contraint points corresponding to the basis variables for the dual problem and the locations of the constraint points corresponding to the local maxima and minima.
2) A contour map of either the frequency response of the filter or the error function.
3) Plots of the frequency response along the band edges.
4) Perspective views of either the frequency response or the error function or both.

Certain plots are frequently generated in other steps of the algorithm; but, for convenience, they have all been grouped into one step. These plots proved to be essential to the successful development of the algorithm. They are still a useful part of the output of the final iteration of a design.

The sixth step in the algorithm is the generation of data for and the solution of the linear programming problem. An example of how the filter design problem can be formulated as a linear programming problem is presented in Subsection 3.1.3. However, the details required to implement this general formulation as a specific computer program depend strongly on the particular linear programming software package being used and so the details of this process are not presented here. The method by which the simplex algorithm arrives at a solution to a linear programming problem is described in the literature [3.19, 20] and is unimportant for the purposes of this discussion.

A rule for termination is an essential part of every algorithm. Two conditions for terminating this filter design algorithm have already been stated. The algorithm is terminated if, first, δ_k (the maximum absolute value of the error on the set of constraint points for iteration k) is not greater than δ_{k-1} or, secondly, the set of constraint points corresponding to the basis variables for iteration k is the same as the set for iteration $k-1$. These two conditions are sufficient but there is another very useful condition for termination. If N_k represents the maximum absolute value of the error over the entire frequency plane for iteration k, then the relative error defined by

$$E_R = \frac{N_k}{\delta_k} - 1$$

is a measure of how close the current approximation is to a best approximation. Small values of E_R indicate that the current approximation is close to a best approximation. It was pointed out in Subsection 3.1.2 that this algorithm is a maximizing method because δ_k (a lower bound on δ^*, the maximum absolute value of the error for a best approximation) is guaranteed to increase from one iteration to the next but N_k (an upper bound on δ^*) is not guaranteed to decrease. In fact, N_k usually fluctuates quite a bit. But because

$$\delta_k \leqq \delta^* \leqq N_k$$

there is little to be gained by continuing the algorithm if N_k is very close to δ_k. Values of E_R as low as 10^{-6} have been obtained by letting the algorithm run until it converges. However, quite good results can be obtained if the algorithm is stopped when E_R is on the order of 10^{-3}. The actual value used to terminate the algorithm depends entirely upon whether a best approximation is needed or whether a "good" approximation will suffice. A "good" approximation that is designed this way can always be made better simply by calculating more iterations. In addition to these three conditions, the algorithm can also be terminated after a specified number of iterations.

3.2.5. Examples and Comparisons

This subsection contains some examples of filters designed by the algorithms described in the two preceding subsections, i.e., the straightforward application of linear programming and the new algorithm. Factors affecting the performance of both techniques are described. These factors include: the effect of solving the primal rather than the dual linear programming problem, the effect of the number of constraint points, and the effect of the number of independent variables. In addition, some comments on the computational complexity and capabilities of each algorithm are presented.

As mentioned in Subsection 3.1.2, if the filter design problem is formulated as a primal linear programming problem, then the number of rows in the constraint matrix is much larger than the number of columns. But, if the same problem is formulated as a dual linear programming problem, then the number of columns is much larger than the number of rows and the solution is achieved more efficiently. Using the straightforward linear programming approach, it was found that if the primal form was used then the largest filter that could be handled consistently by the linear programming software package has a 7×7 unit sample response or 25 independent unit sample response coefficients.

Table 3.1. Comparison of primal and dual forms

Size	Primal form			Dual form		
	Rows	Columns	Time [min]	Rows	Columns	Time [min]
5 × 5	378	14	1.2	14	392	0.13
7 × 7	918	26	16.2	26	828	1.7
9 × 9	1274	42	50.8	42	1316	19.5

One filter with a 9×9 unit sample response was successfully designed. The number of constraint points that could be conveniently handled was approximately 600 for a total of about 1200 constraint equations. However, when the problem was reformulated as a dual linear programming problem, it was found that filters with 9×9 unit sample responses (41 independent coefficients) could be handled with up to several thousand constraint points. Some specific examples showing this effect are given in Table 3.1 which shows the time required for the simplex algorithm to obtain the optimal solution for problem of various sizes in primal and dual form. The total time required to design the filter includes this time plus the time required to generate input data, set up the problem, and produce the output plots. These examples were chosen because the constraint matrices for the primal and dual problems are approximately the same size. The physical problems solved in each case are not the same. It should be clear that, with the particular linear programming software package used, it is much better to formulate the filter design problem as a dual problem rather than as a primal problem.

If the filter design problem is formulated as a dual linear programming problem then the number of columns in the constraint matrix is equal to two times the number of constraint points. The number of rows is equal to the number of independent unit sample response coefficients plus one. The solution obtained is a best approximation on the set of constraint points. As the number of constraint points (and therefore the number of columns in the constraint matrix) increases, the solution gets closer to the desired solution which is a best approximation over the appropriate subset of the area $-\pi \leq \mu \leq \pi$, $-\pi \leq \nu \leq \pi$. The use of a finite set of constraint points to obtain an approximation to a function over this area is called discretization and the errors introduced by this are called discretization errors [3.6].

In order to demonstrate the effect of discretization on the resulting solution, a low-pass filter was designed four different ways. The filter has a 7×7 unit sample response and it was only assumed that the unit sample response and the frequency response were to be real so there are 25 independent unit sample response coefficients. The pass-band edge is at $\varrho = \sqrt{\mu^2 + v^2} = 1.5$ and the stop-band edge is at $\varrho = 2.5$. The method used to determine the locations of the constraint points is described in Subsection 3.2.3. The first three solutions were obtained using the straightforward linear programming approach. These are shown in Figs. 3.7 through 3.9. [Note: The vertical scale in all of the figures like Figs. 3.7 through 3.9 is linear and the data have been shifted and scaled (based on the maximum and minimum values) to fill the range from zero to one.] The new algorithm was used to obtain the fourth solution which is shown in Fig. 3.10. The initial approximation for this fourth solution was the first one of the other three solutions.

Figures 3.7 through 3.9 indicate that the choice of constraint points has a small effect on the shape of the frequency response. Larger variations in the shape of the frequency response have been observed in other examples. The frequency response shown in Fig. 3.10 is clearly one where the error function is nearly constant along the edges $-\pi \leqq \mu \leqq \pi$, $v = \pi$ and $\mu = \pm \pi$, $0 \leqq v \leqq \pi$. Situations like this one led to

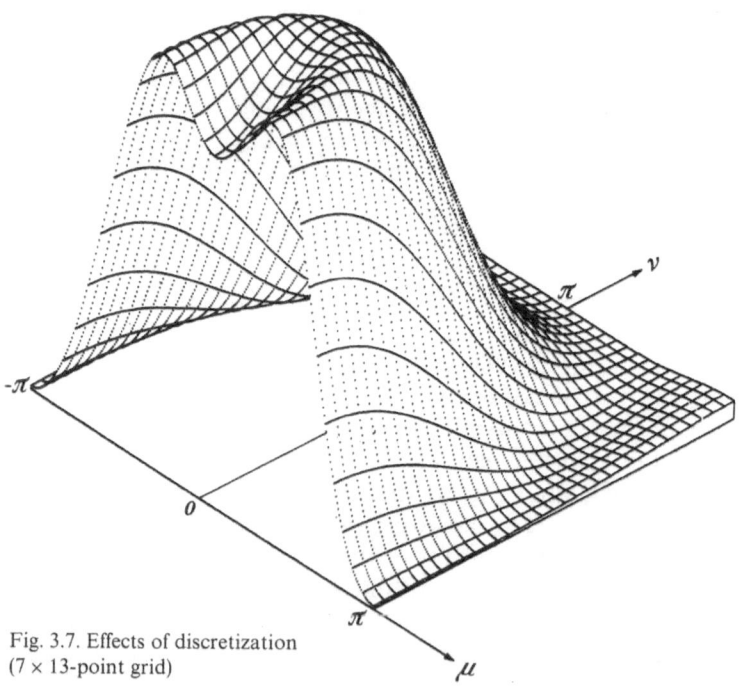

Fig. 3.7. Effects of discretization (7 × 13-point grid)

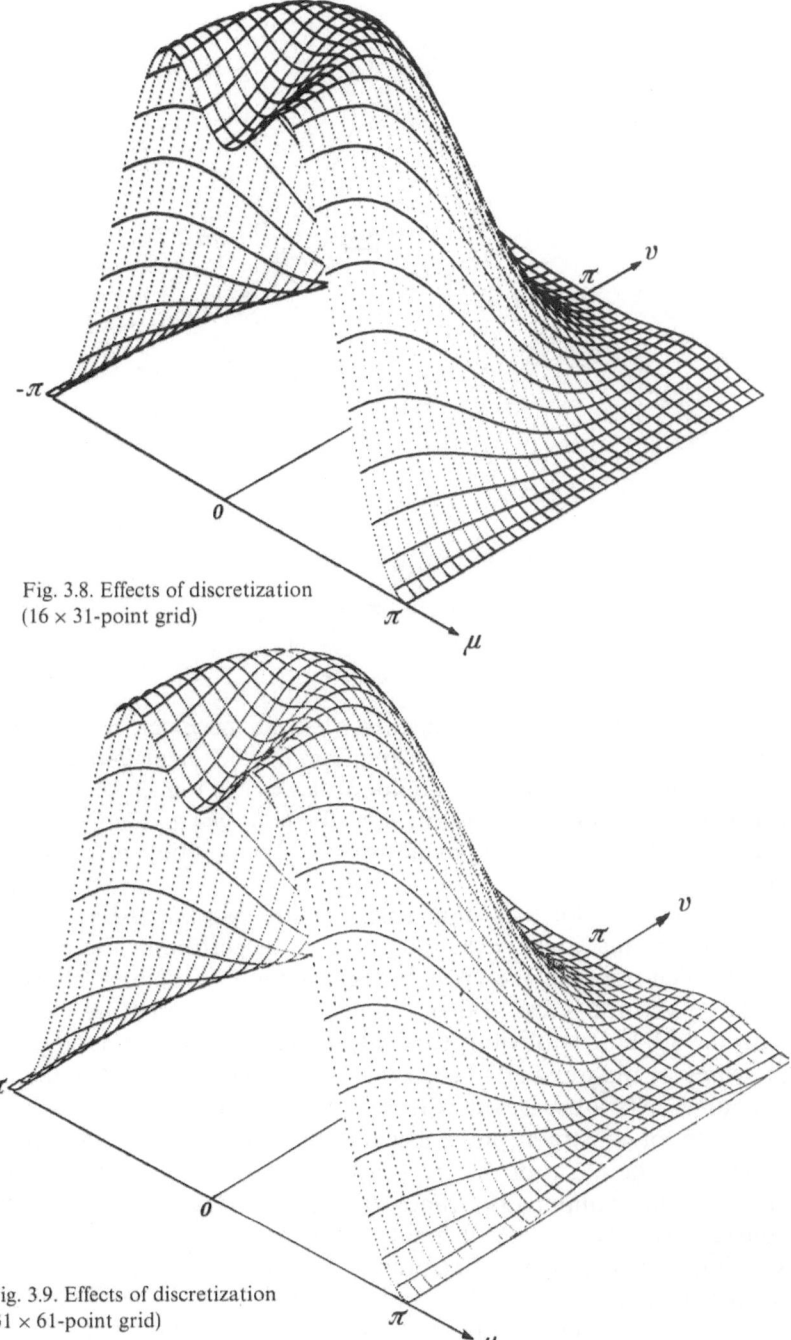

Fig. 3.8. Effects of discretization
(16 × 31-point grid)

Fig. 3.9. Effects of discretization
(31 × 61-point grid)

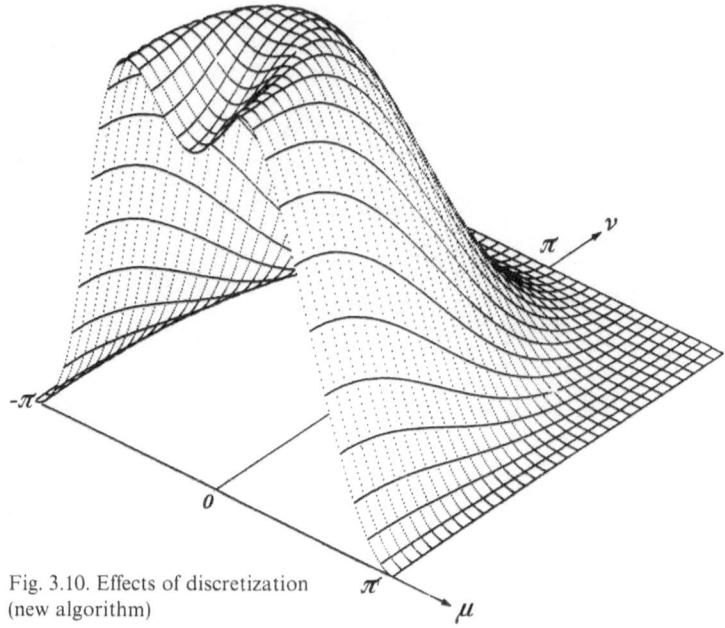

Fig. 3.10. Effects of discretization (new algorithm)

the development of the method described in the preceding subsection for finding local maxima and minima in the error function. It is interesting to note that the solution shown in Fig. 3.10 very nearly satisfies the symmetry contraints $h(n, m) = h(|n|, |m|)$ and $h(n, m) = h(m, n)$ even though no effort was made to force the solution to satisfy these constraints.

The statistics for these four solutions are summarized in Table 3.2. Delta is the maximum absolute value of the error on the constraint points. The norm of the error is the maximum absolute value of the error on the 256×256 grid of frequency response samples calculated with the FFT. It is obvious from the table that the computation time increases as the number of columns in the constraint matrix increases and that delta increases as the set of constraint points gets denser. As the set of constraint points becomes denser, the approximation gets better (i.e., the norm of the error decreases) and the relative error decreases. The computation time required for the new algorithm to converge is only 22 percent longer than the computation time for the solution with 1311 constraint points. The smallest of the 13 iterations of the new algorithm used only 47 constraint points while the largest used 79 constraint points. It is interesting to note that after only three iterations (3.61 min) of the new algorithm the relative error was

Table 3.2. Effects of discretization

	Straightforward linear programming			New algorithm
Constraint point grid size	7×13	16×31	31×61	
Band-edge constraint points	$11 + 18$	$25 + 41$	$47 + 78$	
Constraint points (total)	70	358	1311	
Delta	0.086920	0.093992	0.094595	0.094641
Norm of error	0.11700	0.096651	0.94750	0.094642
Relative error	3.46×10^{-1}	2.83×10^{-2}	1.64×10^{-3}	2.86×10^{-6}
Computation time [min]	1.34	3.62	10.95	13.32
Number of iterations				13

Relative error = norm/delta − 1.

1.28269×10^{-3} just slightly less than that achieved by the largest straightforward linear programming solution.

Another filter designed with the new algorithm is shown in Fig. 3.11. In this case the unit sample response is 9×9 (41 independent coefficients). the pass-band edge is at $\varrho = 1.0$ and the stop-band edge is at $\varrho = 1.5$. The algorithm converged in 12 iterations with delta equal to 0.115725 and with the norm of the error equal to 0.115726. The total computation time was 31.75 min.

Figure 3.12 shows the locations of the points that would have been constraint points if another iteration had been calculated. This figure illustrates two important points. First, the ridges in the frequency response cause too many local maxima and minima to be selected. There are 42 variables in the basis for this problem and so one would expect to find approximately this number of local maxima and minima. The number of local maxima and minima actually found in this case is 109. As pointed out before, this does not cause any theoretical problem but it does lengthen the total computation time. The second important point to observe is that the constraint points corresponding to the basis variables have a tendeny to occur in pairs (or sometimes triplets) in the final solution and for that matter in earlier iterations. This has been observed in nearly all of the examples that have been done. No explanation for this behavior has been discovered.

As in the last example, no effort was made to force the solution to satisfy the symmetry constraints $h(n, m) = h(|n|, |m|)$ and $h(n, m) = h(m, n)$. The goal in this case was simply to find a best approximation to the function

$$F(\mu, v) = \begin{cases} 1 & \text{for} \quad \sqrt{\mu^2 + v^2} \leq 1.0 \\ 0 & \text{for} \quad \sqrt{\mu^2 + v^2} \geq 1.5 \, . \end{cases}$$

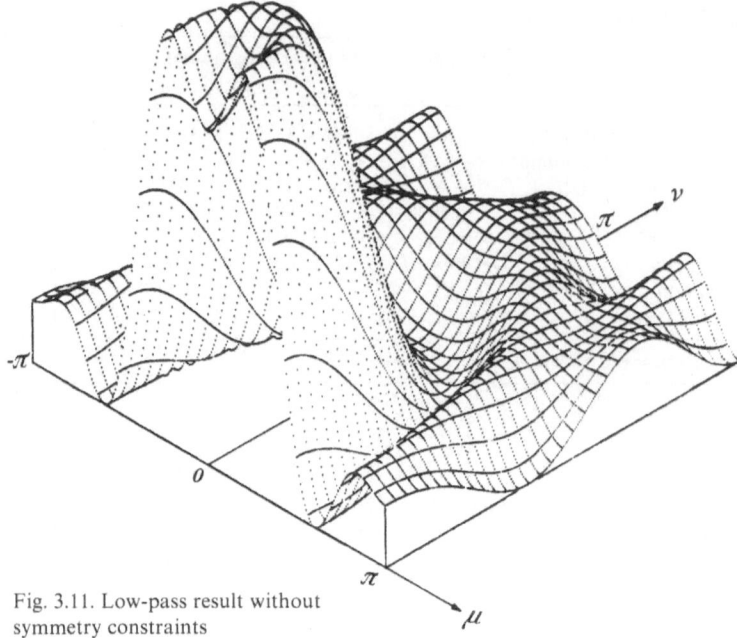

Fig. 3.11. Low-pass result without
symmetry constraints

The resulting solution, however, does very nearly satisfy the symmetry constraints. No proof that this must always be the case for (low-pass) filters like these is available but this has been observed in several other examples. If the symmetry constraints are assumed beforehand, then the solution is as shown in Fig. 3.13. For all practical purposes it is identical to Fig. 3.11. In this case there are only 15 independent unit sample response coefficients and the algorithm converged in 18 iterations with delta equal to 0.115726 and the norm of the error equal to 0.115727. The total computation time was 17.45 min. It is interesting to note that the algorithm required six more iterations but 14.3 fewer minutes of computation time to arrive at an optimal solution with the symmetry constraints imposed.

The effect of the size of the unit sample response on the resulting filter is demonstrated by the following four examples which were designed with the new algorithm. The ideal filter is the two-dimensional differentiator whose frequence response is given by

$$F(\mu, v) = \frac{1}{2\pi^2}(\mu^2 + v^2)$$

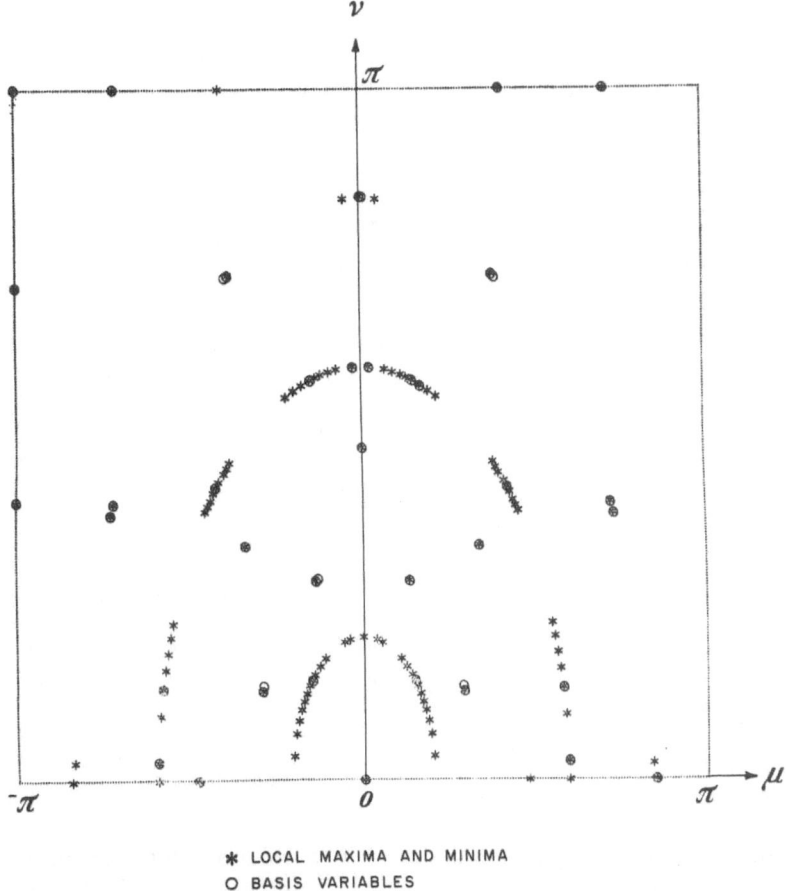

* LOCAL MAXIMA AND MINIMA
O BASIS VARIABLES

Fig. 3.12. Constraint point locations for Fig. 3.11

and the symmetry conditions $h(n, m) = h(|n|, |m|)$ and $h(n, m) = h(m, n)$ were assumed. One quadrant of the frequency response of each filter is shown in Figs. 3.14 through 3.17 and the statistics for each design are summarized in Table 3.3. In each case, the algorithm was allowed to run until it converged. The values of the relative error, which are on the order of 10^{-4}, are one or two orders of magnitude greater than the relative errors generally obtained when designing low-pass filters. No explanation of this is available. Nor can the fact that ridges were never observed in the error function be explained. In every iteration of each example, all of the local maxima and minima in the error function occurred at isolated points. It is interesting to note that the 13×13 example took

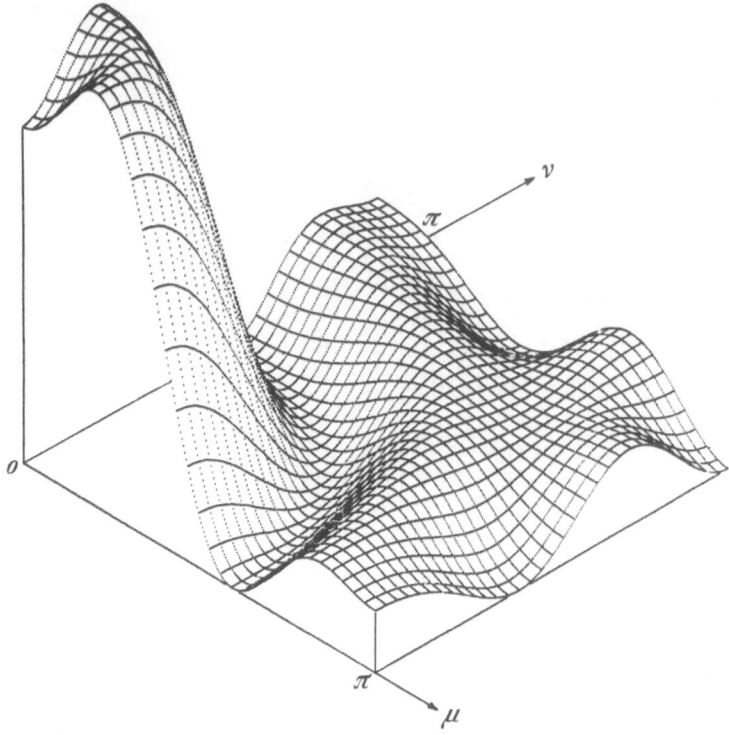

Fig. 3.13. Low-pass result with symmetry constraints

one more iteration than did the 15×15 example, but the total computation time and the time per iteration increased as the size increased. The value of delta decreased by a factor of 2.82 in going from a filter with 10 independent coefficients to one with 45 independent coefficients while the total computation time increased by a factor of 8.22 and the time per iteration increased by a factor of 2.32.

The examples just given indicate that linear programming can be used quite successfully to design two-dimensional nonrecursive digital filters. In theory, of course, the simplex algorithm can provide a solution for every linear programming problem. In practice, however, one must worry about the computational complexity of the algorithm. An indication of the computational complexity of the algorithm is given by the scheme of tolerances that must be used in the actual implementation of the algorithm because infinite-precision arithmetic is unattainable. There are basically two types of required tolerances, accuracy tolerances and significance tolerances. Accuracy tolerances allow for a certain amount of self-checking which permits early termination of the calculation if a serious loss of accuracy is detected. Significance tolerances are imposed so that numbers that are very close to zero will be set to zero in the calculations. This can also lead to early termination of the calculation if no "nonzero" elements of the constraint matrix can be found when needed in the calculation.

The simplex algorithm is itself iterative in nature and the objective function is guaranteed to decrease (or increase in the case of a maximization problem) at every iteration. Furthermore, the algorithm is guaranteed to converge to a solution in a finite number of iterations. However, because of the nature of the algorithm, the number of iterations that are actually required for a specific problem is totally unpredictable. Consequently, one problem may take longer to solve than another problem even though both problems have exactly the same number of rows and columns in the constraint matrix. In the case of filter design problems, the fact that one filter of a certain size can be designed in a specified time is no indication that any other filter of the same size, with the same number of constraint points, can also be designed.

The same comments obviously apply to the computational complexity of the new design algorithm because a linear programming problem is solved in the course of each iteration of the algorithm. The complexity is compounded by the fact that the local maxima and minima in the error function must be found at each iteration. The problems encountered in finding these points have already been discussed. The actual solution that is obtained at each iteration obviously depends on the choice of constraint points but it is very difficult to investigate this relationship. Furthermore, the number of iterations that must be computed is again unpredictable. The maximum absolute value of the error on the set of constraint points must increase from one iteration to the next but the size of the increase may be very small indeed. The algorithm cannot be terminated in this case, however, because the size of the increase may again be significant after several itera-

Table 3.3. Examples of differentiators

	Figure 3.14	Figure 3.15	Figure 3.16	Figure 3.17
Unit sample response size	7×7	13×13	15×15	17×17
Independent coefficients	10	28	36	45
Delta	0.055345	0.026919	0.022752	0.019636
Norm of error	0.055359	0.026933	0.022767	0.019649
Relative error	2.55×10^{-4}	5.09×10^{-4}	6.28×10^{-4}	7.05×10^{-4}
Computation time [min]	8.68	33.55	38.55	71.27
Number of iterations	9	24	23	32
Average time per iteration	0.964	1.398	1.675	2.227

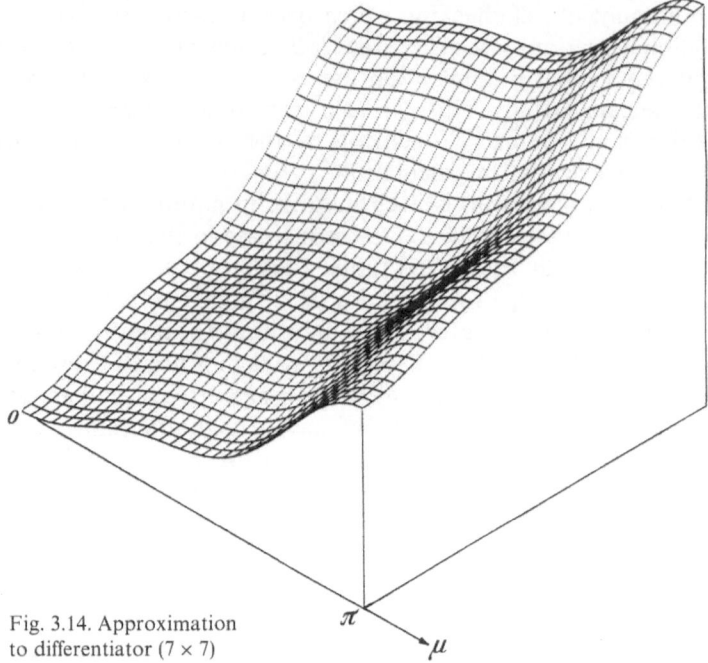

Fig. 3.14. Approximation
to differentiator (7 × 7)

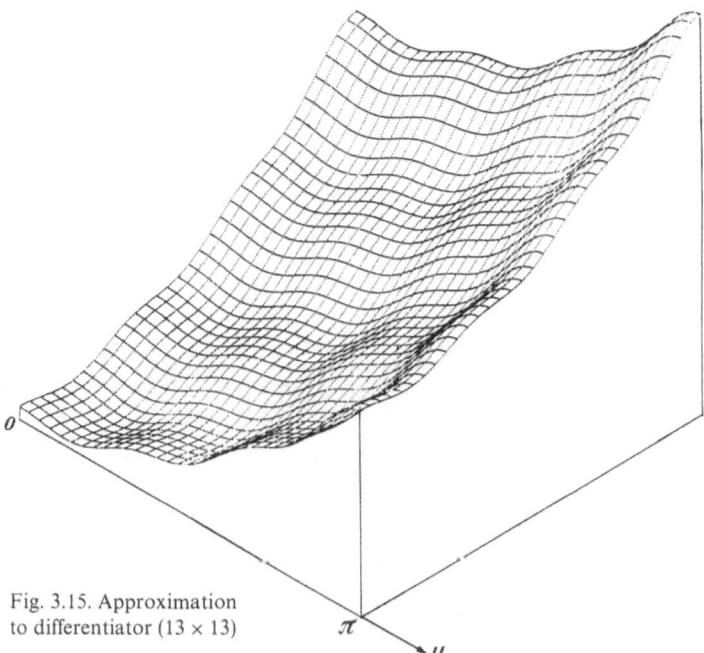

Fig. 3.15. Approximation
to differentiator (13 × 13)

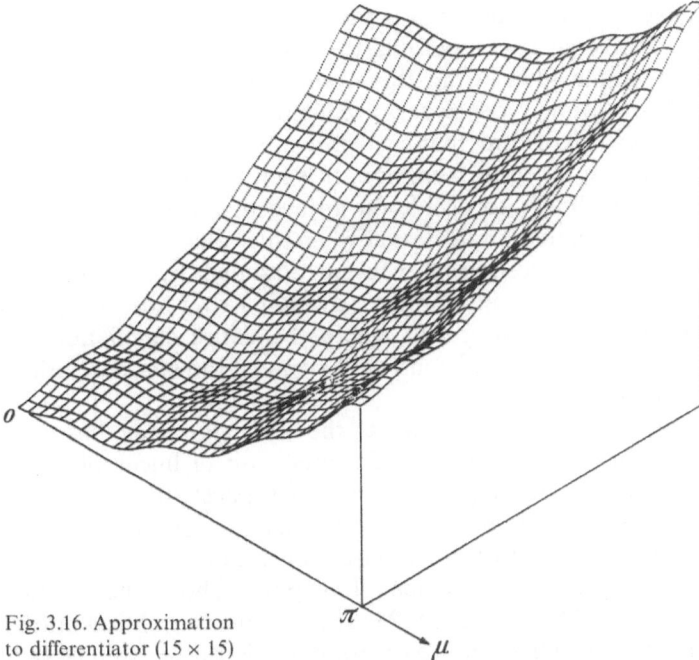

Fig. 3.16. Approximation
to differentiator (15 × 15)

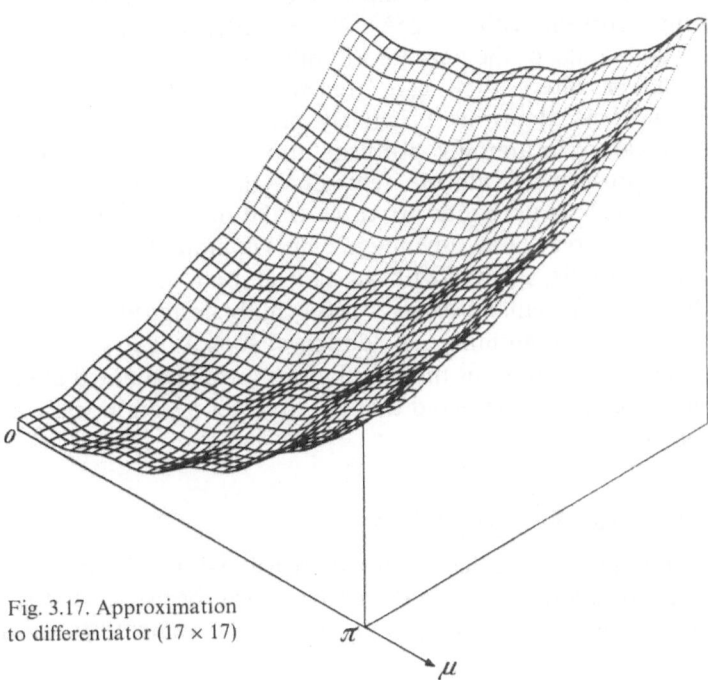

Fig. 3.17. Approximation
to differentiator (17 × 17)

tions have elapsed. Sometimes, the time required for the new algorithm to converge is comparable to or less than the time required to design the same filter using the straightforward linear programming approach on a fairly dense set of constraint points. Sometimes, however, the new algorithm requires more time to produce only a small improvement in the approximation. It is felt that the best way to utilize the new algorithm is to decide on an acceptable value of the relative error and terminate the calculation when this value is reached. If a better approximation is desired, it can be obtained by performing more iterations. This is a considerable advantage over the straightforward linear programming approach where the entire calculation must be repeated on a denser set of constraint points if the approximation is not good enough.

In addition to this advantage, the new algorithm has other advantages over the straightforward application of linear programming. To begin with, the new algorithm computes a best approximation to the desired frequency response over a 256×256-point grid in the frequency domain instead of just over the set of constraint points used in solving the linear programming problem. Two points should be mentioned in this regard. First, the size 256×256 is not a theoretical limitation; it is a practical limitation. Best approximations on denser sets can be generated by incorporating a larger FFT. The algorithm remains conceptually unchanged. Second, it would be impossible to solve a single linear programming problem with $2 \times 256 \times 256 = 131072$ constraint equations. Another advantage of the new algorithm is that it can be used to design slightly larger filters than can otherwise be designed. This is so because the new algorithm effectively provides a means of breaking a large linear programming problem into a set of small linear programming problems. Unfortunately, since the time per iteration and the total number of iterations increases as the size of the filter increases, this is not that great of an advantage unless computation time is not an issue. For example, a filter with an 11×11 unit sample response (61 independent coefficients) was never successfully designed using straightforward linear programming. The design of such a filter was attempted with the new algorithm but the calculations were terminated after a few iterations when it was observed that the time per iteration was approximately six minutes.

3.3. Summary and Conclusions

Four techniques for designing two-dimensional nonrecursive digital filters were considered. The first three techniques, the use of window functions, the frequency sampling technique, and the straightforward application of linear programming, are direct extensions of the corresponding techniques for designing one-dimensional filters. Of these three, only the last one has the potential for designing optimal filters. The fourth technique is a new algorithm that was developed specifically for the two-dimensional filter design problem. This new algorithm is superior to the straightforward application of linear programming because a) "good" approximations can generally be obtained rapidly and these approximations can always be made better simply by performing additional iterations, b) best approximations can be obtained on extremely dense sets of points, and c) the new algorithm provides a way to break up one large linear programming problem into a series of smaller problems. Unfortunately, the new algorithm is not as computationally efficient as was hoped.

One of the most important insights gained from this study is that the general two-dimensional filter design problem is considerably harder than the one-dimensional filter design problem. There are many reasons for this. Probably the most important one is that the set of functions used in the approximation does not satisfy the Haar condition (Subsection 3.1.2). As a result, it is impossible to state a general interpolation formula analogous to the one-dimensional Lagrange interpolation formula [3.27–29, 32]. Furthermore, because the set of functions does not satisfy the Haar condition, the question of the uniqueness of the best approximation becomes a very complicated issue [3.33] unlike the one-dimensional case where the best approximation is unique. Finally, the alternation theorem, one of the most important theorems relating to one-dimensional approximation, cannot be applied to the multi-dimensional case because of the fact that functions of more than one variable generally do not satisfy the Haar condition.

Another reason why the two-dimensional problem is harder than the one-dimensional problem is because one generally must deal with larger arrays of data in the two-dimensional case than in the one-dimensional case. As a result, storage requirements and computation time are generally much greater in the two-dimensional case. Consider, for example, the problem of eavaluating the frequency response of a filter. In the one-dimensional case, the rule of thumb is to evaluate the frequency response on a grid containing 20 times the number of points in the unit sample response of the filter. Thus, the frequency response of a filter whose unit sample response is 100 samples long would be evaluated at about 2000 points. Throughout the course of this work, the frequency responses of filters were evaluated on a 256×256-point grid for a total of 65536 samples of the frequency response.

Several of the one-dimensional filter design algorithms rely on peak picking, i.e., on finding local maxima and minima in the error function.

This is also an essential part of the new algorithm that was developed for designing two-dimensional filters. However, several phenomena occur exclusively in the two-dimensional case and make peak picking considerably more difficult than in the one-dimensional case. The first of these phenomena is the presence of ridges in the error function. This obviously could not occur in the one-dimensional case and, as mentioned in Subsection 3.2.4, this makes the peak picking problem much harder than it would otherwise be. The second phenomenon affecting peak picking is the presence of nearly flat edges (either band edges or the edges at $\mu = \pm \pi$ and $v = \pi$) in the error function. Finding local maxima and minima along such edges is a one-dimensional problem. However, it is not as easy as in the one-dimensional case where the ripple in the error function is at least as great as the maximum absolute value of the error on the last set of constraint points. In the two-dimensional case, the error can be very nearly constant along such edges. It is very difficult to find local maxima and minima along an edge when this occurs.

It is considerably harder to visualize the frequency response of a two-dimensional filter than it is to visualize the frequency response of a one-dimensional filter. In the one-dimensional case, all that is required is a simple graph. In the two-dimensional case, however, the frequency response is a surface and so it is essential to have the capability for displaying projections of surfaces. If the frequency response of a filter does not contain a high degree of symmetry, it may be necessary to display the surface from several different vantage points. Even with this, though, it is frequently helpful to have a contour map of the frequency response and/or error function and also a display showing the locations of the local maxima and minima in the error function. Producing such displays is obviously considerably more complex and time-consuming than drawing a simple graph which can supply all of this information about the frequency response of a one-dimensional filter.

The final reason why the two-dimensional filter design problem is harder than the one-dimensional filter design problem deals specifically with linear programming. For a given number of independent unit sample response coefficients, the two-dimensional problem nearly always involves more constraint points than the one-dimensional problem. This is true if the straightforward linear programming approach is used simply because the dense set of constraint points must cover an area rather than just a line. In the case of iterative algorithms for solving the one-dimensional filter design problem, the number of constraint points required at each iteration is equal to one more than the number of independent unit sample response coefficients and this does not change from iteration to iteration. The number of constraint points required at each iteration of the new algorithm for designing two-dimensional filters is equal to one more than the number of independent unit sample response coefficients plus the number of local maxima and minima in the current error function. This latter number varies quite a bit in most cases (in all cases except the differentiators that were designed) depending on the presence or absence of ridges and flat band edges in the error function. The increased number of constraint points tends to increase the amount of time that is required to obtain a solution.

The question immediately arises: How much harder than the two-dimensional filter design problem would the three-dimensional filter design problem be? All of the theory in Subsection 3.1.2 relating to the filter design problem is phrased specifically in terms of the two-dimensional filter design problem. This was done strictly for notational convenience. All of the results which apply to the two-dimensional case apply with equal force to the 3-, 4-, or n-dimensional case. As in the two-dimensional case, the set of functions used in the approximation would in general not satisfy the Haar condition, problems with ridges and flat band edges would occur (although in a more generalized sense), and the n-dimensional z-transform would present problems not found in the one-dimensional case. Problems in these areas would get no worse in going from two dimensions to n dimensions. However, there are several problems which would get worse. To begin with, it would be extremely difficult, if not totally impossible, to visualize the frequency response of the filter in any convenient form. Second, the problems relating to storage requirements and computation time that are caused by having large arrays of data would be magnified considerably. Finally, if linear programming were used, the problems caused by the number of constraint points for a given unit sample response size would also be magnified.

Acknowledgement. The work reported here was performed at Lincoln Laboratory, a center for research operated by M.I.T., with the support of the Department of the Air Force. I would like to express my appreciation to Dr. B. GOLD, Dr. E. M. HOFSTETTER, and Professors T. S. HUANG, D. H. STAELIN, A. V. OPPENHEIM and K. STEIGLITZ for the assistance which each offered to the completion of this work.

The material in this chapter is based on a thesis entitled "Two-Dimensional Nonrecursive Digital Filters" submitted to the Department of Electrical Engineering at the Massachusetts Institute of Technology on 4 May 1973 in partial fulfillment of the requirements for the degree of Doctor of Philosophy.

128   J. G. Fiasconaro

References

3.1. H. D. Helms: IEEE Trans. Audio Electroacoust. AU-**16**, 336 (1968).
3.2. L. R. Rabiner: IEEE Trans. Commun. Technol. COM-**19**, 188 (1971).
3.3. J. F. Kaiser: "Digital Filters". In *System Analysis by Digital Computer*, ed. by F. F. Kuo and J. F. Kaiser (J. Wiley, New York, 1966).
3.4. L. R. Rabiner, B. Gold, C. A. McGonegal: IEEE Trans. Audio Electroacoust. AU-**18**, 83 (1970).
3.5. L. R. Rabiner: IEEE Trans. Audio Electroacoust. AU-**20**, 280 (1972).
3.6. E. W. Cheney: *Introduction to Approximation Theory* (McGraw-Hill, New York, 1966).
3.7. O. Herrmann: Electron. Letters **6**, (1970).
3.8. E. M. Hofstetter, A. V. Oppenheim, J. Siegal: "A New Technique for the Design of Nonrecursive Digital Filters", Proc. Fifth Annual Princeton Conference on Information Sciences and Systems (1971).
3.9. T. W. Parks, J. H. McClellan: IEEE Trans. Circuit Theory CT-**19**, 189 (1972).
3.10. T. S. Huang: IEEE Trans. Audio Electroacoust. AU-**20**, 158 (1972).
3.11. J. L. Shanks, S. Treitel, J. H. Justice: IEEE Trans. Audio Electroacoust. AU-**20**, 115 (1972).
3.12. E. I. Jury: *Theory and Application of the Z-Transform Method* (John Wiley, New York, 1964).
3.13. B. Gold, C. M. Rader: *Digital Processing of Signals* (McGraw-Hill, New York, 1969).
3.14. D. C. Hanscomb: "Functions of Many Variables". In *Methods of Numerical Approximation*, ed. by D. C. Hanscomb (Pergamon Press, New York, 1966).
3.15. J. R. Rice: *The Approximation of Functions*, Vol. II (Addison-Wesley, Reading, Mass., 1969).
3.16. J. R. Rice: *The Approximation of Functions*, Vol. I (Addison-Wesley, Reading, Mass., 1964).
3.17. E. Ya. Remes: *General Computational Methods of Tchebycheff Approximation* (Kiev, 1967) (Atomic Energy Commission Translation 4491).
3.18. E. L. Stiefel: "Numerical Methods of Tchebycheff Approximation". In *On Numerical Approximation*, ed. by R. E. Langer (The University of Wisconsin Press, 1959).
3.19. W. W. Garvin: *Introduction to Linear Programming* (McGraw-Hill, New York, 1960).
3.20. W. A. Spivey, R. M. Thrall: *Linear Optimization* (Holt, Rinehart and Winston, New York, 1970).
3.21. T. S. Huang: IEEE Trans. Audio Electroacoust. AU-**20**, 88 (1972).
3.22. A. Papoulis: *Systems and Transforms with Applications in Optics* (McGraw-Hill, New York, 1968).
3.23. L. R. Rabiner: "Processing of Two-Dimensional Signals". In *Digital Signal Processing* by L. R. Rabiner and B. Gold (unpublished).
3.24. J. V. Hu, L. R. Rabiner: IEEE Trans. Audio Electroacoust. AU-**20**, 249 (1972).
3.25. J. V. Hu: "Frequency Sampling Design of Two-Dimensional Finite Impulse Response Digital Filters", M.I.T. S.M. Thesis, E.E. Dept. (1972).
3.26. IBM System/360 Scientific Subroutine Package (360A-CM-03X), Version III, Programmer's Manual.
3.27. W. E. Milne, W. Arntzen, N. Reynolds, J. Wheelock: "Mathematics for Digital Computers, Vol. 1: Multivariate Interpolation", WADC Technical Report 57–556 (1958) ASTIA Document No. AD 131033.

3.28. H. C. Thacher, Jr., W. E. Milne: J. Soc. Indust. Appl. Math. **8**, 33 (1960).
3.29. R. B. Guenther, E. L. Roetman: Math. of Computation **24**, 517 (1970).
3.30. N. M. Brenner: "Three Fortran Programs that Perform the Cooley-Tukey Fourier Transform", Technical Note 1967–2, Lincoln Laboratory, M.I.T. (July, 1967).
3.31. L. R. Rabiner, R. W. Schafer, C. M. Rader: IEEE Trans. Audio Electroacoust. AU-**17**, 86 (1969).
3.32. R. W. Hamming: *Numerical Methods for Scientists and Engineers* (McGraw-Hill, New York, 1962).
3.33. T. J. Rivlin, H. S. Shapiro: Comm. Pure Appl. Math. **13**, 35 (1960).

Further References with Titles

Y. Kamp, J. P. Thiran: Maximally flat nonrecursive two-dimensional digital filters. IEEE Trans. Circuits and Systems CAS-**21**, No. 3, 437–449 (1974).
Y. Kamp, J. P. Thiran: Chebyshev approximation for two-dimensional nonrecursive digital filters. IEEE Trans. Circuits and Systems CAS-**22**, No. 3, 208–218 (1975).

4. Two-Dimensional Recursive Filtering

R. R. READ, J. L. SHANKS, and S. TREITEL

With 33 Figures

4.1. Introduction

Definition of a Recursive Filter in Terms of the z-Transform

A two-dimensional digital recursive filter is characterized by the two-dimensional transfer function

$$H(z_1, z_2) = \frac{A(z_1, z_2)}{B(z_1, z_2)} = \frac{\displaystyle\sum_{i=1}^{M_a}\sum_{j=1}^{N_a} a_{ij} z_1^{i-1} z_2^{j-1}}{\displaystyle\sum_{k=1}^{M_b}\sum_{l=1}^{N_b} b_{kl} z_1^{k-1} z_2^{l-1}}. \tag{4.1}$$

The a_{ij} and b_{kl} are constants. Without loss of generality, b_{11} can be set to 1. Given the transform of the input data array, $D(z_1, z_2)$, the transform of the output of the filter $H(z_1, z_2)$ is

$$R(z_1, z_2) = H(z_1, z_2) \cdot D(z_1, z_2). \tag{4.2}$$

Solving for $R(z_1, z_2)$, we obtain

$$R(z_1, z_2) = \left[\sum_{i=1}^{M_a}\sum_{j=1}^{N_a} a_{ij} z_1^{i-1} z_2^{j-1}\right] D(z_1, z_2)$$
$$- \left[\sum_{\substack{k=1 \\ k \cdot l \neq 1}}^{M_b}\sum_{l=1}^{N_b} b_{kl} z_1^{k-1} z_2^{l-1}\right] R(z_1, z_2), \tag{4.3}$$

since

$$b_{11} = 1$$

and where $R(z_1, z_2)$ is the z-transformation of the output of the filter. Each output of the filter, r_{mn}, can be expressed in terms of the previously calculated values of r.

Causality

In our notation, z_1 and z_2 represent unit delays in the horizontal and vertical directions respectively. Equation (4.3) defines a filter that recurses in the positive m and n directions. It is possible to obtain a recursive filter that recurses in either of the other three directions by normalizing with respect to the other corner coefficients to the B array. This situation is illustrated in Fig. 4.1 (where for simplicity we have set $M_a = M_b = p - 1$ and $N_a = N_b = q - 1$). The recursive filter that recurses in the positive m and n direction is called "causal". Unless otherwise specified, we will consider a recursive filter specified by (4.1) to be causal.

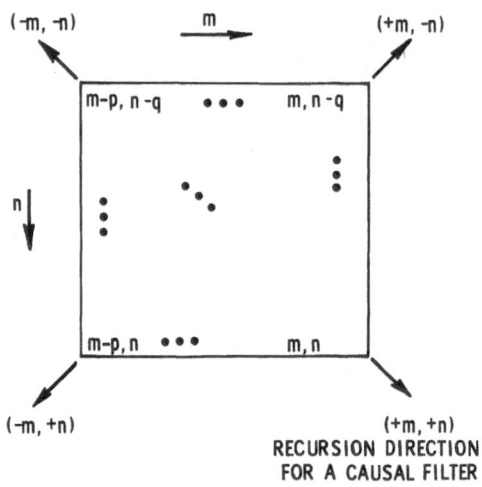

Fig. 4.1. Four directions of recursion. (After HUANG)

Unit Pulse Response

The unit pulse response of a two-dimensional recursive filter is the output of the filter when the input is given by the matrix specified by U, where the u_{11} element is equal to 1 and all of the rest are zeroes. Alternatively, the unit pulse response of a recursive digital filter can be thought of as the two-dimensional power series expansion of the function of (4.1).

To apply a recursive digital filter to two-dimensional data, first the filter must be properly designed; secondly, the filter must be applied to the data. There are two major problems in the design of a recursive filter: approximation and stability. The approximation problem involves choosing the filter coefficients so that the filter meets an appropriate design criterion. Usually this specification is a desired frequency-domain response. Occasionally the specification involves choosing the filter coefficients so that the unit pulse response of the digital filter approximates a given unit pulse response.

Absolute Stability

The stability problem amounts to constraining the filter coefficients so that the impulse response of the filter should satisfy

$$\sum_{m=0}^{\infty} \sum_{n=0}^{\infty} |h_{mn}| < \infty , \qquad (4.4)$$

where h_{mn} specifies the unit pulse response of the filter. If a filter is unstable, any noise, including round-off errors to computation, will propagate to the output and be amplified.

Initial Conditions

Initial conditions for a causal filter are often set to zero before the filter encounters the data (see Fig. 4.2, where again for simplicity we have set $M_a = M_b = p - 1$ and $N_a = N_b = q - 1$). When the filter encounters the data, the result is roughly equivalent to the excitation of a one-dimensional filter with a step function. If $|h_{mn}| \simeq 0$ for $m, n > L$, and L is small compared to the dimensions of the input and output data arrays, then the effects of the step-like excitation at the beginning of

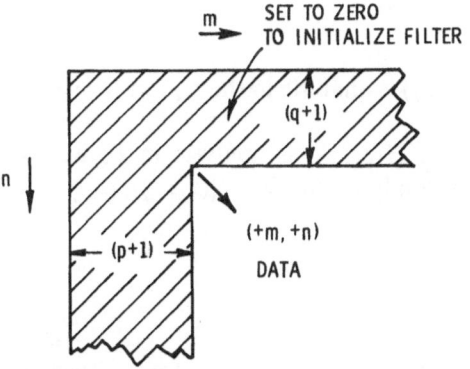

Fig. 4.2. Initial conditions for a causal filter. (After HUANG)

the filter operation will be minimal. Thus, in some cases it is not sufficient simply to insure stability in terms of the (4.4), but also a rapidly decaying unit pulse response is required.

4.2. Stability

The stability of a two-dimensional recursive filter is determined by the coefficients of the denominator $B(z_1, z_2)$ in (4.1). The general criterion for stability is given by Shank's Theorem [4.10].

4.2.1. Theorem 1 (SHANKS)

Given that $B(z_1, z_2)$ is a polynomial in (z_1, z_2), for the coefficients of the expansion of $1/B(z_1, z_2)$ in positive powers of z_1 and z_2 to converge absolutely it is necessary and sufficient that $B(z_1, z_2)$ not be zero for $|z_1|$ and $|z_2|$ simultaneously less than or equal to 1.

Stated another way, the theorem says that if there are any values of z_1 and z_2 for which $B(z_1, z_2)$ is zero and for which z_1 and z_2 are simultaneously less than or equal to 1 in magnitude, then the filter $1/B(z_1, z_2)$ will be unstable. If there are no such values, then the filter will be stable.

Proof[1] [4.1]: Let

$$H(z_1, z_2) = \frac{A(z_1, z_2)}{B(z_1, z_2)} = \sum_{m=0}^{\infty} \sum_{n=0}^{\infty} h_{mn} z_1^m z_2^n . \qquad (4.5)$$

We want to show that

$$\sum_{m=0}^{\infty} \sum_{n=0}^{\infty} |h_{mn}| < \infty$$

if and only if $H(z_1, z_2)$ is analytic in the region $D = \{(z_1, z_2); |z_1| \leq 1 \cap |z_2| \leq 1\}$.

The "if" part: If $H(z_1, z_2)$ is analytic in D, we can find $\varepsilon > 0$ such that $H(z_1, z_2)$ is analytic in $D_1 = \{(z_1, z_2); |z_1| < 1 + \varepsilon \cap |z_2| < 1 + \varepsilon\}$, which implies that

$$\sum_m \sum_n h_{mn} z_1^m z_2^n$$

---

[1] This proof is valid for the case that $B(z_1, z_2)$ is a polynomial in z_1 and z_2. For the case that $B(z_1, z_2)$ is a power series, i.e., contains a nonfinite number of coefficients, a proof is given by JUSTICE and SHANKS [4.9].

is absolutely convergent in D_1. Therefore,

$$\sum_m \sum_n |h_{mn}| < \infty .$$

The   only if' part: If

$$\sum_m \sum_n |h_{mn}| < \infty$$

then by the M test,

$$\sum_m \sum_n h_{mn} z_1^m z_2^n$$

is absolutely convergent in D, which implies in turn that $H(z_1, z_2)$ is analytic in D.   Q.E.D.

Root Maps

In order to use this theorem to test for stability, in general, we need to find the continuum of (z_1, z_2) values for which $B(z_1, z_2)$ equals zero. The region of these zeroes can be estimated in the following way. Values can be assigned to the variable z_1, and the roots of B are then found in terms of z_2. For stability, all the roots of z_2 must be greater than 1 in magnitude if the magnitude of z_1 is less than 1.

Example 1 (Unstable)

Consider the filter $H(z_1, z_2) = 1/B(z_1, z_2)$, where

$$B(z_1, z_2) = 1 - 1.5z_1 + 0.6z_1^2 - 1.2z_2 + 1.8z_1 z_2$$
$$- 0.72z_1^2 z_2 + 0.5z_2^2 - 0.75z_1 z_2^2 + 0.25z_1^2 z_2^2 .$$

Letting $z_1 = \hat{z}_1$, we get

$$B(\hat{z}_1, z_2) = (1 - 1.5\hat{z}_1 + 0.6\hat{z}_1^2)$$
$$+ (-1.2 + 1.8\hat{z}_1 + 0.72\hat{z}_1^2)z_2 \qquad (4.6)$$
$$+ (0.5 - 0.75\hat{z}_1 + 0.25\hat{z}_1^2)z_2^2 .$$

Equation (4.6) is quadratic in z_2 with complex coefficients. Thus, we can compute the roots and plot them as a function of their location in the z_1 plane. Figure 4.3 shows a contour map of the magnitude of one of the

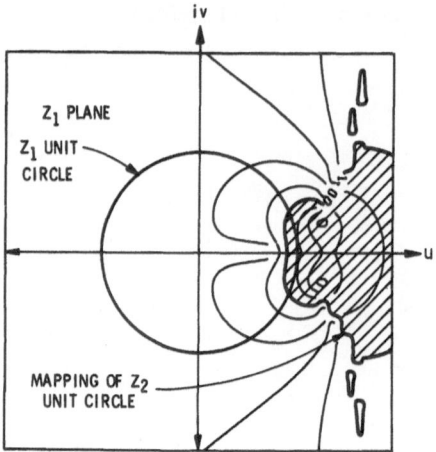

Fig. 4.3. Mapping of z_2 unit circle onto z_1 unit circle for unstable filter

z_2 roots as a function of complex $z_1 = u + iv$. The z_2 root chosen is the one with the minimum magnitude, since we are interested in finding those roots for which $|z_2| \leq 1.0$. The contour interval is 0.1 and the darkest contour corresponds to the magnitude of $z_2 = 1$. The shaded area is for $|z_2| < 1$ and is therefore the mapping of the interior of the z_2 unit circle onto the z_1 plane for $B(z_1, z_2) = 0$. In this case, the filter is unstable, since the shaded area intersects the z_1 unit circle.

Example 2 (Stable)

Figure 4.4 shows the mapping of the z_2 unit circle onto the z_1 plane for the polynomial

$$B(z_1, z_2) = 1 - 1.2z_2 + 0.5z_2^2 - 1.5z_1$$
$$+ 1.8z_1 z_2 - 0.75z_1 z_2^2 + 0.6z_1^2 \qquad (4.7)$$
$$- 0.72z_1^2 z_2 + 0.29z_1^2 z_2^2 = 0 .$$

As before, the shaded area represents the mapping of the z_2 unit circle onto the z_1 plane. In this case the z_2 unit circle does not intersect the z_1 unit circle and the filter $H(z_1, z_2) = 1/B(z_1, z_2)$ is stable. Figures 4.3 and 4.4 illustrate the fact that the root maps are symmetric with respect to this u axis, therefore it is only necessary to plot the upper half of the z plane.

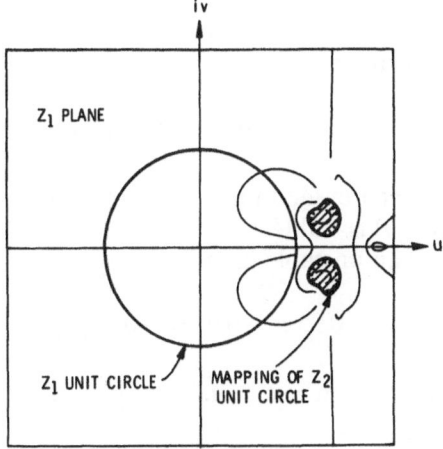

Fig. 4.4. Mapping of z_2 unit circle onto z_1
unit circle for stable filter

Stability Conditions for First-Order Filter

In the special case that the denominator polynomial is first order in
both directions, the z_1 plane $\leftrightarrow z_2$ plane transformation becomes bilinear.
and specific limits can be set on the filter coefficients. Let $H(z_1, z_2)$
$= 1/B(z_1, z_2)$, where $B(z_1, z_2) = 1 + b_{21}z_1 + b_{12}z_2 + b_{22}z_1z_2$. The filter
$F(z_1, z_2)$ will be stable if and only if following conditions hold:
   Condition 1:

$$\left| \frac{1}{b_{12}} \right| > 1.$$

Condition 2:

$$\left| \frac{1 + b_{12}}{b_{21} + b_{22}} \right| > 1.$$

Condition 3:

$$\left| \frac{1 - b_{12}}{b_{21} - b_{22}} \right| > 1. \tag{4.8}$$

These conditions have been derived in an unpublished note by SHANKS
(1969). A proof using Theorem 2 above was shown by HUANG [4.1].

Examples

Let us consider the filter

$$H(z_1, z_2) = 1/(1 - 0.7z_1 - 0.5z_2 + 0.3z_1z_2). \qquad (4.9)$$

The parameters given in (4.9) meet all of the conditions of (4.8). Therefore, the filter of (4.9) is stable.

An example of an unstable filter is given by

$$H(z_1, z_2) = 1/(1 - 0.95z_2 - 0.95z_1 + 0.5z_1z_2). \qquad (4.10)$$

This filter satisfies Conditions 1 and 3 of (4.8) but does not meet Condition 2.

Testing for stability using Theorem 1 involves finding the roots of the polynomial an infinite number of times; in practice, a large number of points suffice. There is a way to reduce the dimensionality of the testing procedure. This technique is based on the following theorem due to HUANG.

4.2.2. Theorem 2 (HUANG)

A causal filter with a z-transform $H(z_1, z_2) = 1/B(z_1, z_2)$, where B is a polynomial, is stable if and only if: 1) the map of $\partial d_1 \equiv (z_1; |z_1| = 1)$ in the z_2 plane, according to $B(z_1, z_2) = 0$, lies outside $d_2 \equiv (z_2; |z_2| \leq 1)$; and 2) no point in $d_1 = (z_1; |z_1| \leq 1)$ maps into the point $z_2 = 0$ by the relation $B(z_1, z_2) = 0$.

Proof:

We want to establish that the stability conditions of Theorems 1 and 2 are equivalent. It is obvious that the stability conditions of Theorem 1 imply those of Theorem 2. So we proceed to show the implication in the other direction.

The two-variable polynomial $B(z_1, z_2) = 0$ defines an algebraic function [4.2] $z_2 = f(z_1)$. We first modify the unit-circle contour in the z_1 plane to exclude any singular points of f inside the contour, resulting in a modified contour $\partial d'_1$, as shown in Fig. 4.5. We use d'_1 to denote the closed region enclosed by $\partial d'_1$. A point $z_1 = z_1^0$ is called a singular point of $z_2 = f(z_1)$, if $B(z_1^0, z_2) = 0$, considered as an equation in z_2, has multiple (finite or infinite) roots.

According to the theory of algebraic functions, in d'_1 the function $z_2 = f(z_1)$ has a number of branches, each of which is holomorphic. Therefore, from the maximum-modulus theorem, the maximum of $|f(z_1)|$ over d'_1 occurs on $\partial d'_1$, and the minimum of $|f(z_1)|$ over d'_1 can occur in

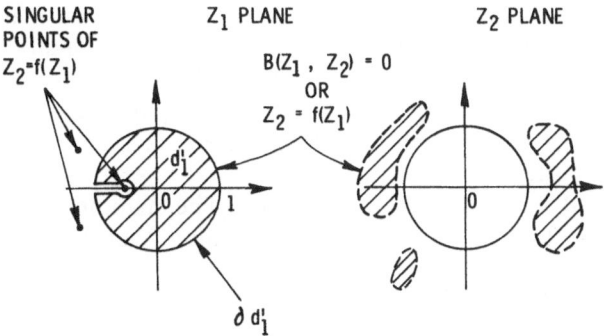

Fig. 4.5. Mapping $z_2 = F(z_1)$. (After HUANG)

the interior only if the minimum is zero. However, condition 2) of Theorem 2 says $f(z_1)$ is never zero in d_1'. Therefore, the minimum of $f(z_1)$ occurs on $\partial d_1'$, i.e.,

$$|f(d_1')| \geqq \min |f(\partial d_1')| \qquad (4.11)$$

which implies that if $|f(\partial d_1')| > 1$, then $|f(d_1')| > 1$, i.e., to ensure that $f(d_1')$ lies outside the unit disk $d_2 = (z_2, |z_2| \leqq 1)$, it is sufficient to ensure that $f(\partial d_1)$ lies outside d_2.

We are almost there, but not quite. What we really want to show is that if $|f(\partial d_1)| > 1$, then $|f(d_1)| > 1$, where $d_1 \equiv (z_1, |z_1| \leqq 1)$ and $\partial d_1 \equiv (z_1; |z_1| = 1)$. Since the detour in $\partial d_1'$ can be any path leading from ∂d_1 to the singular point, what is left to show is simply that $|f(s)| > 1$, where s is the singular point. But since each branch of $z_2 = f(z_1)$ is continuous at $z_1 = s$, and since $|f(s + \varepsilon e^{j\theta})| > 1$ for arbitrarily small ε and any θ, we have $|f(s)| > 1$.   Q.E.D.

Although the conditions of Theorem 2 are stated in terms of mapping the z_1 plane unit circle onto the z_2 plane, the theorem can also be stated by mapping the z_2 plane unit circle onto the z_1 plane. This can be written as Corollary 2.

Corollary 2. A causal filter with a z-transform $H(z_1, z_2) = 1/B(z_1, z_2)$, where B is a polynomial, is stable if and only if: 1) the map of $\partial d_2 \equiv (z_2; |z_2| = 1)$ in the z_1 plane, according to $B(z_1, z_2) = 0$, lies outside $d_1 \equiv (z_1; |z_1| \leqq 1)$; and 2) no point in $d_2 = (z_2; |z_2| \leqq 1)$ maps into the point $z_1 = 0$ by the relation $B(z_1, z_2) = 0$.

This corollary can be illustrated by considering the root map in Fig. 4.4. In Fig. 4.4, the mapping of the z_2 unit circle is, of course, the continuous contours traced by the roots of $B(z_1, |z_2| = 1) = 0$. In this example, Condition 1 of Corollary 2 is satisfied. Condition 2 of Corollary 2 is also satisfied, since the interior of the z_2 unit circle maps into the interior of the mapping of the z_2 unit circle. Both conditions are met and the filter is stable. Condition 1 of Corollary 2 insures that the map of the z_2 unit circle does not intersect the z_1 unit disk. Condition 2 further insures that the map of the z_2 unit disk intersects no point in the z_1 unit disk.

In Fig. 4.3, Condition 2 is violated. The mapping of the z_2 unit circle intersects the z_1 unit circle and the filter is unstable.

Test Procedure

To test stability using Theorem 2, we have to map $\partial d_1 = (z_1; |z_1| = 1)$ into the z_2 plane according to $B(z_1, z_2) = 0$, and see whether the image lies outside $d_2 \equiv (z_2; |z_2| \leq 1)$. Also, we have to solve $B(z_1, 0) = 0$ to see whether there is any root with magnitude smaller than 1.

There are two techniques for using this theorem as the basis for a stability test that will yield a result in a finite number of steps. One involves the use of the Hurwitz test, and is explained by HUANG [4.1]. The other involves use of the Schur-Cohn matrix, and is given by ANDERSON and JURY [4.3].

HUANG's *Method Using the* HURWITZ *Test.* The conditions of Theorem 2 can be restated as in ANDERSON and JURY:

$B(z_1, z_2) \neq 0$ for $|z_1| \leq 1 \cap |z_2| \leq 1$, if and only if the following two conditions hold:

$$B(z_1, 0) \neq 0, |z_1| \leq 1 , \tag{4.12}$$

$$B(z_1, z_2) \neq 0, |z_1| = 1, |z_2| \leq 1 . \tag{4.13}$$

Checking of (4.12) is straightforward because $B(z_1, 0)$ is a single variable polynomial and there are a number of tests for determining whether its zeros all lie outside the unit circle. One group of tests relies on replacement of $B(z_1, 0)$ by another polynomial through a bilinear transformation which then must be checked for Hurwitz character. (A polynomial is Hurwitz if all its zeros have negative real parts.) There are, of course, many tests for checking the Hurwitz property (see, e.g., [4.4]).

ANDERSON *and* JURY *Method Using the Schur-Cohn Matrix.* The second technique involves use of the Schur-Cohn matrix [4.5, 6, 8]; this matrix is square, Hermitian, of size equal to the degree of $B(z_1, 0)$, and with elements which are simple functions of the coefficients of $B(z_1, 0)$. The matrix is negative definite if, and only if, $B(z_1, 0)$ has all its zeros in $|z_1| > 1$. Of course, the negative definiteness can be established by examining the signs of the leading principal minors of the matrix.

Other tests may be found in [4.7, 11–12]. These tests include a computationally efficient one involving the recursive construction of a finite set of polynomials, each member of the set having lower degree than the preceding, and with $B(z_1, 0)$ being the first member of the set. Examination of some of the coefficients of these polynomials quickly yields the information whether (4.12) is satisfied.

The check for Condition (4.13) is more complicated than the test for Condition (4.12), but can be performed in a finite number of steps. As with the test for Condition (4.12), there are two techniques, one based on the work of ANSELL and described in HUANG [4.1], and another based on the use of the Schur-Cohn matrix, which is presented in ANDERSON and JURY [4.3]. Both methods are lengthy to derive. Interested readers should refer to the above-mentioned references.

4.3. Design

As pointed out earlier, there are two major problems in the design of a recursive filter: approximation and stability. The theoretical foundations for stability tests of two-dimensional recursive filters have been presented. Now we will consider the problem of choosing the coefficients of a recursive filter and the techniques for making that recursive filter stable.

There are many ways to choose the coefficients for a two-dimensional recursive filter. We will deal with three methods:
1) Extension of one-dimensional techniques.
2) Choosing the coefficients of the filter so that the unit pulse response approximates a desired unit pulse response.
3) Choosing the filter coefficients by approximation techniques so that the frequency response of the filter approximates a desired frequency response.

4.3.1. Extensions of One-Dimensional Techniques

Many frequency domain recursive filter design techniques are available for the one-dimensional case [4.13, 14]. Much of the work in one-dimensional frequency domain synthesis has been in the design of the low-pass and band-pass type filters. Some of the most useful techniques for designing two-dimensional recursive filters involve extensions of one-dimensional approaches.

In one dimension, the design parameters are often the cutoff frequencies, or the points of transition between the "passband" and the "stopband". In two dimensions the transition is a boundary separating the pass and reject regions. The shape of the boundary depends on the designer's filter requirements and the number of coefficients he is willing to have in the filter.

Separable Filters

The simplest case occurs for a separable filter, that is,

$$F(z_1, z_2) = F_1(z_1)F_2(z_2).$$ (4.14)

and the two-dimensional filter takes the form of two one-dimensional filters in cascade. The four-quadrant frequency domain symmetry associated with the separable filter is occasionally an undesirable feature.

   The general symmetry condition for a filter with real coefficients is symmetry about a line through the two-dimensional frequency plane origin. Two techniques are given below that procedure separable filters with real coefficients, and do so without the four quadrant (in the two-dimensional frequency plane) symmetry associated with (4.14).

The Rotation of One-Dimensional Filters

Suppose we design a low-pass or band-pass filter in one dimension. Such a filter can also be viewed as a two-dimensional filter that varies in one dimension only. In terms of the two-dimensional Laplace variables [4.15] (s_1, s_2) such a filter will be written

$$F(s_1, s_2) = F_2(s_2).$$ (4.15)

The frequency response of this filter can be found by setting $s_1 = i2\pi f_1$ and $s_2 = i2\pi f_2$. Thus, the response varies only in the f_2 frequency direction. However, suppose we were to perform a transformation, or mapping, of the (s_1, s_2) plane onto an (s'_1, s'_2) plane such that the passband of $F(s_1, s_2)$ maps into a suitable region in the (s'_1, s'_2) plane. In particular, let us consider rotating the axes of the (s_1, s_2) plane.
Technique. The (s_1, s_2) axes can be rotated by an angle β by means of the transformation

$$s_1 = s'_1 \cos \beta + s'_2 \sin \beta,$$ (4.16)

$$s_2 = s'_2 \cos \beta - s'_1 \sin \beta.$$ (4.17)

Rotation of the (s_1, s_2) axes also causes the two-dimensional frequency axes (f_1, f_2) to be rotated. Consider a filter with the transfer function

$F(s_1, s_2)$. Using (4.16) and (4.17), we can rotate the (f_1, f_2) axes by an angle β to a new position (f'_1, f'_2). Conversely, this has the effect of rotating the "pass" region by an angle $(-\beta)$ with respect to the new frequency axes.

For example, consider the filter

$$F(s_1, s_2) = \frac{1}{s_2 + \alpha}. \tag{4.18}$$

The amplitude response of this varies in the f_2 direction only. Applying the transformation Eq. (4.16) and (4.17) we get

$$F'(s'_1, s'_2) = \frac{1}{s'_2 \cos \beta - s'_1 \sin \beta + \alpha}. \tag{4.19}$$

This filter varies in both frequency directions.

We can design two-dimensional bandpass filters whose response pattern has a particular angle with respect to the (f_1, f_2) frequency axes. First we would design a suitable one-dimensional low-pass or bandpass filter in the one-dimensional s plane. Expressing this filter as a function of either s_1 or s_2, we would use either (4.16) or (4.17) to rotate this filter some desired angle into the (s'_1, s'_2) plane. This resultant form $F'(s'_1, s'_2)$ describes a continuous two-dimensional filter.

It is still necessary to produce the equivalent two-dimensional discrete filter. For this purpose we can use the bilinear z transform [4.14]. We have

$$s'_1 = \frac{2(1 - z_1)}{\varDelta_1 (1 + z_1)}, \tag{4.20}$$

$$s'_2 = \frac{2(1 - z_2)}{\varDelta_2 (1 + z_2)}, \tag{4.21}$$

where $\varDelta_1$ and $\varDelta_2$ are the sample intervals. Applying the bilinear equations to the $F'(s'_1, s'_2)$ given in (4.19), we get

$$F(z_1, z_2) = \frac{\varDelta_1 \varDelta_2}{2} \frac{(1 + z_1)(1 + z_2)}{(1 + b_{21} z_1 + b_{12} z_2 + b_{22} z_1 z_2)} \tag{4.22}$$

where

$$b_{21} = \left(\frac{\alpha \Delta_1 \Delta_2}{2} + \Delta_1 \cos\beta + \Delta_2 \sin\beta \right) / D , \qquad (4.23)$$

$$b_{12} = \left(\frac{\alpha \Delta_1 \Delta_2}{2} - \Delta_1 \cos\beta - \Delta_2 \sin\beta \right) / D , \qquad (4.24)$$

$$b_{22} = \left(\frac{\alpha \Delta_1 \Delta_2}{2} - \Delta_1 \cos\beta + \Delta_2 \sin\beta \right) / D , \qquad (4.25)$$

and

$$D = \frac{\alpha \Delta_1 \Delta_2}{2} + \Delta_1 \cos\beta - \Delta_2 \sin\beta .$$

These equations can be simplified by letting $\Delta_1 = \Delta_2 = \Delta$. In (4.18) the constant α represents the cutoff frequency (3 dB down) in rad/unit [4.16][2]. Let us set

$$\alpha = 2\pi f_c f_N ,$$

where f_c is cutoff frequency as a fraction of the Nyquist frequency, and f_N is the Nyquist frequency in cycles per unit. Since $f_N = 1/2\Delta$,

$$\alpha = \frac{\pi f_c}{\Delta} .$$

Thus we get

$$b_{21} = [\pi f_c + 2(\cos\beta + \sin\beta)]/D$$

$$b_{12} = [\pi f_c - 2(\cos\beta + \sin\beta)]/D$$

$$b_{22} = [\pi f_c - 2(\cos\beta - \sin\beta)]/D$$

$$D = [\pi f_c + 2(\cos\beta - \sin\beta)] .$$

---

[2] The unit depends on the problem. In the case of seismic records, the unit in one direction would be feet, and in the other direction, time. In the case of a photograph, both units would be spatial.

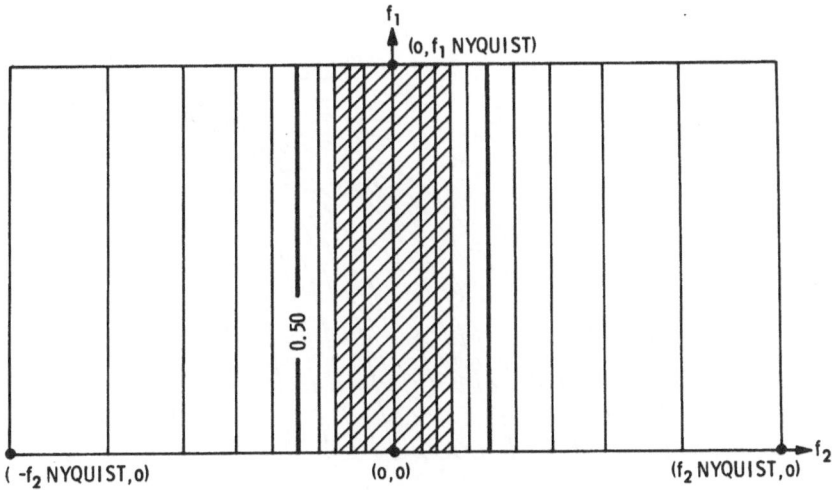

Fig. 4.6. Two-dimensional amplitude response of $(1 + z_2)/(1 - 0.6186 z_2)$

Examples. Let us consider a filter $F(s) = 1/(s + \alpha)$, where α is chosen to give a cutoff frequency at 0.15 Nyquist. Thus, $f_c = 0.15$ and for $\beta = 0°$, we get

$$F(z_1, z_2) = \frac{\Delta^2}{2} \frac{(1 + z_1)(1 + z_2)}{1 - 0.6186 z_2 + z_1 - 0.6186 z_1 z_2}$$

$$= \frac{\Delta^2}{2} \frac{1 + z_2}{1 - 0.6186 z_1}$$

Figure 4.6 shows a contour map of the amplitude response of this filter, normalized to a peak response of 1.0. The contour interval is 0.1. The pass region, for which the amplitude response is greater than 0.7, is shaded. Since $\beta = 0$, this response does not vary along the f_1 frequency axis. Along the f_2 frequency axis, the filter is low-pass. Note that the cutoff frequency is 15% of the Nyquist frequency, which agrees with the design parameter $f_c = 0.15$.

For $\beta = 15°$, we obtain

$$F(z_1, z_2) = \frac{\Delta^2}{2} \frac{(1 + z_1)(1 + z_2)}{1 - 0.3229 z_2 + 0.6455 z_1 - 0.6773 z_1 z_2}.$$

Figure 4.7 shows a contour map of the amplitude response of this filter. The contour interval is again 0.1. As required, the amplitude response

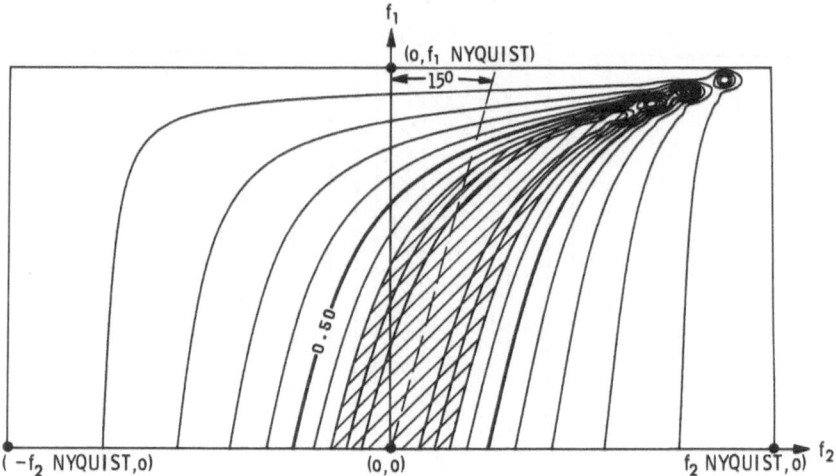

Fig. 4.7. Amplitude response of filter rotated 15°

has been rotated by 15° from the f_2 axis. However, the contour lines are no longer straight, and tend to be distorted at the higher (f_1, f_2) frequencies. This effect is caused by the bilinear (z_1, z_2) transform. When the bilinear Eqs. (4.20) and (4.21) are applied to (4.19), the numerator of $F(z_1, z_2)$ will contain the factors $(1 + z_1)(1 + z_2)$. The factor $(1 + z_1)$ causes $F(z_1, z_2)$ to be zero for all frequencies for which f_1 is equal to the Nyquist frequency. Similarly, the factor $(1 + z_2)$ causes $F(z_1, z_2)$ to go to zero at the f_2 Nyquist frequencies.

Figure 4.8 shows the amplitude response of a filter rotated to 45°. The filter equation is

$$F(z_1, z_2) = \frac{(1 + z_1)(1 + z_2)}{1 + 0.1428z_2 + 0.1428z_1 - 0.7144z_1z_2}.$$

We can also "rotate" the bandpass type of filter. Figure 4.9 shows the amplitude response of a bandpass filter. This is a bilinear transform of an eighth-order Butterworth filter [4.17]. Figure 4.10 exhibits the response of the same filter rotated 45°, again using the bilinear transform.

In order to obtain a rotation through a negative angle, we can filter the data in a different manner. Suppose $D(z_1, z_2)$ is an $M \times N$ input and $F(z_1, z_2)$ is the filter which produces the response shown in Fig. 4.10. Instead of applying $F(z_1, z_2)$ in the normal way, as described by (4.3), let us start filtering at the input point $d_{1,N}$, and filter backwards in the

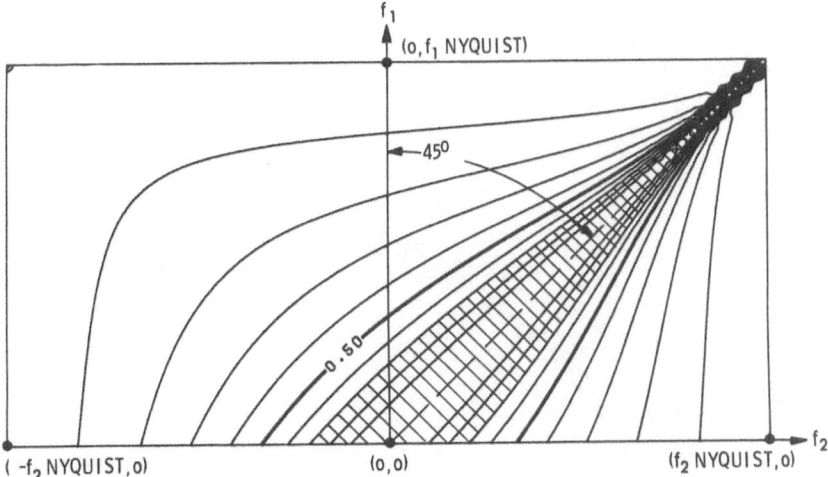

Fig. 4.8. Amplitude response of filter rotated 45°

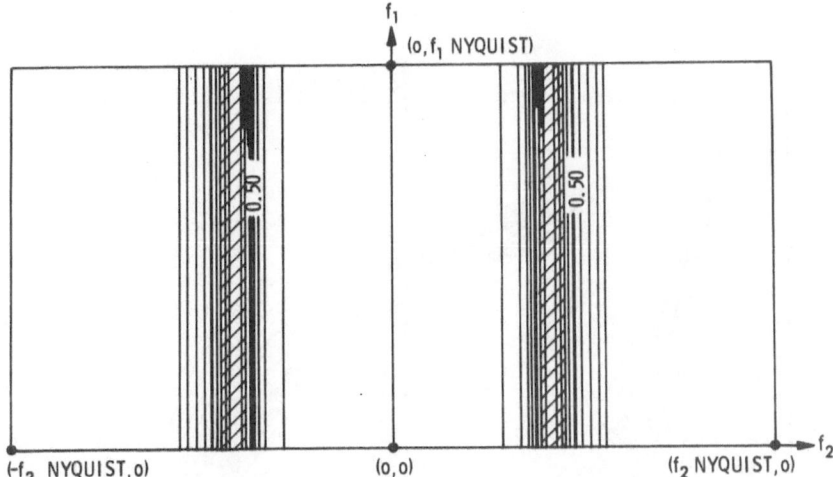

Fig. 4.9. Amplitude response of band-pass filter

y direction. This is equivalent to applying the filter

$$G(z_1, z_2) = F(z_1, 1/z_2).$$

This filter has the amplitude response curve shown in Fig. 4.11, with the bandpass characteristic rotated by $-45°$.

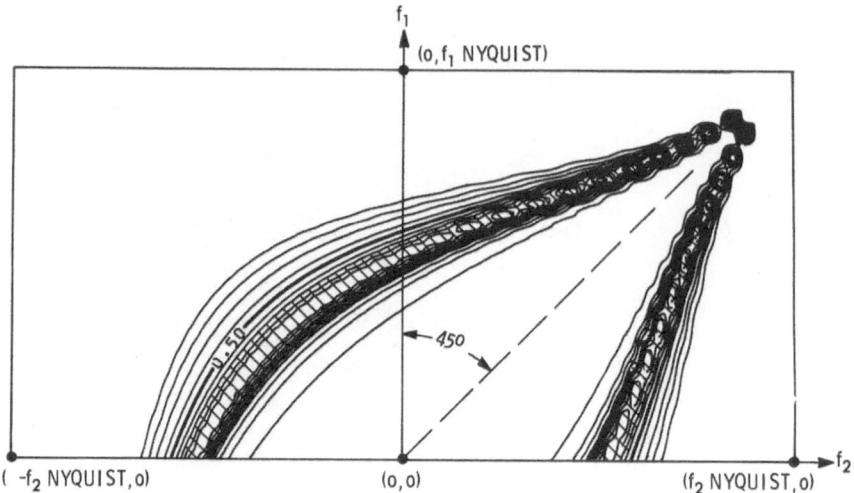

Fig. 4.10. Amplitude response of band-pass filter rotated 45°

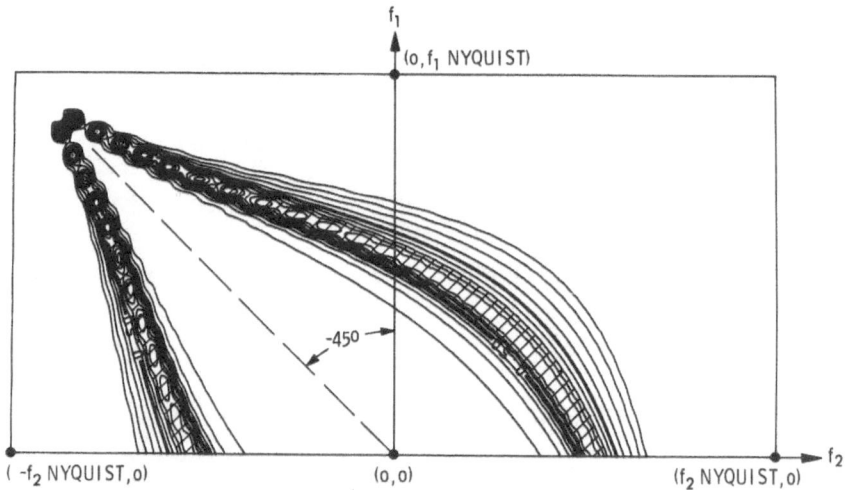

Fig. 4.11. Amplitude response of band-pass filter rotated − 45°

We can also combine these rotated filters to obtain other filter shapes. Consider the filter with an amplitude response shown in Fig. 4.12. This is a low-pass filter rotated 45°. Suppose we filter the data with this filter in cascade with the filter described by Fig. 4.11. The overall amplitude response will be the product of the two amplitude responses. The amplitude response of the cascade filters is shown in Fig. 4.13.

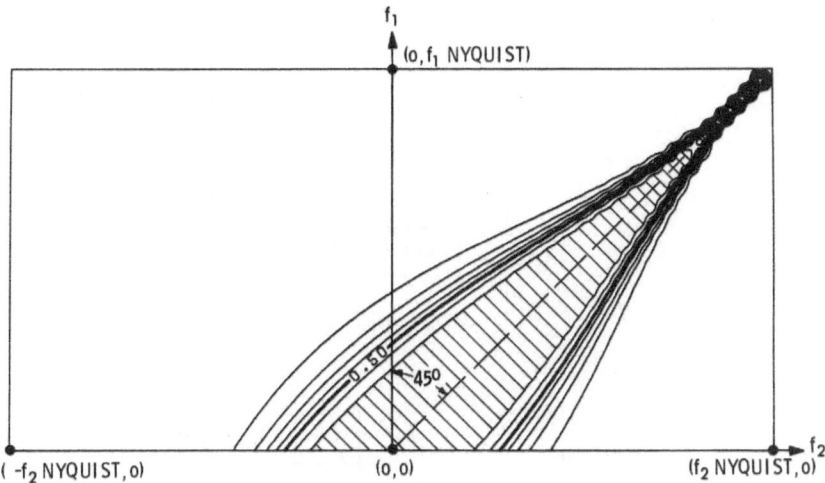

Fig. 4.12. Amplitude response of low-pass filter rotated 45°

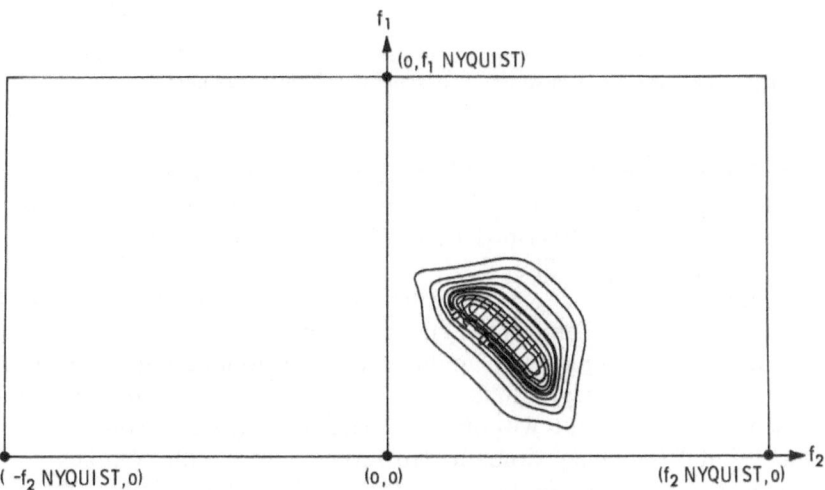

Fig. 4.13. Amplitude response of cascade response of filters from Figs. 4.11 and 4.12

Shifting Two-Dimensional Low-Pass Filters[3]

Another very useful technique for designing two-dimensional recursive filters uses a filter shifting scheme. If a two-dimensional recursive filter

---

[3] This material constitutes work by one of the authors (RRR).

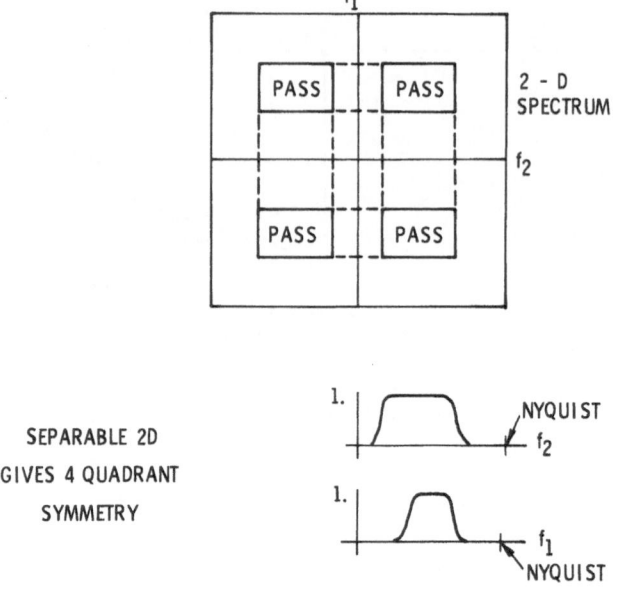

Fig. 4.14. Amplitude response of two band-pass one-dimensional filters operated in cascade to produce a four-quadrant, two-dimensional band-pass filter

is real (has real coefficients) and separable, then it follows that the filter has four-quadrant symmetry in the two-dimensional frequency domain. Such filters can be described by (4.14). This symmetry is indicated in Fig. 4.14 for bandpass filters.

Suppose, however, we allow the one-dimensional filters of (4.14) to have complex coefficients. The one-dimensional filters $F_1(z_1)$ and $F_2(z_2)$ will then not have the familiar symmetry about the zero frequency origin. Applying these filters in cascade produces a separable two-dimensional filter, but without any particular symmetry in the two-dimensional frequency domain. However, a filter with complex coefficients is not often useful in processing real data. If we combine the real parts of two complex one-dimensional filters in cascade, we get

$$\mathrm{Re}\{F_1\}\,\mathrm{Re}\{F_2\} = \tfrac{1}{2}(F_1 + F_1^*)\tfrac{1}{2}(F_2 + F_2^*)$$
$$= (F_1 F_2 + F_1 F_2^* + F_2 F_1^* + F_1^* F_2^*)/4 , \tag{4.26}$$

where $\mathrm{Re}\{\cdot\}$ indicates taking the real part of the coefficients, and * denotes the original filter using the complex conjugate of the coefficients. The filter described by (4.26) is real and separable.

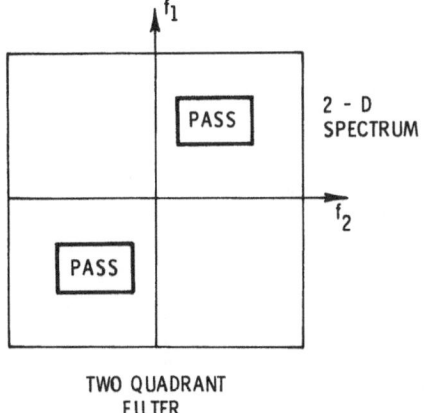

Fig. 4.15. Amplitude response of nonseparable two-dimensional band-pass filter

If we apply the complex one-dimensional filters in cascade and take the real part of the result, we get

$$\mathrm{Re}\{F_1 F_2\} = \tfrac{1}{2}(F_1 F_2 + F_1^* F_2^*).$$ (4.27)

This filter has real coefficients, is not separable, and has the desired symmetry about any line through the two-dimensional frequency plane origin.

This shows that generally it is possible to derive a filter with real coefficients with a pass region in two opposite quadrants only, as shown in Fig. 4.15. The technique for doing this is given below.

Technique. Now we will show how to derive the two-quadrant filter using the one-dimensional recursive filters F_1 and F_2 characterized by

$$F_1(z_1) = \frac{A_1(z_1)}{B_1(z_1)}, \quad F_2(z_2) = \frac{A_2(z_2)}{B_2(z_2)},$$ (4.28)

where the A_1 and B_1 polynomials are given by

$$A_1(z_1) = \sum_{i=1}^{N_1} a_i z_1^{i-1}$$

$$B_1(z_1) = \sum_{j=1}^{M_1} b_j z_1^{j-1}$$

with similar definitions for A_2 and B_2.

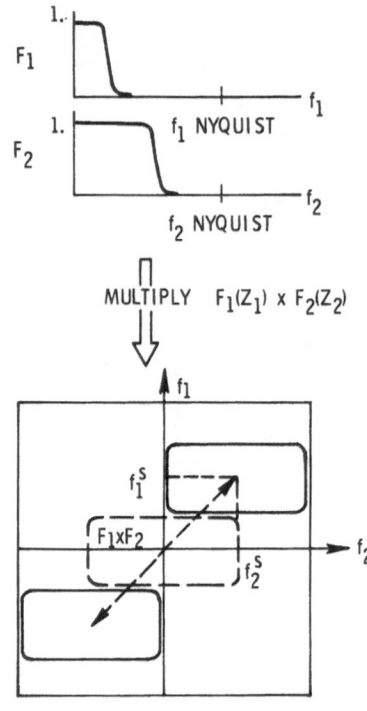

Fig. 4.16. Illustration of nonseparable band-pass filter produced by shifting the cascade combination of two one-dimensional low-pass filters

SHIFTED LOW-PASS $F_1 \times F_2$ PRODUCES
TWO-QUADRANT FILTER

The technique for deriving the two-quadrant filter involves starting with two real one-dimensional low-pass filters $F_1(z_1)$ and $F_2(z_2)$. We then shift these two filters to desired locations in the one-dimensional f_1 and f_2 frequency planes. The resulting filters will have complex coefficients but can be combined according to (4.27) to produce the desired two-quadrant band-pass filter. Note that the extraction of the real part means that the filter, in general, is no longer separable. Thus, if we are given a low-pass, one-dimensional filter F_1, a filter F_2, and an angle θ_1 corresponding to the frequency shift (or rotation in the z_1 plane) of the filter F_1 and an angle θ_2 corresponding to the shift of filter F_2, the filter can be split as indicated in Fig. 4.16 by the following procedure:

Substitute

$$z_1 = z_1' \, e^{-j\theta_1}$$

$$z_2 = z_2' \, e^{-j\theta_2}$$

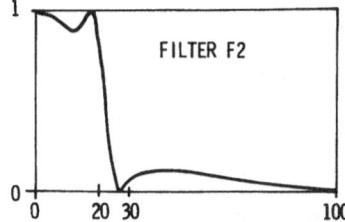

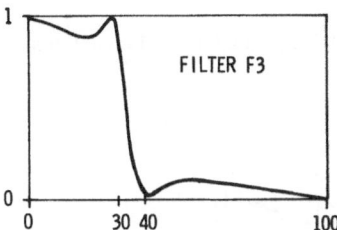

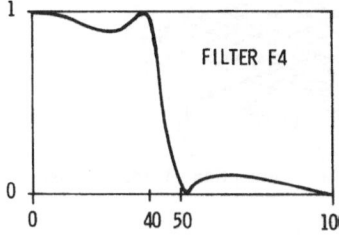

Fig. 4.17. Amplitude response of one-dimensional filters used in producing two-dimensional band-pass filters

in F_1 and F_2, respectively. This gives the filter's complex coefficients $\hat{A}_1, \hat{B}_1, \hat{A}_2$, and $\hat{B}_2$ and shifts their individual frequency responses in their frequency planes. Denote these filters by $\hat{F}_1$ and $\hat{F}_2$ then

$$\hat{F}_1 \hat{F}_2 + \hat{F}_1^* \hat{F}_2^* = \frac{2\,\mathrm{Re}\,\{\hat{A}_1 \hat{B}_1 \hat{A}_2 \hat{B}_2\}}{|\hat{B}_1|^2 |\hat{B}_2|^2}. \tag{4.29}$$

This resulting filter has a separable denominator containing even powers of z_1 and z_2. The form of the denominator guarantees stability if the original filters F_1 and F_2 are stable since the zeros of the $\hat{B}$'s are rotated versions of the zeros of the B's.

Examples. Equation (4.29) can easily be implemented on a digital computer. As an example of this technique, consider three third-order elliptic function filters with a passband variation of 0.1, and a reject band variation

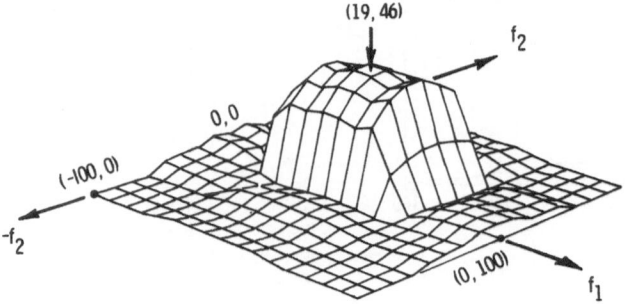

Fig. 4.18. Two-dimensional amplitude response of the combination of filters $F2$ and $F3$ shifted to $(19, 46)$

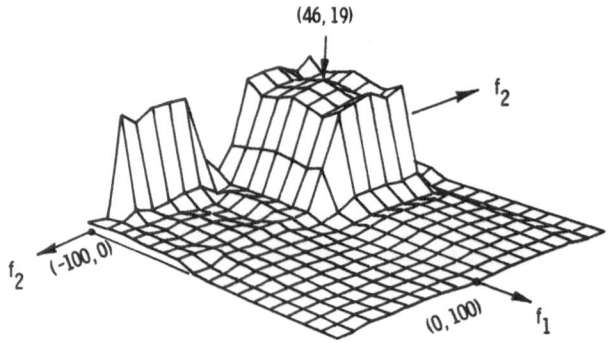

Fig. 4.19. Two-dimensional amplitude response of the combination of one-dimensional filters $F2$ and $F3$ shifted to $(46, 19)$

of 0.1. The transition zone is 10 Hz for each filter with a Nyquist frequency of 100 Hz. Filter F_2 has a passband from zero to 20 Hz, filter F_3 has a passband from zero to 30 Hz, and filter F_4 has a passband from zero to 40 Hz. These filters are illustrated by Fig. 4.17.

The results of applying the frequency shifting technique are pictured in Figs. 4.18–4.20.

As can be seen from the figures, the technique is especially useful in deriving two-dimensional recursive filters with square passbands or reject bands located anywhere in the two-dimensional frequency spectrum. It is also apparent from (4.29) that the denominator coefficients form a separable polynomial, and this implies fewer computational steps in the calculation of each output point. Equation (4.29) implies that the response in each individual quadrant is not completely independent of the response in other quadrants. The response outside the passband in quadrant 1 (associated with the first term on the left-hand side) affects

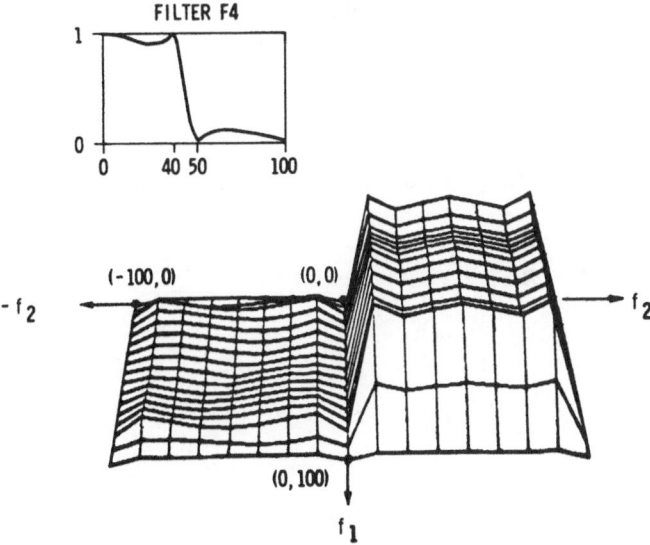

Fig. 4.20. Two-dimensional amplitude response of the combination of filter $F4$ with itself and shifted to $(50, 50)$

the passband of the symmetrical portion in quadrant 3. However, this effect is usually rather small, provided that the response of the filter is negligible outside the pass regions. Note that the response in each individual quadrant, as indicated by (4.29), is additive. If the pass regions in the shifted responses happen to overlap, then instead of an amplitude response in the pass region equal to 1, the value in the passband where the response in the two quadrants overlap would be 2.

4.3.2. Synthesis in the Space Domain

Suppose we know the two-dimensional unit pulse response for the discrete filter we wish to use on the data. We could use two-dimensional convolution or the two-dimensional Fourier transform to apply this filter. However, it might also be possible to approximate the unit pulse response with a two-dimensional recursive filter. In some cases the recursive filter can be more efficient than convolution or filtering in the frequency domain.

Method

The following is an extension of a synthesis technique in one dimension [4.17–19]. Suppose the desired impulse response is d_{ij}, for $i = 1, 2,$

$3, \ldots, M_d$ and $j = 1, 2, 3, \ldots, N_d$. That is

$$D(z_1, z_2) = \sum_{i=1}^{M_d} \sum_{j=1}^{N_d} d_{ij} z_1^{i-1} z_2^{j-1} . \qquad (4.30)$$

Let the approximating two-dimensional recursive filter be

$$F(z_1, z_2) = \frac{A(z_1, z_2)}{B(z_1, z_2)} , \qquad (4.31)$$

where

$$A(z_1, z_2) = \sum_{i=1}^{M_a} \sum_{j=1}^{N_a} a_{ij} z_1^{i-1} z_2^{j-1}$$

$$B(z_1, z_2) = \sum_{i=1}^{M_b} \sum_{j=1}^{N_b} b_{ij} z_1^{i-1} z_2^{j-1}$$

$$b_{11} = 1.0$$

M_a, N_a, M_b, N_b = arbitrary (but fixed) parameters .

From (4.31), $F(z_1, z_2) B(z_1, z_2) = A(z_1, z_2)$. Multiplication of the z polynomials is equivalent to convolution of the arrays,

$$a_{mn} = \sum_{i=1}^{M_b} \sum_{j=1}^{N_b} b_{ij} f_{m-i+1, n-j+1} . \qquad (4.32)$$

The coefficients a_{mn} are defined by (4.32) over the integers $m = 1, 2, \ldots, M_a, n = 1, 2, \ldots, N_a$. Outside this region, the a_{mn} are zero. Let us define the set of integer pairs

$$I_a = \{(i, j) : 1 \leq i \leq M_a, 1 \leq j \leq N_a\} .$$

Furthermore, let us define $\hat{I}_a$ as the set of all other values of (m, n) greater than zero.

$$\hat{I}_a = \{(m, n) : m > 0, n > 0, (m, n) \notin I_a\} .$$

Thus, for $(m, n) \in I_a$, the values of a_{mn} are yet to be determined. For $(m, n) \in \hat{I}_a$, $a_{mn} \equiv 0$. Therefore, using (4.32) and the fact that $b_{11} = 1.0$, we get

$$f_{mn} = - \sum_{\substack{i=1 \\ i \cdot j \neq 1}}^{M_b} \sum_{j=1}^{N_b} b_{ij} f_{m-i+1,n-j+1} \quad \text{for} \quad (m, n) \in \hat{I}_a .$$

If we judiciously choose the b_{ij}, the f_{mn} will approximate the desired response d_{mn}. Thus we may write

$$d_{mn} \cong - \sum_{\substack{i=1 \\ i \cdot j \neq 1}}^{M_b} \sum_{j-1}^{N_b} b_{ij} d_{m-i+1,n-j+1} \quad \text{for} \quad (m, n) \in \{\hat{I}_a \cap I_d\}, \qquad (4.33)$$

where (m, n) must be a member of the intersection of the set $\hat{I}_a$ and the set

$$I_d = \{(m, n): 1 \leq m \leq M_d, 1 \leq n \leq N_d\} .$$

Let us define an error e_{mn} that can be added to the right side of (4.33) to produce equality.

$$d_{mn} = e_{mn} - \sum_{\substack{i=1 \\ i \cdot j \neq 1}}^{M_b} \sum_{j=1}^{N_b} b_{ij} d_{m-i+1,n-j+1} \quad \text{for} \quad (m, n) \in \{\hat{I}_a \cap I_d\} .$$

Therefore, we can put d_{mn} inside the summation to get

$$e_{mn} = \sum_{\substack{i=1 \\ i \cdot j \neq 1}}^{M_b} \sum_{j=1}^{N_b} b_{ij} d_{m-i+1,n-j+1} \quad \text{for} \quad (m, n) \in \{\hat{I}_a \cap I_d\} .$$

Now that we have defined the error, let us choose the b_{ij} such as to minimize the mean square error,

$$\overline{e^2} = \sum_{(m,n)} \sum \left[\sum_{(i,j)} b_{ij} d_{m-i+1,n-j+1} \right]^2 , \qquad (4.34)$$

for $(i, j) \in I_b, (m, n) \in \hat{I}_a \cap I_d\}$.

Except for b_{11}, which is 1.0, the b_{ij} are not yet specified. By differentiating (4.34) with respect to the b_{ij} and setting the resultant equations equal to zero, we can find those b_{ij} that minimize the mean square error. This will give $(M_b \cdot N_b - 1)$ equations of the type

$$\sum_{\substack{i=1 \\ i \cdot j \neq 1}}^{M_b} \sum_{j=1}^{N_b} b_{ij} \Phi_{klij} = \Phi_{kl}, \quad \text{for} \quad \begin{aligned} k &= 1, 2, 3, ..., M_b, \\ l &= 1, 2, 3, ..., N_b, \end{aligned} \qquad (4.35)$$

$$\text{but} \quad \dot{k} \cdot l \neq 1,$$

where

$$\Phi_{klij} = \sum_{(m,n)} d_{m-l+1,n-j+1} d_{m-k+1,n-l+1} \, , \tag{4.36}$$

$$\Phi_{k1} = - \sum_{(m,n)} d_{mn} d_{m-k+1,n-j+1}, \quad (m,n) \in \{\hat{I}_a \cap I_d\} . \tag{4.37}$$

Thus, (4.35) describes a set of $(M_b \cdot N_b - 1)$ simultaneous linear equations with $(M_b \cdot N_b - 1)$ unknowns. The solutions of (4.35) give us the b_{ij} that minimize the mean square error.

Having computed the denominator $B(z_1, z_2)$ of the filter, we now compute the numerator $A(z_1, z_2)$. One way to compute $A(z_1, z_2)$ is to compute those coefficients that minimize the mean square difference between the coefficients of $F(z_1, z_2) = A(z_1, z_2)/B(z_1, z_2)$ and the coefficients of the desired response $D(z_1, z_2)$. This is a discrete Wiener filtering problem in two dimensions [4.20]. It consists of finding the optimum filter $A(z_1, z_2)$ given an input $1/B(z_1, z_2)$ and a desired output $D(z_1, z_2)$.

In a simpler, but less accurate, method we can convolve the B array with the D array to get the A array. Since $A(z_1, z_2)/B(z_1, z_2) \simeq D(z_1, z_2)$, we compute the coefficients a_{mn} from $A(z_1, z_2) = B(z_1, z_2) \cdot D(z_1, z_2)$ for $(m, n) \in I_a$. If $B(z_1, z_2)$ has been chosen well, the coefficients of $B(z_1, z_2)$ will be comparatively small for $(m, n) \in \hat{I}_a$.

Example

To illustrate the method, consider the impulse response of a two-dimensional recursive filter, $D(z_1, z_2) = A'(z_1, z_2)/B'(z_1, z_2)$. Both A' and B' are 3×3 arrays. The coefficients of the A' and B' arrays are shown in Table 4.1, under Columns A' and B'. The first 20×20 values of the impulse response were used as the desired response. M_a, M_b, N_a, and N_b were set at three. The coefficients given by (4.36) and (4.37) were computed and (4.35) was solved to find the denominator coefficients. The numerator coefficients were found by two-dimensional convolution of $B(z_1, z_2)$ with $D(z_1, z_2)$. The recursive coefficients computed by the program are shown in Table 4.1 under Columns A and B.

The results show very good agreement with the coefficients of the original filter. The problem of choosing the proper values for the order of the numerator and denominator arrays is a difficult one even in the one-dimensional case. Here the ¨proper¨ orders were chosen because it was known that the desired unit pulse response originated from a filter with 3×3 numerator and denominator. In a more general case where this information is not available, trial and error offers the most workable solution.

Table 4.1. Test of space domain synthesis

Term	A'	A	B'	B
Constant	1.0	1.0	1.0	1.0
z_1	3.0	3.0000005	-1.2	-1.1999995
z_1^2	2.0	2.0000016	0.5	0.4999996
z_2	2.0	2.0000002	-1.5	-1.4999998
$z_1 z_2$	4.0	4.0000014	1.8	1.7999989
$z_1^2 z_2$	-1.0	-0.9999976	-0.75	-0.7499992
z_2^2	-1.0	-0.9999995	0.6	0.5999998
$z_1 z_2^2$	2.0	2.0000002	-0.72	-0.7199994
$z_1^2 z_2^2$	1.0	0.9999998	0.29	0.2899996

Note: A', B' are the actual coefficients. A, B, are the estimated coefficients.

4.3.3. Generalized Approximation and Stabilization

Approximation

The general approximation problem for designing two-dimensional recursive filters is a nonlinear minimization problem. It involves choosing the coefficients of the A and B arrays so that the filter response closely approximates a desired filter response in the frequency domain. One common error criterion is the mean square norm. There are many algorithms available to perform such a minimization. For example, see [4.21, 22, 28].

Stabilization Procedures

Unless constraints are placed on the filter coefficients, the resulting coefficients may produce an unstable filter. There are two possible stabilization procedures which are described below. Both methods operate on the denominator polynomial to alter its phase characteristics. The methods are intended to produce a filter whose amplitude response is similar to the original unstable filter.

Double Planar Least Squares Inverse. The first method uses the properties of the planar least squares inverse. Let us suppose that we have some given array C. We would like to find an array P such that C convolved with P approximates the unit pulse array U. That is,

$$C * P \simeq U . \tag{4.38}$$

The symbol $*$ denotes convolution.

Definition of a Planar Inverse. In general, it will not be possible to make $C * P$ exactly equal to U. In actuality, $C * P$ will be equal to some other array, G. If we choose P such that the sum of the squares of the elements of $U - G$ is minimized, we call P a *least squares inverse* of C [4.25]. In the case that C, P, G, and U are two-dimensional arrays, we call P a planar least squares inverse (PLSI) of C. In the planar case, P is found from U and C by a two-dimensional Wiener technique described by WIGGINS [4.20].

The Minimum-Phase Conjecture. As a group, PLSI's have interesting properties. One property, in particular, which we shall use is described by the following conjecture:

Conjecture A:

Given an arbitrary, real, finite matrix C, any planar least squares inverse of C is minimum phase. This conjecture states that given a matrix C from which we find a PLSI matrix P, then P must be minimum phase.

This is an important conjecture because it implies that the filter $F(z_1, z_2) = 1/P(z_1, z_2)$ must be stable if P is a PLSI.

Unfortunately we do not have a proof for this conjecture, which is an extension of a one-dimensional theorem [4.25]. However, a large number of numerical cases have been tested and no counterexamples have been found [4.24].

Example. As an example of the PLSI, consider the array A in Fig. 4.21a. Figure 4.21b shows the (3×3) PLSI of A. The root map of this PLSI is shown in Fig. 4.21c. Since the mapping of the z_2 unit circle (shaded part) does not intersect the z_1 unit circle, then the PLSI must be minimum phase.

Suppose that a two-dimensional recursive filter $F(z_1, z_2) = 1/B(z_1, z_2)$ has been designed, only to find that it is unstable, i.e., that the denominator of the filter is not minimum phase. How can the coefficients be altered in order to produce a stable filter?

Let us denote the denominator matrix of this unstable filter by B. We can compute a planar least squares inverse, B', or B. According to Conjecture A, B', the PLSI of B, is minimum phase. Suppose that we compute a PLSI of B', and call it $\hat{B}$. Now $\hat{B}$ is the inverse of the inverse, or the "double inverse" of B. Intuitively, $\hat{B}$ and B have some characteristics in common. Moreover, $\hat{B}$ is itself a PLSI. Hence, it is minimum phase, while B is not. Therefore, the filter $\hat{F}(z_1, z_2) = 1/\hat{B}(z_1, z_2)$ is stable.

For example, consider the filter $F(z_1, z_2) = 1/B(z_1, z_2)$, where

$$H = \begin{bmatrix} 1.0 & -1.20002759 & 0.40002239 \\ -1.00003018 & 1.70007079 & -0.65005088 \\ 0.40002035 & -0.7005488 & 0.25004387 \end{bmatrix}. \tag{4.39}$$

$$\begin{bmatrix} 1.0 & -1.2 & 0.4 \\ -1.2 & 1.7 & -.65 \\ .4 & -.7 & .25 \end{bmatrix}$$

(a)   Input Matrix $\underline{A}$.

$$\begin{bmatrix} 1.0 & .87742 & .39488 \\ .90441 & .65133 & .22754 \\ .39818 & .21555 & .04207 \end{bmatrix}$$

(b)   (3 x 3)   PLSI of $\underline{A}$.

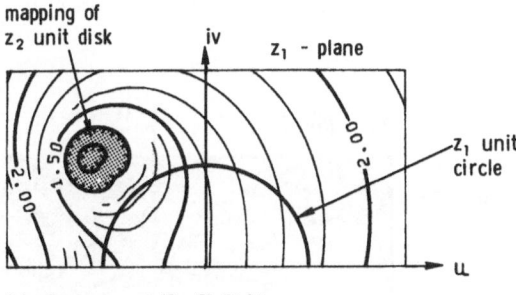

(c)   Root map of (3 x 3) PLSI.

Fig. 4.21a–c. Root map of PLSI of array A

This filter is unstable, as illustrated in Fig. 4.22, which shows a perspective view of the unit pulse response of $F(z_1, z_2)$. Note that the unit pulse response tends to grow larger and larger with increasing values of the spatial coordinates x and y. This instability can be verified by plotting the root map of B. This map is shown in Fig. 4.23. Since a portion of the dark area (mapping of the z_2 unit disk) falls inside the z_1 circle, B is not minimum phase.

In order to produce a stable filter we first compute a PLSI, B', of B. For this example, B' is an 8×8 array. We then compute a 3×3 PLSI, which we shall call $\hat{B}$. The evaluation of $\hat{B}$ yields

$$\hat{B} = \begin{bmatrix} 1.11567 & -1.23469 & 0.38431 \\ -1.0932 & 1.68765 & -0.6821 \\ 0.37577 & -0.65178 & 0.29767 \end{bmatrix}. \tag{4.40}$$

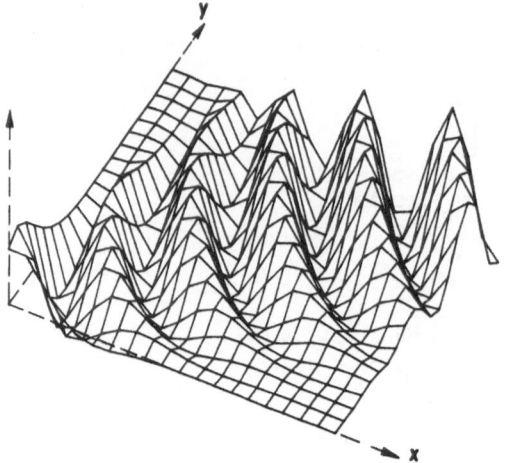

Fig. 4.22. Two-dimensional pulse response of an unstable filter

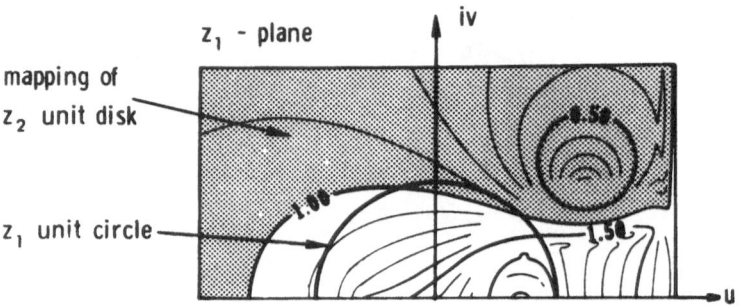

Fig. 4.23. Root map of an unstable filter

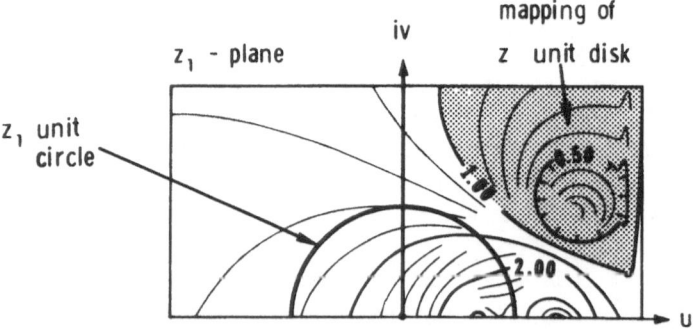

Fig. 4.24. Root map of an DPLSI of unstable denominator

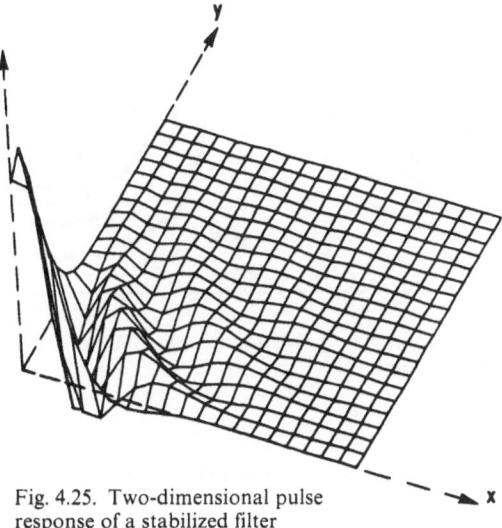

Fig. 4.25. Two-dimensional pulse
response of a stabilized filter

The root map of $\hat{B}$ is shown in Fig. 4.24, which shows that $\hat{B}$ is minimum phase. Hence $\hat{F}(z_1, z_2) = 1/\hat{B}(z_1, z_2)$ should be stable; this conclusion is confirmed by inspection of Fig. 4.25, which shows the unit pulse response of $\hat{F}(z_1, z_2)$. This response is clearly seen to decay as an x and y increase.

  Comments about Amplitude and Phase. In what sense are B and $\hat{B}$ similar? We have already seen that $\hat{B}$ is minimum phase, while B does not have this property. On the other hand, $\hat{B}$ is the double PLSI of B. Hence, we conjecture that $\hat{B}$ must in fact be the approximate minimum-phase version of B. In other words, we would expect the amplitude spectra of B and $\hat{B}$ to be roughly equal. Figure 4.26 shows the two-dimensional amplitude spectra of B and $\hat{B}$ in the form of contour plots. The method has produced a stable filter by operating on the denominator of an unstable filter. The amplitude spectra of the denominator arrays are similar but their phase spectra have been altered to produce stability.

The Two-Dimensional Discrete Hilbert Transform. The second approach for stabilizing two-dimensional recursive filters is called the Hilbert transform method [4.26]. This method, like the previous one, is based on the extension of a one-dimensional technique.

  One-Dimensional Case. It is well-known [Ref. 4.27, p. 248], that for a minimum-phase sequence $a(i)$, the phase $\theta(\exp j\omega)$ and the log magnitude of the amplitude spectrum, $\log |A(\exp j\omega)|$, are related through the Hilbert transform,

$$\theta(\exp j\omega) = -\frac{1}{2\pi} \int_0^{2\pi} \log |A(\exp j\Omega)| \cot \frac{\omega - \Omega}{2} \, d\Omega, \qquad (4.41)$$

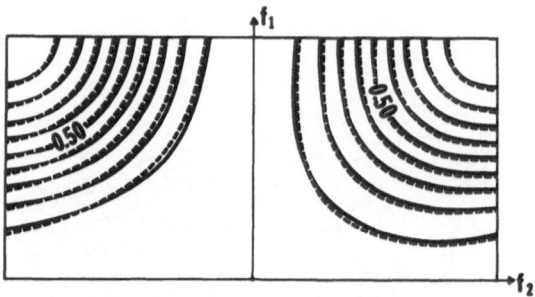

Fig. 4.26. Comparison of amplitude spectra of nonminimum-phase B (dashed lines) and minimum-phase $\hat{B}$ (solid lines)

where (4.41) results from the fact that $a(i)$ $(i = 0, 1, \ldots, M-1)$ is minimum phase if, and only if, the inverse z-transform of $\log[A(z)]$ is causal. Since the Hilbert transform relates the real and imaginary parts of causal functions, application of the Hilbert transform to $\log|A(\exp j\omega)|$ [Eq. 4.41)] yields the imaginary part of $\log[A(z)]$, which is the phase of $A(z)$.

A discrete approximation to (4.41) using the trapezoidal rule can be implemented using the discrete Fourier transform [4.26]. This procedure for obtaining the approximate minimum-phase spectrum from knowledge of the amplitude spectrum of a mixed or minimum-phase sequence $a(i)$ is given by (4.42).

$$\theta(i) = -jDFT\{\text{sgn}(i) \cdot IDFT[\log|A(i)|]\}, \tag{4.42}$$

Here $A(i)$ is the ith component of the amplitude spectrum, and $\theta(i)$ is the ith component of the phase spectrum. DFT and IDFT represent the operation of taking the discrete Fourier transform and inverse discrete Fourier transform. The function $\text{sgn}(i)$ is the finite discrete signum function of the same length as the discrete Fourier transform operations. A block diagram summarizing the procedure required to obtain the minimum-phase version of a mixed or maximum-phase pulse is shown in Fig. 4.27.

Two-Dimensional Case. By using the two-dimensional version of the discrete Fourier transform and by deriving a special two-dimensional signum function, this technique can be extended to two dimensions [4.26]. Thus, given a two-dimensional amplitude spectrum $A(i_1, i_2)$ of a causal sequence, the minimum phase spectrum $\theta(i_1, i_2)$ is calculated by the equation

$$\theta(i_1, i_2) = -j \, \text{DFT}(\text{sgn}(i_1, i_2) \cdot \text{IDFT}[\log A(i_1, i_2)]). \tag{4.43}$$

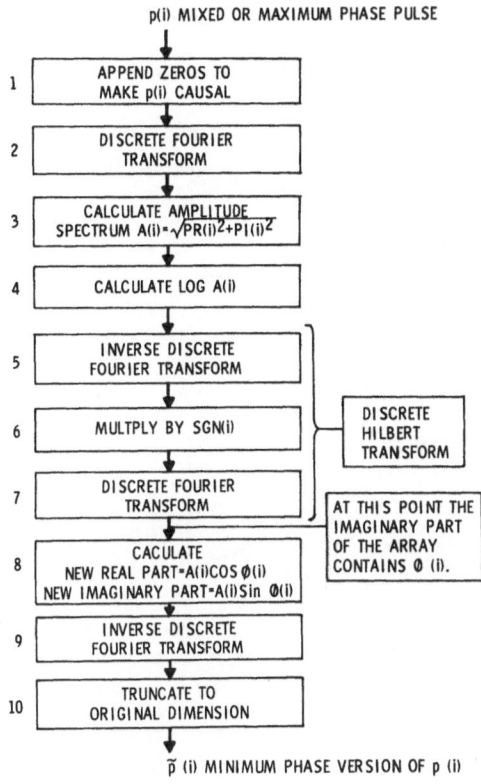

p(i) MIXED OR MAXIMUM PHASE PULSE

1. APPEND ZEROS TO MAKE p(i) CAUSAL

2. DISCRETE FOURIER TRANSFORM

3. CALCULATE AMPLITUDE SPECTRUM $A(i) = \sqrt{PR(i)^2 + PI(i)^2}$

4. CALCULATE LOG A(i)

5. INVERSE DISCRETE FOURIER TRANSFORM

6. MULTPLY BY SGN(i)

7. DISCRETE FOURIER TRANSFORM

DISCRETE HILBERT TRANSFORM

8. CACULATE NEW REAL PART=A(i)COS Ø(i) NEW IMAGINARY PART=A(i)Sin Ø(i)

AT THIS POINT THE IMAGINARY PART OF THE ARRAY CONTAINS Ø (i).

9. INVERSE DISCRETE FOURIER TRANSFORM

10. TRUNCATE TO ORIGINAL DIMENSION

$\bar{p}$ (i) MINIMUM PHASE VERSION OF p (i)

Fig. 4.27. Procedure for obtaining a minimum-phase version of a pulse

The procedure for using (4.43) to obtain the minimum-phase version of a mixed or maximum-phase array is the same as in Fig. 4.27 except that all of the operations are performed in two dimensions.

A Comparison to the Double Planar Least Squares Inverse Method

Example 1. The first example is a comparison between Hilbert stabilization procedure and the double-least-squares approach. We assume a two-dimensional recursive filter whose denominator is the polynomial $B(z_1, z_2)$, with a coefficient array given by

$$B = \begin{bmatrix} 1.0 & -0.75 & 0.9 \\ 1.5 & -1.2 & 1.3 \\ 1.2 & 0.9 & 0.5 \end{bmatrix}. \tag{4.44}$$

The root map of the array is given in Fig. 4.28a. It shows that the image of the z_2 unit disk intersects the z_1 unit disk in the z_1-plane. This means that the filter with the denominator $B(z_1, z_2)$ is unstable. Figure 4.28b

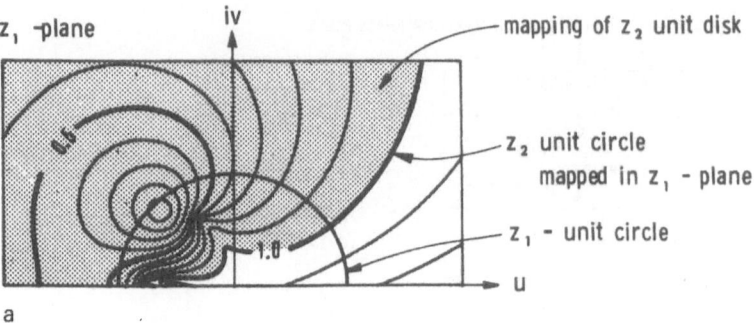

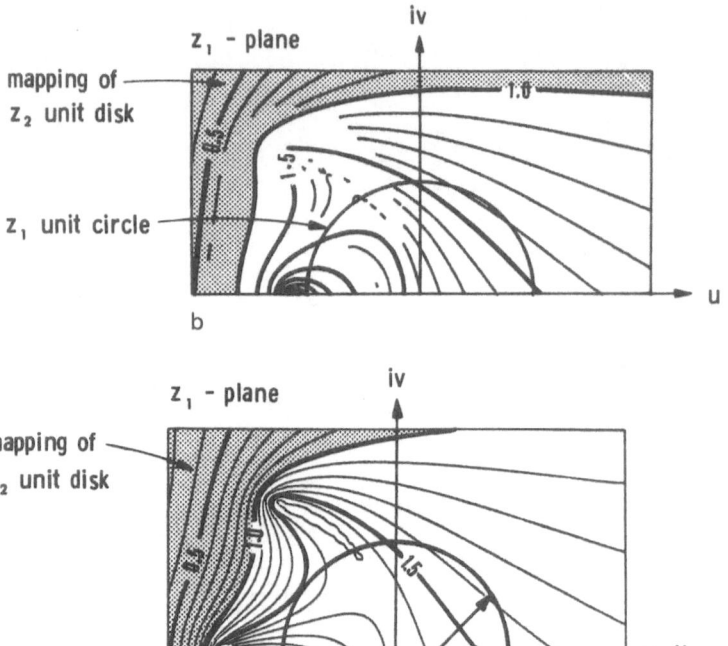

Fig. 4.28a–c. Root maps. (a) A nonminimum-phase pulse, (b) a version stabilized by the double least squares techniques, and (c) a version stabilized by the Hilbert transform technique

shows the root map of G,

$$G = \begin{bmatrix} 6.761 & -1.248 & 1.258 \\ 3.156 & -0.602 & 1.027 \\ 0.706 & 0.209 & 0.368 \end{bmatrix} \qquad (4.45)$$

that results from applying the double-least squares stabilization technique to the array B. It is noted that the z_2 unit disk mapped in the z_1-plane does not intersect the z_1 unit disk. Thus, we conclude that the recursive filter with denominator given by $G(z_1, z_2)$ is stable. Figure 4.28c is the root map of the two-dimensional polynomial whose coefficients are given by

$$H = \begin{bmatrix} 2.46 & -0.619 & 0.881 \\ 1.29 & -0.427 & 0.770 \\ 0.263 & 0.325 & 0.212 \end{bmatrix}. \tag{4.46}$$

This array results from applying the Hilbert transform algorithm to the given coefficient array B.

The root map for $H(z_1, z_2)$ shows that the image of the z_2 unit disk does not intersect the z_1 unit disk in the z_1-plane. Hence we conclude that the recursive filter with denominator given by the array H is minimum-phase. The differences in the shapes of the contours of Fig. 4.28b and c imply that the two-dimensional spectra of H and G differ appreciably. Amplitude spectra of B, H, and G are contoured in Fig. 4.29a–c, respectively. The squared error for the amplitude spectrum is defined by

$$e(B, G) = \sum_{i_1=0}^{N_1-1} \sum_{i_2=0}^{N_2-1} [AB(i_1, i_2) - AG(i_1, i_2)]^2, \tag{4.47}$$

where AB and AG are the amplitude spectra of B and G, respectively. The error for $e(B, H)$ is defined similarly. For the spectra shown in Fig. 4.29, $e(B, H)$ is 18.5 times smaller than $e(B, G)$.

From Fig. 4.27 it is clear that before truncation, H is derived directly from the inverse discrete Fourier transform of AH and from the phase resulting from the discrete Hilbert transform procedure. This result is then truncated to the size of the original array B. If the discrete Hilbert transform stabilization procedure yielded the *precise* minimum-phase array, all the elements in the inverse transform of AH would be zero beyond the dimensions of the original array B. In practice the element magnitudes of H beyond the boundaries defined by the original array B are small compared to element magnitudes within these boundaries, but they are not zero. These elements are nonzero because the discrete Hilbert transform method, implemented using the discrete Fourier transform, is an approximation to the integral transform [4.26]. The Hilbert procedure may lead to arrays which are not minimum-phase. Substantial numerical work has shown that this situation is unlikely to arise often; in most cases the method yields a minimum-phase version of an array with an amplitude spectrum which closely approximates the amplitude spectrum of the original array.

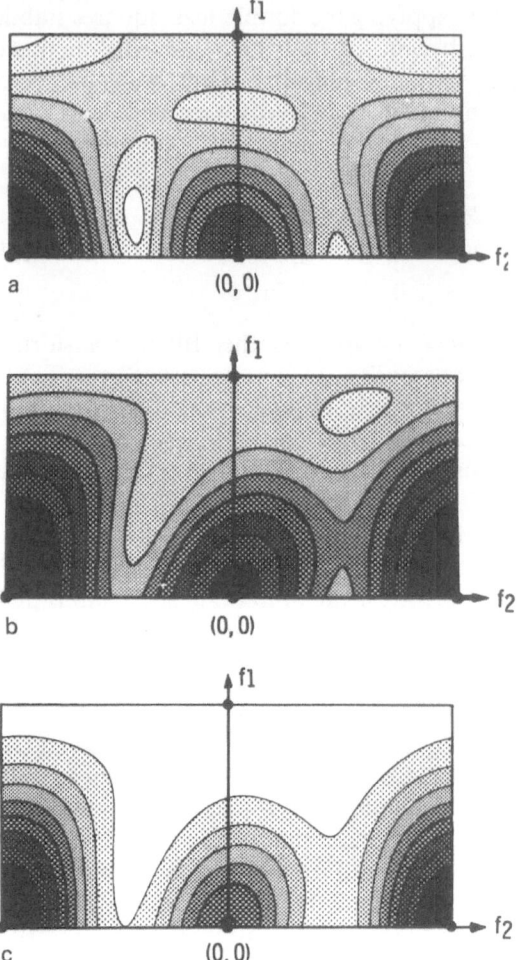

Fig. 4.29a–c. Comparison of amplitude spectra after stabilization techniques. (a) Contours or original spectrum, (b) after Hilbert stabilization technique, and (c) after double least squares technique

Example 2. The polynomial $B(z_1, z_2)$, whose coefficients are given by the 5×5 array

$$B = \begin{bmatrix} 1.00 & 1.50 & -1.90 & -0.80 & 1.10 \\ 1.40 & 2.10 & -2.60 & -1.10 & 1.50 \\ -1.80 & -2.40 & 3.30 & 1.30 & -1.60 \\ -0.70 & -0.90 & 1.10 & 0.50 & -0.80 \\ -0.90 & 1.30 & -1.60 & -0.60 & 1.00 \end{bmatrix} \qquad (4.48)$$

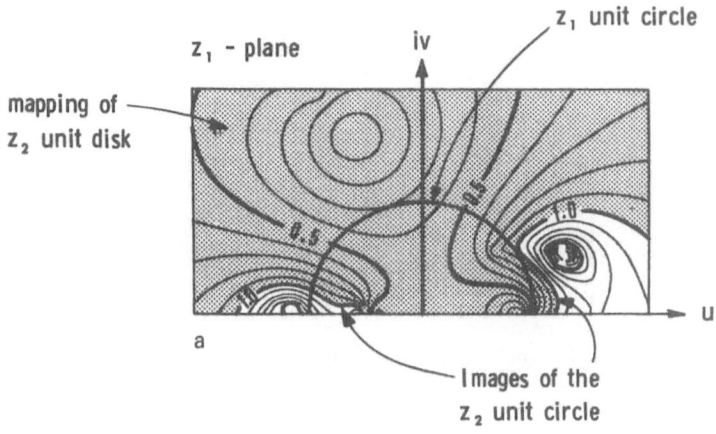

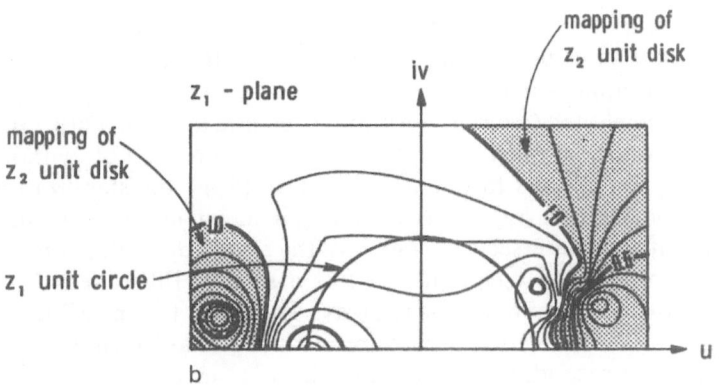

Fig. 4.30a and b. Root maps. (a) A nonminimum-phase pulse and (b) after the Hilbert stabilization technique

has the root map shown in Fig. 4.30a. As in the previous example, the z_2 unit disk mapped in the z_1-plane intersects the z_1 unit disk. Thus, a recursive filter with denominator $B(z_1, z_2)$ is unstable. Application of the Hilbert stabilization procedure leads to the minimum-phase array H,

$$H = \begin{bmatrix} 4.76 & -0.891 & -3.43 & 0.851 & 0.847 \\ -0.392 & 0.206 & 0.305 & 0.0787 & -0.0187 \\ -3.04 & 0.721 & 2.94 & -0.753 & -0.853 \\ 0.204 & -0.469 & -0.503 & 0.238 & 0.165 \\ 0.586 & -0.416 & -0.704 & 0.312 & 0.211 \end{bmatrix} \tag{4.49}$$

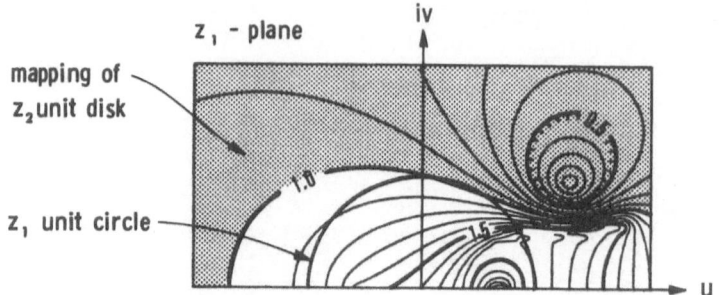

Fig. 4.31. Root map after applying the Hilbert stabilization technique

whose root map is shown to Fig. 4.30b. The mapped z_2 unit disk does not intersect the z_1 unit disk, and the recursive filter with denominator $H(z_1, z_2)$ is stable. Note that the disturbance at approximately $z_1 = (-0.5, 0)$ in Fig. 4.30a is reflected with respect to the z_1 unit circle in Fig. 4.30b to approximately $z_1 = (-2, 0)$.

A Counter-Example and Comments. Although leading to excellent spectral approximations, the Hilbert stabilization procedure is not without drawbacks. When applied to the array given in (4.39), the stabilization procedure produces an array whose root map is shown in Fig. 4.31. Note that the z_1 unit circle intersects the image of the z_2 unit disk after application of the stabilization technique. In this case the Hilbert technique failed to produce a minimum-phase array. There are two possible explanations: first, the theoretical procedure does not guarantee a minimum-phase result, and second, numerical difficulties are associated with this example. The root map of Fig. 4.23 shows that the 1.00 contour remains very close to the periphery of the z_1 unit disk in the zone where penetration into this unit disk occurs. Figure 4.31 shows that the Hilbert technique has reduced, but not eliminated, the extent of this penetration. The phenomenon may be related to numerical difficulties encountered when working with one-dimensional arrays whose z-transform zeroes are close to unity in magnitude.

4.4. Implementation

Once the two-dimensional recursive filter is designed and stabilized, it must be applied to the data. As given in (4.3), it is possible to derive a recursive algorithm that describes the recursion relationship for a filter recursing in the positive m and n directions.

4.4.1. Phase Response of Two-Dimensional Recursive Filters

The recursive filtering algorithm of (4.3) produces a filter with a "one-quadrant" impulse response. If the input to the filter is a unit pulse at some grid point, the filter responds in the positive m and n directions only. Due to its one-quadrant response, the recursive filter cannot have a zero-phase response. However, by the proper combination of two or more recursive filters, we can produce a filtering operation which has zero phase or other useful symmetries.

Zero-Phase Cascade

A zero-phase two-dimensional filter has the particular property that its unit pulse response is symmetric about the (x, y) origin along any radius through the origin. We can perform zero-phase filtering with recursive filters by first filtering an input array with a recursive filter, starting at one corner of the input array. We then take the resultant array and, starting at the diagonally opposite corner, filter the result in the opposite direction with the same recursive filter [4.10]. This equivalent to filtering with the filter

$$G(z_1, z_2) = F(z_1, z_2) \cdot F(1/z_1, 1/z_2), \qquad (4.50)$$

where $F(z_1, z_2)$ is the one-quadrant recursive filter[4].

Setting $z_1 = \exp(j\omega)$ and $z_2 = \exp(j\omega_2)$ in (4.50) gives us the two-dimensional frequency response of $G(z_1, z_2)$. Since $F(e^{j\omega_1}, e^{j\omega_2})$ is the complex conjugate of $F(e^{-j\omega_1}, e^{-j\omega_2})$, $G(e^{j\omega_1}, e^{j\omega_2})$ will be real for any value of (ω_1, ω_2). Therefore, the filter $G(z_1, z_2)$ has zero-phase response. The amplitude response of $G(z_1, z_2)$ will be the square of the amplitude response of $F(z_1, z_2)$. That is, $|G(\omega_1, \omega_2)| = |F(\omega_1, \omega_2)|^2$.

Zero-Phase Sum

We can also get zero-phase response by filtering an input array forward and backward with the same recursive filter and adding the results. This is equivalent to filtering with

$$H(z_1, z_2) = F(z_1, z_2) + F(1/z_1, 1/z_2). \qquad (4.51)$$

Like $G(z_1, z_2)$, $H(z_1, z_2)$ has zero-phase response. However, the amplitude responses are different. The amplitude response of $H(z_1, z_2)$ is

$$|H(\omega_1, \omega_2)| = 2 \, \mathrm{Re} \, \{F(\omega_1, \omega_2)\}. \qquad (4.52)$$

---

[4] Use of this technique requires that the computer output array be sufficiently large to allow output values to decay to reasonably small numbers.

A filtering operation can be made symmetric with respect to the x axis (or y axis) by filtering forward and reverse in the x (or y) direction. For example, if we relate z_1 with the x axis and z_2 with the y axis, $G(z_1, z_2) = F(z_1, z_2) \cdot F(1/z_1, z_2)$ and $H(z_1, z_2) = F(z_1, z_2) + F(1/z_1, z_2)$ are symmetric about the x axis.

Finally, the filtering operations

$$G(z_1, z_2) = F(z_1, z_2) \cdot F(1/z_1, z_2) \cdot F(z_1, 1/z_2) \cdot F(1/z_1, 1/z_2) \qquad (4.53)$$

and

$$H(z_1, z_2) = F(z_1, z_2) + F(1/z_1, z_2) + F(z_1, 1/z_2) + F(1/z_1, 1/z_2) \quad (4.54)$$

are symmetric with respect to both the x and y axes, as well as having zero-phase response.

4.4.2. State Variable Implementation[5]

An implementation that is different from the direct method of (4.3) can be derived by introducing a state-variable array. This alternate method of realizing a recursive algorithm can result in savings in computer storage. The state-variable array is designated by $W(z_1, z_2)$. This is indicated in (4.55)

$$\frac{R(z_1, z_2)}{D(z_1, z_2)} = \frac{A(z_1, z_2) W(z_1, z_2)}{B(z_1, z_2) W(z_1, z_2)}. \qquad (4.55)$$

To see how the introduction of this array affects implementation, the identifications given in (4.56) and (4.57) are made,

$$D(z_1, z_2) = B(z_1, z_2) W(z_1, z_2), \qquad (4.56)$$

$$R(z_1, z_2) = A(z_1, z_2) W(z_1, z_2). \qquad (4.57)$$

Equations (4.56) and (4.57) imply a two-step operation. First, the B array operates recursively on the D array to generate the state-variable array, W; then the A array operates on the W array to produce the output array, R.

---

[5] This material constitutes work by one of the authors (RRR).

Equation (4.56) yields (4.58). Here it is assumed the normalization steps indicated in (4.3) have been carried out.

$$\left(1 + \sum_{\substack{k=1 \\ k \cdot l \neq 1}}^{M_b} \sum_{l=1}^{N_b} b_{kl} z_1^{k-1} z_2^{l-1}\right) W(z_1, z_2) = D(z_1, z_2), \tag{4.58}$$

$$W(z_1, z_2) = D(z_1, z_2) - W(z_1, z_2) \sum_{\substack{k=1 \\ k \cdot l \neq 1}}^{M_b} \sum_{l=1}^{N_b} b_{kl} z_1^{k-1} z_2^{l-1}. \tag{4.59}$$

Recursive Equations

From (4.59) the recursive nature of the algorithm is clear. The inverse transform of (4.59) is given in (4.60).

$$w_{mn} = d_{mn} - \sum_{\substack{k=1 \\ k \cdot l \neq 1}}^{M_b} \sum_{l=1}^{N_b} b_{kl} w_{m-k+1, n-l+1}. \tag{4.60}$$

Equation (4.60) describes the evolution of the state-variable array. The output array R is generated from the state-variable array according to (4.61), i.e.

$$r_{mn} = \sum_{i=1}^{M_a} \sum_{j=1}^{N_a} a_{ij} w_{m-i+1, n-j+1}. \tag{4.61}$$

Note that the state-variable array provides the coupling between the input and the output arrays. Each new state depends only on the input plus previous states. Likewise, each new output depends only on the state variables. The implementation of this algorithm is illustrated in Fig. 4.32.

Comments on Implementation

From 4.32 it would appear that an additional two-dimensional array has been introduced that would actually require more computer storage than the algorithm described in Fig. 4.33, which corresponds to the direct implementation of (4.3). This would be true if all of the arrays involved in the computation were kept in the main storage of the computer. However, for processing large two-dimensional arrays, this is very often impractical. A more usual situation for large data arrays would be to keep the two-dimensional data on an external bulk storage device and transfer it into the computer for processing one column or row at a time. For the sake of illustration, we will consider the data

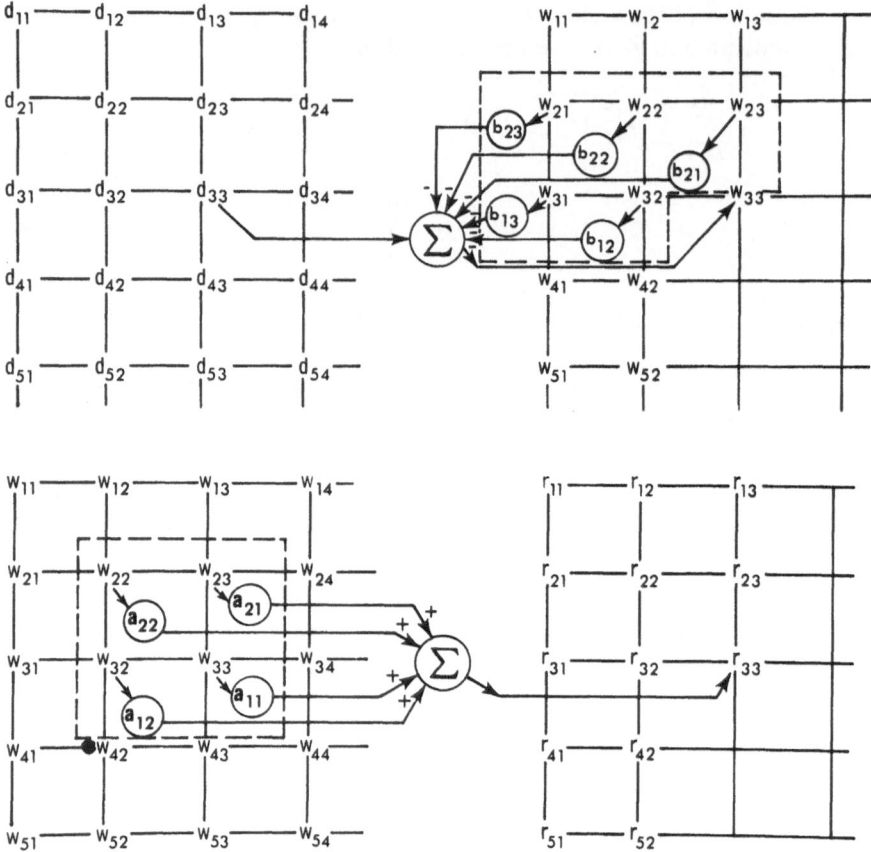

Fig. 4.32. The implementation of a two-dimensional recursive filter using a state-variable array

to be transferred in one column at a time. Under this assumption, Fig. 4.32 shows that to generate one column of output, one column of data would be input and the state-variable array updated using this column, as indicated by (4.60). Then a column of output would be generated by operating on the W array according to (4.61). Using this technique, the number of columns of the W array required in the main storage of the computer at any time would be the maximum of N_a or N_b. From Fig. 4.33 it can be seen that direct implementation of the algorithm requires $N_a + N_b$ columns of data in the main storage at one time. In situations where the processor has limited main storage, this factor can be significant.

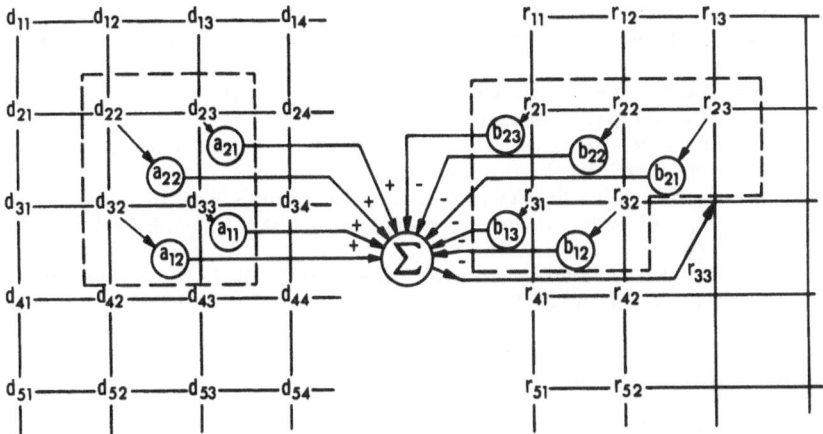

Fig. 4.33. The direct implementation of a two-dimensional recursive filter

References

4.1. T. H. HUANG: IEEE Trans. Audio Electroacoust. AU-**20**, 158 (1972)
4.2. G. A. BLISS: *Algebraic Functions* (Am. Math. Soc., New York 1933).
4.3. B. D. O. ANDERSON, E. D. JURY: IEEE Trans. Audio Electroacoust. AU-**21**, 366 (1973).
4.4. H. G. ANSEL: IEEE Trans. Circuit Theory CT-**11**, 214 (1964)
4.5. A. COHN: Math. J. **14**, 110 (1922).
4.6. M. FUJIWARA: Math. J. **24**, 160 (1926).
4.7. E. I. JURY: *Theory and Application of the z-transform Method* (J. Wiley, New York, 1964)
4.8. S. BARNETT: *Matrices in Control Theory* (Van Nostrand- Reinhold, London, 1971)
4.9. J. H. JUSTICE, J. L. SHANKS: IEEE Trans. Automatic Control AC-**18**, 284 (1973).
4.10. J. L. SHANKS, S. TREITEL, J. H. JUSTICE: IEEE Trans. Audio Electroacoust. AU-**20**, 115 (1972).
4.11. K. J. ÅSTROM: *Introduction to Stochastic Control Theory* (Academic Press, New York, 1970).
4.12. E. I. JURY: IEEE Trans. Circuit Theory CT-**11**, 292 (1964).
4.13. C. M. RADER, B. GOLD: Proc. IEEE **55**, 149 (1967).
4.14. R. M. GOLDEN, J. F. KAISER: Bell Syst. Tech. J. **43**, 1533 (1964).
4.15. V. A. DITKIN, A. P. PRUDNIKOV: *Operational Calculus in Two Variables* (English translation by D. M. G. Wishart) (Pergamon Press, New York, 1962).
4.16. R. G. BROWN, J. W. NILSSON: *Introduction to Linear Systems Analysis* (J. Wiley, New York, 1962), pp. 121–125.
4.17. J. L. SHANKS: Geophysics **32**, 33 (1967).
4.18. C. S. BURRUS, T. W. PARKS: IEEE Trans. Audio Electroacoust. AU-**18**, 137 (1970).
4.19. T. W. PARKS, C. S. BURRUS: "Applications and extensions of Prony's method to parameter identification and digital filtering", presented at 5th Princeton Conf. Inform. Sci. Syst. (March 1971).
4.20. R. A. WIGGINS: "On factoring the correlations of discrete multivariable stochastic processes", MIT Sci. Rep. 9 of Contract AF 19(604) 7378, pp. 127–152, 1965. Also, Ph. D. Thesis, Dept. of Geology and Geophysics, MIT (March 1965).
4.21. R. FLETCHER, M. J. D. POWELL: Computer J. **6**, 163 (1963).

4.22. System/360 Scientific Subroutine Package (360-CM-03X) Version III, Programmer's Manual, Docmt. H20-0205-3 IBM Data Proc. Div., White Plains, New York, U.S.A. (1968).
4.23. H. G. ANSELL: IEEE Trans. Circuit Theory CT-11, 214 (1964).
4.24. T. S. HUANG: Private communication (March 1971).
4.25. E. A. ROBINSON: *Statistical Communication and Detection* (Hafner Publishing Company, New York, 1967), pp. 173–174.
4.26. R. READ, S. TREITEL: IEEE Trans. Geoscience Electronics GE-11, 153 and 205 (1973).
4.27. B. GOLD, C. M. RADER: *Digital Processing of Signals* (McGraw-Hill Book Co., New York, 1969).
4.28. D. E. DUDGEON: "Two-dimensional recursive filtering", Ph. D. dissertation, Dept. of Electr. Engg., MIT (May 1974).

Further References with Titles

N. K. BOSE, P. S. KAMAT: Algorithm for stability test of multi-dimensional filters. IEEE Trans. Acoustics, Speech, Signal Proc. ASSP-22, No. 5 (1974).
N. K. BOSE, E. I. JURY: Positivity and stability test for multi-dimensional filters (discrete-continuous). IEEE Trans. Acoustics, Speech, Signal Proc. ASSP-22, No. 3 (1974).
J. M. COSTA, A. N. VENETSANOPOULOS: Design of circularly symmetric two-dimensional recursive filters. IEEE Trans. Acoustics, Speech. Signal Proc. ASSP-22, No. 6 (1974).
E. I. JURY: The theory and applications of the inners. IEEE Proc. 63, No. 7, 1044–1068 (1975).
G. A. MARIA, M. M. FAHMY: On the stability of two-dimensional digital filters. IEEE Trans. Audio Electroacoustics AU-21, 470–472 (1973).
G. A. MARIA, M. M. FAHMY: An L_p design technique for two-dimensional digital recursive filters. IEEE Trans. Acoustics, Speech, Signal Proc. ASSP-22, No. 1 (1974).
R. M. MERSEREAU, D. E. DUDGEON: The representation of two-dimensional sequences as one-dimensional sequences. IEEE Trans. Acoustics, Speech, Signal Proc. ASSP-22, No. 5 (1974).
S. K. MITRA, A. D. SAGAR, N. A. PENDERGRAS: Realizations of two-dimensional recursive digital filters. IEEE Trans. Circuits and Systems CAS-22, No. 3 (1975).
M.-D. NI, J. K. AGGARWAL: Two-dimensional digital filtering and its error analysis. IEEE Trans. Computers C-23, No. 9 (1974).
D. D. SILJAK: Stability for two-variable polynomials. IEEE Trans. Circuits and Systems CAS-22, No. 3 (1975).

5. Image Enhancement and Restoration

B. R. FRIEDEN

With 17 Figures

The aim of collecting data is to gain meaningful information about a phenomenon of interest. Unfortunately, often the phenomenon is not a direct physical observable. Instead, e.g., the data at hand may be a linear superposition of the desired quantities. This linear, and simplest type of information mixing is endemic in the physical sciences, arising in fields as diverse as atmospheric physics and medial diagnostics (see listing in Section 5.1). The common problem confronting workers in these fields is how to "unmix" (or, restore, enhance, de-blur, de-convolve), the data.

A complicating factor is the phenomenon of random noise, which is inevitably added to the data by the very act of collecting it. In some instances, noise is added even before detection, by the nature of the information channel (say, the turbulent atmosphere). It will be seen that noise is the sole obstacle to perfect restoration of the desired data. By contrast, if the collected data is bandwidth-limited, this does not *in itself* rule out a perfect restoration. Furthermore, even in the presence of noise, it is often possible for the restored bandwidth to significantly (by factors of 2 or more) exceed the data bandwidth. Perhaps an accounting of this significant recent accomplishment will be the major point of interest to the reader of this chapter. (See Section 5.12 onward.)

At this point we should quantify the problem. We shall use the language of optical image formation, for personal convenience, but the reader can easily substitute from his own field (see comparative Section 5.1). Also, the notation will be one-dimensional, for simplicity, even though optical images are usually two-dimensional. In this way, we avoid cumbersome double arguments, subscripts, etc., and can more easily focus attention on the salient features of the restoring methods.

In optical image formation, an unknown spatial radiance distribution $o(x)$ called "the object" produces an image irradiance distribution $i(y)$ which is collected as data $i(y_m)$, $m = 1, ..., M$. It is desired to estimate $o(x)$ at a discrete subdivision of points x_n, $n = 1, ..., N$. We designate the estimate $\hat{o}(x_n)$, $n = 1, ..., N$ or as simply $\hat{o}(x)$ for brevity. The connecting relation between image and object is the linear form

$$i(y_m) = \int_{-X}^{X} dx\, o(x)\, s(y_m; x) + n(y_m), \quad m = 1, ..., M, \qquad (5.1)$$

where s designates the point spread function of the overall image-forming system and n denotes noise (random or otherwise). Note the finite limits X. These result from the fact that any optical instrument has a limited field of view; due to vignetting

$$s(y_m; x) = 0 \quad \text{for all } |x| > X. \tag{5.2}$$

Therefore, even if the object $o(x)$ is of infinite extent the light that forms the data $i(y_m)$ only emanates from effectively a limited region of the object. We are interested in estimating, or restoring, $o(x)$ only over this region.

The preceding is not of mere academic importance. It will be seen below, that because the limits X are finite in (5.1), $o(x)$ can be perfectly restored in the absence of noise n, or restored over a bandwidth exceeding that of data $i(y_m)$ in the presence of significant amounts of noise.

In order to facilitate finding $o(x)$ at the discrete subdivision x_n, $n = 1, ..., N$ it has proven convenient [5.1] to replace the integral in (5.1) by an approximating sum, so that now

$$i(y_m) = \sum_{n=1}^{N} w_n\, o(x_n)\, s(y_m; x_n) + n(y_m), \quad m = 1, ..., M. \tag{5.3}$$

Numbers $\{w_n\}$ are input weights (say, due to Simpson's rule) for facilitating accuracy in the approximating sum.

At this point in time, there is no single optimum method for estimating a *general* $o(x)$. The growing body of empirical evidence suggests, in fact, that there is a "best" (however one defines it) restoring method *relative to* the type of a priori information regarding object and noise one might have at hand. Such types of prior information as object shape (whether smooth or sharp, edge-type or impulse-type), or object and noise statistics, prove important in fostering a choice of method.

Other, more peripheral factors, such as the amount of data to be processed, and permissible computer cost, of course also must enter into the choice. These factors must often be balanced against an "optimum" choice based purely on accuracy. The reason is that, at this time, most of the reliably accurate restoring methods for exceeding the data bandwidth in the presence of noise appear too slow to allow their use on extensive objects. This is because most were developed for specifically one-dimensional problems, where such operations as iterative convergence and iterative matrix inversion are not excessively time-consuming.

For these reasons, we shall be content to examine a variety of proposed restoring methods, specifically those that have appeared in the past 10 or so years. The user will still have to decide for himself which one to use!

5.1. Nomenclature and References, for Non-Optical Fields

Most recent progress in solving (5.1) or (5.3) has been made by workers in fields exterior to optics. Exercising a great deal of imagination (and rewarded with doctoral degrees, in many cases) these workers have produced restoring algorithms that are as different as are the fields themselves. Some of the more promising of these methods, toward solution of the image restoration problem, will be described below.

The number and diversity of physical and mathematical problems that can be modeled after (5.1) are astonishingly large. See, e.g., the partial listing given in TWOMEY [5.2]. Likewise is the number of different names given to the problem (or course, at least one for each field). A partial list, by field, giving problem name and one or two representative references, is as follows:

1) Atmospheric physics; indirect sensing; TWOMEY [5.2], YAMAMOTO and TANAKA [5.3],

2) Fourier transform spectroscopy; Fourier inversion of the interferogram; VANASSE and SAKAI [5.4],

3) Geophysics; estimation of power spectrum from autocorrelation function; BURG [5.5], LACOSS [5.6],

4) Medical diagnostics; reconstruction of pictures from their projections: GORDON and HERMAN [5.7],

5) Numerical analysis; inversion of the Fredholm equation of the first kind: PHILLIPS [5.1],

6) Radio astronomy; antenna de-smoothing, de-convolution: SCHELL [5.8], BIRAUD [5.9],

7) Spectroscopy; de-convolution: JANSSON [5.10],

8) Statistics; estimation of probability densities from moment measurements: JAYNES [5.11], WRAGG and DOWSON [5.12].

The reader can probably add to this list.

5.2. Preliminary Mathematics, Notation and Definitions

We have already defined the *physical* quantities: object $o(x)$, its required estimate $\hat{o}(x)$, image $i(y)$ and noise $n(y)$, at (5.1). Some mathematical definitions are as follows.

Given a function f of space coordinate x, its Fourier transform F of spatial frequency (radians/length) ω is defined as

$$F(\omega) = (2\pi)^{-\frac{1}{2}} \int\limits_{-\infty}^{\infty} dx \, f(x) \exp(-j\omega x). \tag{5.4}$$

The Fourier transform operation will sometimes be denoted by F.T. The "power spectrum" $\phi_f(\omega)$ corresponding to spectrum F is defined as the ensemble average

$$\phi_f(\omega) \equiv \langle |F(\omega)|^2 \rangle. \tag{5.4a}$$

In general, functions of space coordinate x will be lower case, as f, with corresponding Fourier transforms upper case, as F. Thus, $o(x)$ and $O(\omega)$ are an F.T. pair, etc. for $i(y)$, $I(\omega)$ and $n(y)$, $N(\omega)$. The exception (there must always be one) to the rule is the F.T. of point spread function s, which will be denoted as $\tau(\omega)$, in conformity with the notation of O'Neill [5.13].

Regarding notation for vectors and matrices, the vector corresponding to $f(x_n)$, $n = 1, \ldots, N$ will be designated F (capitalized), and the matrix corresponding to $s(y_m; x_n)$ will be $[S]$. Matrices and Fourier transforms will never be used together, so there is no problem of confusing one for the other.

With the choice (5.4) of Fourier transform, the inverse relation between f and F becomes

$$f(x) = (2\pi)^{-\frac{1}{2}} \int_{-\infty}^{\infty} d\omega\, F(\omega) \exp(+j\omega x). \tag{5.5}$$

A function $f(x)$ is called "frequency-band limited", or simply "band limited" if there exists sharp cutoff in its Fourier transform (or, spectrum): $F(\omega) = 0$ for $|\omega| > \Omega$. Frequency Ω is then called the "cut-off frequency", and the infinite limits in (5.5) may be replaced by $\pm \Omega$.

Conversely, a function $F(\omega)$ may be called "space-band limited" if a length X exists such that $f(x) = 0$ for $|x| > X$. In ordinary language, $f(x)$ is of limited spatial extent. As an example, we see from (5.1) and (5.2) that $o(x)$ is of limited spatial extent, so that $O(\omega)$ is space band-limited.

It is useful, when working with band-limited functions, to define special functions $\text{Rect}(x)$ and $\text{sinc}(x)$:

$$\text{Rect}(x) \equiv \begin{cases} 1 & \text{for} \quad |x| \leq 1 \\ 0 & \text{for} \quad |x| > 1, \end{cases} \tag{5.6a}$$

$$\text{sinc}(x) \equiv x^{-1} \sin(x). \tag{5.6b}$$

The three concepts (5.4) and (5.6a) and (5.6b) can be combined by observing that F.T. of $\text{Rect}(x) = 2\,\text{sinc}(\omega)$. This serves to display the chief virtue for definitions (5.4)–(5.6): the avoidance of factors 2π in the arguments of functions.

A band-limited function $f(x)$ obeys the "Whittaker-Shannon sampling theorem"

$$f(x) = \sum_{n=-\infty}^{\infty} f(n\pi/\Omega) \operatorname{sinc}(\Omega x - n\pi),$$ (5.7a)

all x

and its nameless (but useful) counterpart

$$F(\omega) = (\pi/\Omega) \sum_{n=-\infty}^{\infty} f(n\pi/\Omega) \exp(-jn\pi\omega/\Omega).$$ (5.7b)

$|\omega| \leq \Omega$

Equation (5.7a) is sometimes called the "perfect interpolation" formula: If f is known at a subdivision of points x spaced by π/Ω, then by (5.7a) it is known everywhere between these points. Points $x_n = n\pi/\Omega$ then take on the meaning of critical points, or independent degrees of freedom, for the function f. Interval π/Ω is often called the "Nyquist interval".

Equations (5.7) may be proved by the use of (5.5) (now with finite limits Ω) combined with the observation that since $F(\omega)$ is non-zero only over a finite interval $|\omega| \leq \Omega$ it may be represented by a Fourier series over that interval. The coefficients of that series turn out, from (5.5), to be precisely $f(n\pi/\Omega)$. Substitution of these into the Fourier series yields (5.7b). Equation (5.7b) is then substituted back into (5.5) and integrated termwise to yield (5.7a).

If $O(\omega)$ is known perfectly over a finite frequency domain $|\omega| \leq \Omega$, the "principal value" or "perfect bandlimited" estimate $\hat{o}_{pb}(x)$ of the object may be formed. This is defined as

$$\hat{o}_{pb}(x) = (2\pi)^{-\frac{1}{2}} \int_{-\Omega}^{\Omega} d\omega\, O(\omega) \exp(+j\omega x).$$ (5.8)

Because the perfect bandlimited estimate is a purely deterministic quantity with a straight-forward meaning, it makes a useful benchmark for comparison with proposed restoring methods.

A mathematical identity which we shall use abundantly below is the Wiener-Khintchine theorem:

$$\int_{-\infty}^{\infty} dx\, f(x)\, g(x) \exp(-j\omega x) = \int_{-\infty}^{\infty} d\omega'\, F(\omega')\, G(\omega - \omega').$$ (5.9)

This is easily derived by substituting for f and g in (5.9) their Fourier integrals (5.5), and performing the x-integration. The identity

$$\int_{-\infty}^{\infty} dx \exp(\pm j\omega x) = 2\pi \delta(\omega) \qquad (5.9a)$$

aids this procedure. $\delta(\omega)$ is the Dirac delta function, defined as

$$\delta(x) = 0 \quad \text{for} \quad x \neq 0,$$

but

$$\int_{-\infty}^{\infty} dx\, \delta(x) = 1. \qquad (5.10)$$

The right-hand side of (5.9) defines the mathematical operation called a "convolution". It is convenient to give this the special notation

$$F \otimes G$$

(convolution). $\qquad (5.11)$

Finally, it is useful to mention the "sifting" property of the Dirac delta function. When $\delta(x)$ is one member of a convolution operation, we may establish via properties (5.10) the result

$$\int_{-\infty}^{\infty} dx'\, f(x')\, \delta(x - x') = f(x). \qquad (5.12)$$

In words, the single value $f(x)$ is sifted from the infinite domain of values $f(x')$ integrated over on the left-hand side. This property and (5.9a) are the main practical uses of the delta function.

5.3. Intuitive Restoring Methods

There are at least two methods of solving for $o(x)$ which immediately appeal to our intuition. The *discrete* problem (5.3) suggests regarding $o(x)$ as a set of unknowns $o(x_n)$ in a system of linear equations. The *continuous* problem (5.1) suggests, to people with engineering backgrounds, the taking of a Fourier transform and subsequent filtering operations. We shall address both these ideas (in Subsection 5.3.1 and Section 5.8), and see where they lead.

5.3.1. Matrix Inversion of the Discrete Imaging Equation

Equation (5.3) actually represents a system of M equations linear in N unknowns $o(x_n)$. Defining a matrix $[S] = \{w_n s(y_m; x_n)\}$, and vectors $\mathbf{I} = \{i(y_m)\}$, $\mathbf{N} = \{n(y_m)\}$ and $\mathbf{O} = \{o(x_n)\}$, (5.3) becomes

$$\mathbf{I} = [S]\mathbf{O} + \mathbf{N}. \tag{5.13}$$

If we make $M = N$, then $[S]$ becomes a square matrix, with an inverse; and $[O]$ may be solved for as

$$\mathbf{O} = [S]^{-1}\mathbf{I} - [S]^{-1}\mathbf{N}. \tag{5.14}$$

Hence, if the noise is zero, (5.14) provides the exact solution to the problem.

However, the presence of even a slight amount of noise in data $\mathbf{I}$ completely ruins this solution, filling it with spurious oscillations. Figure 5.1 from PHILIPPS [5.1] illustrates the situation. The maximum noise value was 0.004 (average value 0.0014), on a scale where the maximum signal (noiseless) image value was 3. This represents an average

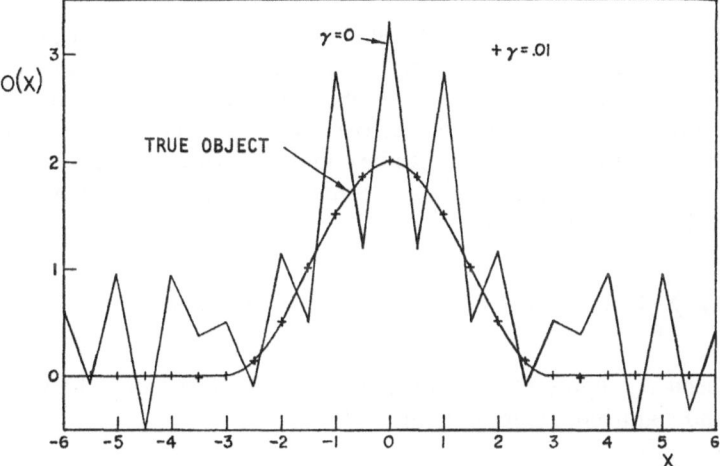

Fig. 5.1. Illustrating the failure of the direct-inverse approach. The true object is the smooth curve. The jagged curve labeled $\gamma = 0$ is the result of using the straightforward matrix-inverse approach indicated in (5.14) (with $N = 0$ assumed). The reason the approach fails is basically because, contrary to the assumption, N is not zero for this case; and no provision is made by the approach for smoothing out its effects. By contrast, Phillips' algorithm (5.22) permits noise-smoothing, through input parameter γ. The points labeled $+$ in the figure result from use of algorithm (5.22) with $\gamma = 0.01$. (After PHILLIPS [5.1])

signal-to-noise value of $3/0.0014 = 2140$, extraordinarily large and virtually impossible to attain experimentally. Nevertheless, Fig. 5.1 shows that the restoring method (5.14) yields a very poor answer.

What is the reason for this idiopathic reaction to noise in the data? Basically, it is because matrix $[S]$, representing the point spread function, is mainly filled with zeroes and small elements near the diagonal, which causes $[S]^{-1}$ to have very large elements. Hence, at points y_m where noise n_m is finite, the error term $[S]^{-1} N$ in (5.14) is very large.

The reason that the error is also inherently *oscillatory* has been described by PHILLIPS [5.1]: The observables $\{i_m\}$ are invariant to the presence of a sinusoid $\sin(kx)$, k large, that is added to the true solution $o(x)$. Invariance stems from cancellation of positive and negative contributions of the $\sin(kx)$ to the superposition operation (5.3). Evidently, only an oscillatory error term like $\sin(kx)$ would cause such cancellation. Hence, in the presence of noise in the data any estimate $\hat{o}(x)$ that is made consistent with (5.3) alone is liable to be in error by a high-frequency, oscillatory term. Evidently, something aside from consistency with (5.3) is required in order to overcome this problem.

5.4. Phillips' Smoothing Method

We may notice in Fig. 5.1 that the superficial oscillations in $\hat{o}(x)$ incur high values for derivative $d^2 o(x)/dx^2$ or, equivalently, for second-differences $\hat{o}_{n+1} - 2\hat{o}_n + \hat{o}_{n-1}$. It seems reasonable, then, that a solution $\hat{o}(x)$ with *small* second-differences over its extent tends to be accurate. At the same time, we want $\hat{o}(x)$ to be consistent with the image inputs via (5.3). A reasonable tack, then, is to trade off consistency for smoothness, and seek the $\hat{o}(x)$ satisfying

$$\gamma \sum_{n=1}^{N} (\hat{o}_{n+1} - 2\hat{o}_n + \hat{o}_{n-1})^2 + \sum_{m=1}^{M} \hat{n}_m^2 = \text{minimum} . \qquad (5.15)$$

The terms $\hat{n}_m$ represent the "mismatch" or "inconsistency" between the left- and right-hand sides of (5.3)

$$\hat{n}_m = i(y_m) - \sum_{n=1}^{N} w_n \, \hat{o}(x_n) \, s(y_m; x_n) . \qquad (5.16)$$

The magnitude of the chosen parameter γ fixes the extent to which smoothness [the first term in (5.15)] is to dominate over data consistency.

5.4.1. Derivation

The solution $\hat{o}(x)$ to problem (5.15) and (5.16) is found by the usual methods of calculus. Setting the partial derivatives $\partial/\partial\hat{n}_m$ of (5.15) equal to zero yields

$$\gamma \sum_{n=1}^{N} (\hat{o}_{n+1} - 2\hat{o}_n + \hat{o}_{n-1})(\partial\hat{o}_{n+1}/\partial\hat{n}_m - 2\partial\hat{o}_n/\partial\hat{n}_m + \partial\hat{o}_{n-1}/\partial\hat{n}_m) \tag{5.17}$$
$$+ \hat{n}_m = 0.$$

But from (5.14) we know the desired partials:
Equations (5.14) may be expressed as (assuming $M = N$)

$$\hat{o}_j = \sum_k \alpha_{jk}(i_k - n_k), \quad \text{with} \quad [S]^{-1} \equiv \{\alpha_{jk}\} \tag{5.18}$$

so that

$$\partial\hat{o}_j/\partial\hat{n}_k = -\alpha_{jk}. \tag{5.19}$$

Hence, (5.17) becomes

$$-\gamma \sum_{n=1}^{N} (\hat{o}_{n+1} - 2\hat{o}_n + \hat{o}_{n-1})(\alpha_{n+1,m} - 2\alpha_{nm} + \alpha_{n-1,m}) + \hat{n}_m = 0. \tag{5.20}$$

This is a linear equation in $\hat{o}$, which can be written as

$$-\gamma[B]\,O + N = 0, \tag{5.21}$$

in terms of a matrix $[B]$ defined in (5.20). Eliminating N from the combined Eqs. (5.13), (5.21) yields the object solution

$$\hat{O} = ([S] + \gamma[B])^{-1} I \tag{5.22}$$

and the noise solution

$$\hat{N} = -\gamma[B]\,\hat{O}. \tag{5.23}$$

5.4.2. Discussion

As a check on the solution, we let $\gamma = 0$, which means *no* smoothing, from (5.15). The results (5.22) and (5.23) mirror this, in that (5.22) becomes just the matrix-inverse solution (5.14) with $N = 0$. More interestingly, as γ is increased solution (5.22) gives more weight to smoothing and less to data consistency, just as (5.15) would require. For γ too large, $\hat{O}$ is even more blurred than is the data I.

Evidently, the user must select a value γ somewhere between these two extremes. This is where estimate (5.23) for $\hat{N}$ enters in. The user generally knows something about the image noise, typically its variance to some accuracy. A *correct* solution (5.23) (i.e., one using about the correct γ), should then yield a noise estimate $\hat{N}$ with about the known variance. Hence, in order to find the right γ the user must generate a few solutions (5.22) and (5.23) based on different γ-values and then select the one whose noise estimate best matches his *a priori* information (say, the variance) about noise.

The points indicated by plus (+) in Fig. 5.1 correspond to use of $\gamma = 0.01$ in restoring formula (5.22). Further discussion on the subject of γ-selection may be found in PHILLIPS [5.1].

5.4.3. Relative Merits for Image-Enhancement Use

The chief advantage of the method is its ability to smooth out noise without unduly blurring the output (provided a proper γ is selected). And, in the limit of zero noise the output $\hat{o}(x)$ is perfect.

Regarding computation time, (5.22) is a closed-form solution requiring but two matrix inversions and one matrix-vector multiplication. These would be definite advantages for a one-dimensional problem.

However, regarding the *two*-dimensional world of image-enhancement, we note that each of the matrices $[S]$, $[B]$ contains $N \times N$ elements, where N is the total number of image inputs. Hence, for a 32×32 matrix of inputs these matrices are 1024 elements on a side! The problem of inverting matrices of this size is formidable (except in special cases), but not inconceivable, with the growing capabilities of computers.

Other disadvantages of the method from the optical standpoint are:

1) The uncertainty in choice of γ.

2) The unconstrained nature of the estimate: e.g., although we know $o(x)$ is positive, output $\hat{o}(x)$ of (5.22) is not necessarily so. Constrained restoring methods will be seen, below, to have important advantages over unconstrained ones.

3) Oscillatory noise in $\hat{o}(x)$ is suppressed, at the expense of smoothing $\hat{o}(x)$ through use of second-difference minimization. However, in many optical cases the object features of interest have *high* second-differences, as in the case of star fields, edges, or line spectra. Therefore, for optical purposes a better criterion than second-difference minimization might well be used. This may be expedited as follows.

5.5. Twomey's Generalized Approach

We may note that PHILLIPS' second-difference sum in (5.15) is quadratic in $\hat{o}(x)$. In seeking to use constraints other than second-difference minimization, TWOMEY [5.2] found that *any* constraint which is *quadratic* in $\hat{o}(x)$ may be used to produce a solution resembling (5.22). Examples might be minimization of third-differences, in

$$\gamma \sum_{n=1}^{N} (\hat{o}_{n+1} - 3\hat{o}_n + 3\hat{o}_{n-1} - \hat{o}_{n-2})^2 + \sum_{m=1}^{M} \hat{n}_m^2 = \min. , \qquad (5.24)$$

minimization of departure from a known curve $p(x)$ (e.g., the image), in

$$\gamma \sum_{n=1}^{N} (\hat{o}_n - p_n)^2 + \sum_{m=1}^{M} \hat{n}_m^2 = \min . \qquad (5.25)$$

or even a linear combination of these. For this general class of constraints, he found a solution

$$\hat{O} = ([S]^T [S] + \gamma [H])^{-1} ([S]^T I + \gamma P) . \qquad (5.26)$$

The derivation is similar to that of (5.22), with the exception of differentiating the new form (5.24) or (5.25), etc., with respect to the $\hat{o}_n$ instead of with respect to $\hat{n}_m$.

In (5.26), the form of matrix H depends on the particular constraints used, and $P = 0$ if a "bias" curve $p(x)$ is not known. For example, with Phillips' criterion (5.15) H is

$$\begin{bmatrix}
1 & -2 & 1 & 0 & 0 & — & — & — \\
-2 & 5 & -4 & 1 & 0 & — & — & — \\
0 & 1 & -4 & 6 & -4 & 1 & — & — \\
0 & 0 & 1 & -4 & 6 & -4 & 1 & — \\
\cdot & \cdot & \cdot & \cdot & \cdot & \cdot & \cdot & \cdot \\
\cdot & \cdot & \cdot & \cdot & \cdot & \cdot & \cdot & \cdot
\end{bmatrix} \qquad (5.27)$$

while for third-difference minimization, H is, see [5.14],

$$\begin{bmatrix}
1 & -3 & 3 & -1 & 0 & 0 & 0 & — & — \\
-3 & 10 & -12 & 6 & -1 & 0 & 0 & — & — \\
3 & -12 & 19 & -15 & 6 & -1 & 0 & — & — \\
-1 & 6 & -15 & 20 & -15 & 6 & -1 & — & — \\
0 & -1 & 6 & -15 & 20 & -15 & 6 & -1 & — \\
0 & 0 & -1 & 6 & -15 & 20 & -15 & 6 & -1 \\
\cdot & \cdot & \cdot & \cdot & \cdot & \cdot & \cdot & \cdot & \cdot
\end{bmatrix} . (5.28)$$

Another useful aspect of solution (5.26) is that matrix $[S]$ does not have to be square, as it has to be in Phillips' solution (5.22). This allows either an overconstrained or an underconstrained problem: either more image data points than unknown object values, or vice versa. An overconstrained situation might, e.g., produce beneficial data averaging in (5.26). TWOMEY [5.14] exemplifies use of the method in a tabular comparison of $\hat{o}(x)$ with other estimates.

5.5.1. Relative Merits for Image-Enhancement Use

The benefits and drawbacks of this method generally follow those of PHILLIPS. In addition, there are the following advantages:

1) Allowance for either an under- or overconstrained problem.

2) Flexibility as to choice of a smoothing criterion. This may alternatively be viewed as a disadvantage, since now the user has even more decisions to make than a mere selection of γ.

5.6. Subsequent Methods

Later work on solving the discrete imaging Eq. (5.3) has presumed the user to have at hand specific statistical information, such as covariance matrices, about $o(x_n)$ and/or $n(y_m)$. See, e.g., STRAND and WESTWATER [5.15]. This kind of information is not, however, usually available to the person with a two-dimensional image enhancement problem, so we shall not further examine these methods.

5.7. Linear or Non-Linear Methods?

Having talked about linear methods of inverting the *discrete* (or matrix) imaging Eq. (5.3), we now address the problem of linearly inverting, or filtering, the *continuous* version (5.1). An enormous amount of research as been done on this problem. The reader may wish to consult the bibliography recently provided by PRATT [5.16], or the extensive list of references in HUANG et al. [5.17].

But, aside from an embarrassment of riches, there is a more basic reason for not taking a global view of the linear field of operations. Put simply, it is that better image-processing methods exist than linear methods. By better, we do not mean speedier (for they are decidedly not) but rather more accurate in output $\hat{o}(x)$. These are the *non*-linear, bandwidth extrapolation schemes due to BURG [5.5], BIRAUD [5.9] and SCHELL [5.8], JANSSON [5.10], and FRIEDEN [5.18], [5.19]. In addition to being nonlinear, all but BURG's method are also iterative.

Justification for the continued use of filtering methods lies in their computational speed, and their application when there already is *enough* bandwidth in the filtered estimate. For example, some sources of image blur, such as image motion, intrinsically have infinite bandwidth, and so do not require an extrapolation thereof. Also, object scenes which contain mainly smooth features have most of their spectrum packed near the origin (DC level) in frequency space.

Regarding speed, the recent advent of the FFT (Fast Fourier Transform) algorithm due to GOOD [5.20], and COOLEY and TUKEY [5.21] has allowed application of filtering methods to extended 2-dimensional images (the order of 64×64 elements). Later extension of the algorithm due to SINGLETON [5.22] allows, in principle, any size array to be FFT-transformed. Numerous examples of the use of the method have been published, e.g., see MCGLAMERY [5.23].

A more minor consideration is the ease with which linear filtering methods may be analyzed (for ultimate resolution, noise propagation, etc.) by analytic means. By contrast, basically trial-and-error methods are needed for analysis of the non-linear methods.

It might further be argued that linear methods are sometimes *optimum*, but this is untrue in the optical case, as we next show.

If, at each y_m-point the signal image (convolution part) and noise in (5.1) are jointly normal, then the best mean-square estimate of the signal *is* linear in $i(y_m)$ (see [Ref. 5.24, p. 256]). Alternatively, the maximum-likelihood estimate of the signal is also linear if signal and noise are independently normal (see [Ref. 5.25, p. 336]). Regarding estimation of the object $o(x)$, since it connects linearly with the signal image (via the deterministic spread function) it also would be linear in the data $i(y_m)$ for these two cases.

However, it is a special property of optical signal images that *they cannot obey normal statistics*. This is because they are always positive quantities. Hence, the basic premise of the above arguments for linear estimation as the optimum procedure are invalid in the optical case. Again, this dovetails with the empirical fact that decidedly *non*-linear methods (as cited above) have proven superior. Moreover, these methods are nonlinear so as to specifically constrain the output to be positive (and non-normal)!

5.8. Filtering Methods Defined

If the point spread function $s(y_m; x)$ in (5.1) is shift-invariant (sometimes called a condition of "stationarity", or "isoplanatism") then $s(y_m; x)$

$= s(y_m - x)$ and a Fourier transform of (5.1) yields

$$I(\omega) = O(\omega)\,\tau(\omega) + N(\omega) \tag{5.29}$$

(see Section 5.2 on notation).

Filtering methods consist of multiplication of spectrum $I(\omega)$ by a chosen filter function $Y(\omega)$,

$$\hat{O}(\omega) \equiv Y(\omega)\,I(\omega) \tag{5.30a}$$

$$= Y(\omega)\,\tau(\omega)\,O(\omega) + Y(\omega)\,N(\omega), \tag{5.30b}$$

with subsequent Fourier transformation to yield an output

$$\hat{o}(x) = (2\pi)^{-\frac{1}{2}} \int_{-\Omega_p}^{\Omega_p} d\omega\,\hat{O}(\omega)\,e^{j\omega x}. \tag{5.30c}$$

Ω_p is a cut-off frequency, which is necessarily finite for various reasons discussed below. Alternatively, this filtering operation can be performed purely by convolution in the space domain, by noting that (5.30a) and (5.30c) are equivalent to [via the identity (5.9)] the single operation

$$\hat{o}(x) = \int_{-\infty}^{\infty} dx'\,i(x')\,y(x - x'). \tag{5.31}$$

Finally, it is useful for later analysis purposes to define a net transfer function

$$\tau_{net}(\omega) \equiv \tau(\omega) \cdot Y(\omega) \tag{5.32}$$

describing the overall image blurring-post filtering operations. Use of (5.32) in (5.30b) discloses that the signal part of output $\hat{o}(x)$ connects with the true $o(x)$ by a net image blurring-post convolution spread function $s_{net}(x)$

$$\hat{o}_{sig}(x) = \int_{-\infty}^{\infty} dx'\,o(x')\,s_{net}(x - x') \tag{5.33}$$

where

$$s_{net}(x) = (2\pi)^{-\frac{1}{2}} \int_{-\infty}^{\infty} d\omega\,\tau(\omega)\,Y(\omega)\,e^{j\omega x}.$$

Every linear filtering method may be analyzed for its resolution and sidelobe properties through $s_{net}(x)$, and so we shall specify $s_{net}(x)$ for each such method described below.

5.9. Inverse Filtering

Equation (5.29) may be formally solved for $O(\omega)$, as

$$O(\omega) = I(\omega)/\tau(\omega) - N(\omega)/\tau(\omega), \quad \text{for} \quad \tau(\omega) \neq 0 .\tag{5.34}$$

Evidentally, at frequencies for which τ is finite and the ratio N/I is sufficiently small the approximation to (5.34)

$$\hat{O}(\omega) = I(\omega)/\tau(\omega)$$
$$\tau(\omega) \neq 0 \tag{5.35}$$

will yield an accurate restoration $\hat{O}$. This is the method of "inverse filtering", sometimes also called "frequency- or transfer function-compensation". In the nomenclature of Section 5.8, this method employs a filter function

$$Y(\omega) = \text{Rect}(\omega/\Omega_p)\, \tau(\omega)^{-1} , \tag{5.36}$$

which has as its aim the net transfer function

$$\tau_{net}(\omega) = \text{Rect}(\omega/\Omega_p) \tag{5.37}$$

and net spread function

$$s_{net}(x) = \text{sinc}(\Omega_p x) . \tag{5.38}$$

Quantity $2\Omega_p$ is the processing bandwidth for the method. Evidently, from the result (5.37) and Eq. (5.33), it is desirable to have Ω_p as large as possible [in the limit $\Omega_p \to \infty$, $s_{net}(x) \propto \delta(x)$]. However, it is the property of optical images that

$$\tau(\omega) = 0 \quad \text{for all} \quad |\omega| \geq \Omega , \tag{5.39}$$

where Ω is proportional to the smallest optical aperture in the overall imaging system. Hence, $Y(\omega)$ in (5.36) is undefined unless $\Omega_p < \Omega$.

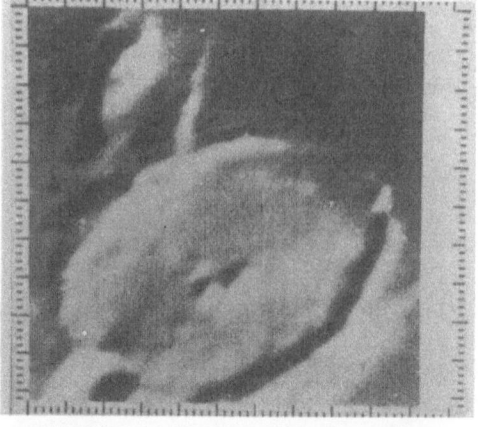

a

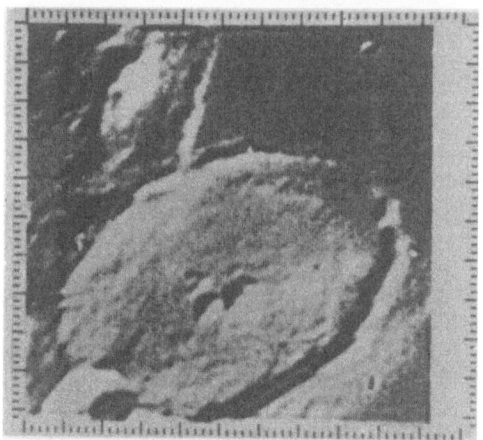

b

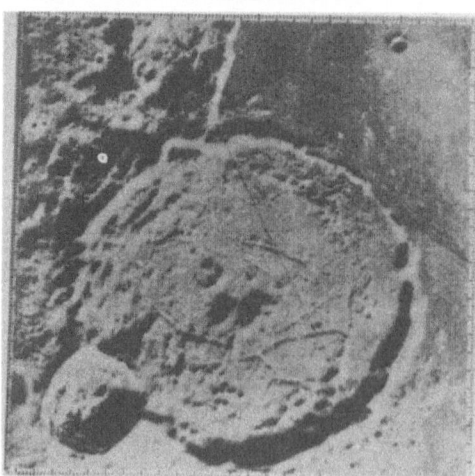

c

Fig. 5.2. (a) Unenhanced picture of the lunar crater Gassendi taken with the 24-in. telescope at the Aerospace Corporation San Fernando Observatory. The blur is due to atmospheric turbulence. (b) Enhanced version of (a), using inverse-filtering. (c) Lunar Orbiter view of Gassendi, providing "ground truth". By inspection, many details of (c) not present in (a) are restored in (b). Long-term exposure through atmospheric turbulence is characterized by roughly a gaussian transfer function, which is not bandlimited and hence very effectively compensated out by inverse filtering. (After O'HANDLEY and GREEN [5.62])

Thus, inverse filtering is limited in signal quality by a net psf (5.38), whose resolution distance (x-coordinate of first zero) x_0 obeys

$$x_0 = \pi/\Omega_p \qquad (5.40)$$

and whose largest sidelobe is about $2(3\pi)^{-1} \simeq 0.22$. However, we wish to emphasize here that (5.40) is *not* the *smallest* possible x_0 over all choices of linear filters over bandwidth $2\Omega_p$. This should become abundantly clear from examples given below. Other problems regarding noise are deferred until later. Finally, the problem of isolated zeroes in $\tau(\omega)$ *within* the processing bandwidth must be confronted; see, e.g., SONDHI [5.26].

For all its problems, inverse filtering has produced the most impressive examples of enhanced *extended* images. See, e.g., Fig. 5.2. In the main, inverse filtering owes its widespread use to the Fast Fourier transform algorithm (see Section 5.7) which, owing to its speed, permits large 2-dimensional matrices of image values to be Fourier transformed. On the other hand, inverse filtering has the following drawbacks:

1) Resolution (5.40) is often insufficient for practical use. This stems from the choice (5.37) for a net transfer function, which is not optimum for resolution. (See Subsection 5.10.1 and Eqs. (5.83), (5.84) for alternatives.)

2) Sidelobes of 0.22 are often unacceptably large insofar as the false detail they produce in output $\hat{o}(x)$.

3) Unphysical (negative) outputs $\hat{o}(x)$ often result over extended regions of x. This ties in with the drawback 2), since a positive-constrained output $\hat{o}(x)$ could not produce spurious negative ripples, or even substantial positive ones in regions where the object is at zero radiance. [The presence of spurious positive ripples would require adjacent negative ones in order to cancel their contribution to the image integral (5.1) and produce zero radiance. See the discussion prior to Section 5.4.]

5.9.1. An Optimum Processing Bandwidth

We have see that, purely on the basis of *resolution* (5.40), it is desirable to have Ω_p as large as possible. On the other hand, since optical systems suffer an attenuation in $\tau(\omega)$ as ω increases, i.e.

$$\tau(\omega) \to 0 \quad \text{as} \quad \omega \to \Omega_p < \Omega \qquad (5.41)$$

algorithm (5.35) has an error term $N(\omega)/\tau(\omega)$ which grows unlimitedly as $\omega \to \Omega_p$. Evidently, there is a possibility for optimizing the choice of Ω_p

through balance of the errors in output $\hat{o}(x)$ due to insufficient resolution and noise propagation. This may be accomplished as follows. (This derivation is not available in the open literature, so we present it in full: it also typifies the beautiful kinds of analysis to which linear methods lend themselves.)

At any x, the error $e(x)$ in $\hat{o}(x)$ due to the use of filter (5.36) in inverse Fourier relation (5.30c) is, using (5.29)

$$
\begin{aligned}
e(x) &\equiv o(x) - \hat{o}(x) \\
&= \int_{-\infty}^{\infty} d\omega \, O(\omega) \, e^{j\omega x} - \int_{-\Omega p}^{\Omega p} d\omega \, [O(\omega) + N(\omega)/\tau(\omega)] \, e^{j\omega x} \\
&= \left(\int_{-\infty}^{-\Omega p} + \int_{\Omega p}^{\infty} \right) d\omega \, O(\omega) \, e^{j\omega x} - \int_{-\Omega p}^{\Omega p} d\omega \, [N(\omega)/\tau(\omega)] \, e^{j\omega x} .
\end{aligned}
\tag{5.42}
$$

Factor $(2\pi^{-\frac{1}{2}})$ is suppressed.

Let us define the optimum Ω_p as the one which minimizes the m.s. error

$$
\varepsilon^2 \equiv \left\langle \int_{-\infty}^{\infty} dx \, |e(x)|^2 \right\rangle .
\tag{5.43}
$$

Combining (5.42–43) yields

$$
\varepsilon^2 = 2 \int_{\Omega p}^{\infty} d\omega \, \phi_0(\omega) + 2 \int_{0}^{\Omega p} d\omega \, \phi_n(\omega)/|\tau(\omega)|^2 ,
\tag{5.44}
$$

using definition (5.4a), the statistical assumptions

$$
\langle N(\omega)^* \, O(\omega') \rangle = 0 \qquad \text{(uncorrelated object and noise)}
$$

$$
\langle O(\omega)^* \, O(\omega') \rangle = \phi_0(\omega) \, \delta(\omega - \omega') \quad \text{(uncorrelated object statistics)}
\tag{5.45}
$$

$$
\langle N(\omega)^* \, N(\omega') \rangle = \phi_n(\omega) \, \delta(\omega - \omega') \quad \text{(uncorrelated noise statistics)}
$$

and the functional assumption of symmetry for power spectra ϕ_0, ϕ_n and for transfer function τ.

We wish to minimize ε^2 in (5.44) through choice of Ω_p. Accordingly, we take

$$
\partial(\varepsilon^2)/\partial\Omega_p = 0 ,
\tag{5.46}
$$

yielding the requirement

$$- \phi_0(\Omega_\mathrm{p}) + \phi_\mathrm{n}(\Omega_\mathrm{p})/|\tau(\Omega_\mathrm{p})|^2 = 0 , \tag{5.47}$$

or

$$|\tau(\Omega_\mathrm{p})|^2 = \phi_\mathrm{n}(\Omega_\mathrm{p})/\phi_0(\Omega_\mathrm{p}) . \tag{5.48}$$

In words, the optimal processing bandwidth is fixed by the frequency for which the M.T.F. (modulus of the transfer function) equals the root noise-to-signal ratio, a rather interesting correspondence. In the particular case of diffraction-limited optics, where

$$\tau(\omega) = \begin{cases} 1 - |\omega|/\Omega & \text{for} \quad |\omega| \leqq \Omega \\ 0 & \text{for} \quad |\omega| > \Omega \end{cases} \tag{5.49}$$

and if also the noise-to-signal is a constant for all frequencies, (5.48) provides an explicit solution for Ω_p:

$$\Omega_\mathrm{p} = \Omega[1 - (\phi_\mathrm{n}/\phi_\mathrm{o})^{\frac{1}{2}}] . \tag{5.50}$$

Hence, the optimum processing bandwidth is reduced from the total optical bandwidth by the extent to which root noise-to-signal departs from zero. Thus, if noise is known to be zero, (5.50) says to inverse filter with full bandwidth; if noise-to-signal is unity, (5.50) says not to inverse filter at all!

5.9.2. Use of Window Functions

For a general object, $\hat{O}(\omega)$ in (5.35) does not go continuously to zero as the limiting processing frequencies $\pm \Omega_\mathrm{p}$ are approached. The resulting abrupt transition from a finite $\hat{O}(\omega)$ to zero in the Fourier integral (5.30c) causes $\hat{o}(x)$ to have highly fluctuating (and misleading) details. For instance, if the true $o(x) = \delta(x)$, then $\hat{O}(\omega) = 1$; but (5.36) truncates the latter at its limits, and produces a restoration $\hat{o}(x)$ which is the $\mathrm{sinc}(\Omega_\mathrm{p} x)$ function, (in)famous for its 0.22-level negative sidelobe.

To reduce this effect, the general approach has been to use a supplemental, purely mathematical function $W(\omega)$ which multiplies $Y(\omega)$ in (5.36) so as to "gently" taper off $\tau_\mathrm{net}(\omega)$ in (5.37) as $\omega \to \Omega_\mathrm{p}$. This results in a new output $\hat{o}_W(x)$ formed as

$$\hat{o}_W(x) = (2\pi)^{-\frac{1}{2}} \int\limits_{-\Omega\mathrm{p}}^{\Omega\mathrm{p}} d\omega \, W(\omega) \, \hat{O}(\omega) \exp(+j\omega x) . \tag{5.51}$$

(See, e.g., [5.27].) The role played by this function, $W(\omega)$, is to make the product $W \cdot \hat{O}$ close to $\hat{O}$ where $\hat{O}$ is finite, and to make product $W \cdot \hat{O}$

smoothly approach zero as $\omega \to \Omega_p$. For these purposes, such *ad hoc* window functions as the elliptical arc $W(\omega) = (1 - \omega^2/\Omega_p^2)^{\frac{1}{2}}$, the "Hanning filter" (see [5.28]) $W(\omega) = \frac{1}{2}[1 + \cos(\pi\omega/\Omega_p)]$, etc., have been used with modest success. See also Subsection 5.11.1 for the derivation of an optimum window function based on a "sharpness" constraint.

That such window functions will fulfill their aim of smoothing out high-frequency details in $\hat{o}(x)$ is indisputable. This can be shown by noting that the new restoration $\hat{o}_W(x)$ in (5.51) relates back to the old (inverse-filter) one $\hat{o}(x)$ through

$$\hat{o}_W(x) = \int_{-\infty}^{\infty} dx' \, \hat{o}(x') \, w(x - x'). \tag{5.52}$$

Hence, provided $w(x)$ is any positive function, $\hat{o}_W(x)$ must be smoother than $\hat{o}(x)$.

The drawback to this approach is that it cannot discriminate between significant and superficial portions of $\hat{o}(x)$, and so it blindly smooths out both [as is evident in (5.52)]. Hence, the price paid is an overall blurring of the entire picture; an effect which is acceptable, however, in certain cases.

With the use of a window function, the net psf connecting $\hat{o}_W(x)$ with true object $o(x)$ becomes

$$s_{net}(x) = (2\pi)^{-\frac{1}{2}} \int_{-\Omega_p}^{\Omega_p} d\omega \, W(\omega) \, e^{j\omega x}, \tag{5.53}$$

as compared with (5.38) for the simple inverse-filter case. The dependence of $s_{net}(x)$ in (5.53) upon $W(\omega)$ is precisely that in optics of the amplitude diffraction pattern upon the pupil function. This allows us to use all the rectangular apodisers of JACQUINOT and DOSSIER [5.29] for window functions. [These apodisers have as their aim the suppression of sidelobes in the point amplitude pattern, exactly our aim here. The price paid is less resolution in $s_{net}(x)$.]

On the other hand, the same optical correspondence also allows us to use for $W(\omega)$ in (5.53) the "extrapolating pupil" functions of FRIEDEN [5.30], whose aim is to produce *extra* resolution (the exact opposite aim of apodisation). This possibility is discussed in Subsection 5.13.1. Another resolution-enhancing $W(\omega)$ is defined in Subsection 5.10.1.

5.10. Other ad hoc Linear Methods

By an *ad hoc* linear method we mean any one which gives an output $\hat{o}(x)$ which better resembles $o(x)$ than does the raw image $i(x)$, if there is no noise in the data $i(y_m)$. These methods do not have noise suppression built in as, for example, a Wiener filter method does.

There are a multitude of interesting methods of this type, some of which are briefly described next.

5.10.1. DC-Suppression

It is, perhaps, not widely known that the simple expedient of zeroing a finite band of image spectral values $I(\omega)$ at frequencies centered about the origin will result in an enhanced output; see [5.31]. The image of a point which is processed in this way will show a decreased first-zero position (for example, asymptotic to $\frac{1}{4}$ the value for an unprocessed, diffraction-limited image), as the zeroed bandwidth is increased toward Ω. Figure 5.3 shows the tendency for this situation, which may be described by the window function

$$W(\omega) = \begin{cases} 0 & \text{for} \quad |\omega| \leq \omega_0 \\ \tau(\omega) & \text{for} \quad \omega_0 \leq |\omega| \leq \Omega \end{cases} \tag{5.54}$$

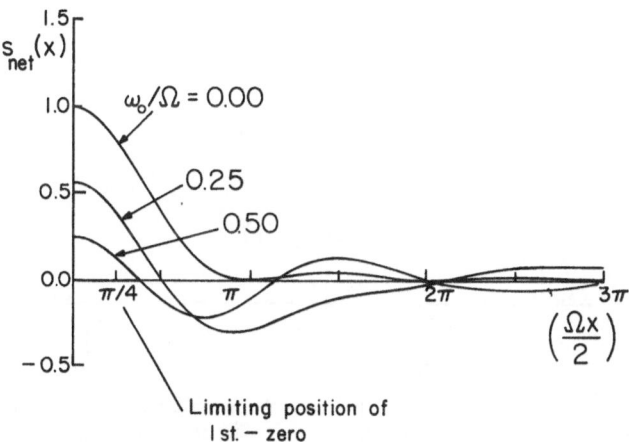

Fig. 5.3. Spread function $s_{net}(x)$ produced by zeroing a finite band of frequencies $|\omega| \leq \omega_0$ in the image. The imagery is here characterized by diffraction-limited optics, of cut-off frequency Ω. $s_{net}(x)$ is shown for different values ω_0/Ω. It is apparent from the curves that resolution is gained, at the expense of growing sidelobes, as ω_0 is increased toward Ω. This kind of tradeoff is common to all linear restoring methods. (We thank our student R. WAGNER for providing the curves)

and

$$s_{net}(x) = \left(\int_{-\Omega}^{-\omega_0} + \int_{\omega_0}^{\Omega} \right) dw\, \tau(\omega)\, e^{j\omega x} \qquad (5.55)$$

$$= s(x) - \text{sinc}(\omega_0 x) \otimes \text{sinc}^2(\Omega x/2)$$

by use of identity (5.9).

5.10.2. Suppression of Image-Motion by Convolving Mask

In the case of linear image motion, the image degradation is $s(x)$ = Rect(x/L). Swindell [5.32] has shown that the simple mask function $m(x)$ shown in Fig. 5.4 will, when convolved with $s(x)$ for this case, produce a net spread function $s_{net}(x)$ having two narrow spikes of arbitrary width and arbitrary separation. Spread function $s_{net}(x)$ is, as shown next, the output spread function connecting estimate $\hat{o}(x)$ with unknown $o(x)$: Assuming shift-invariance for s in (5.1), we have [see

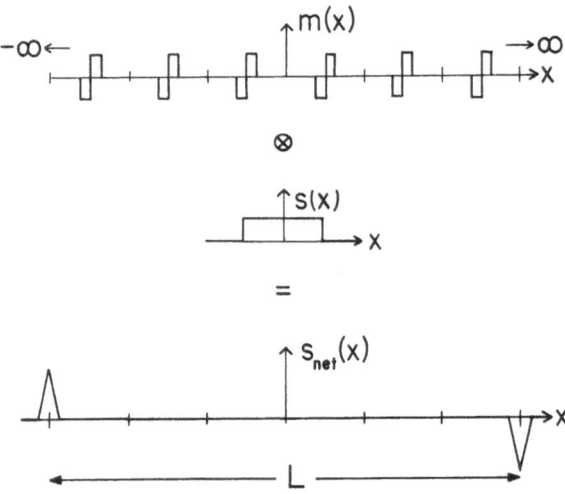

Fig. 5.4. Illustrating the principle by which deconvolving mask $m(x)$ reduces by an arbitrary amount the presence of image motion in an image. The convolution of $m(x)$ with the degradation $s(x)$ produces two arbitrarily narrow spikes, with a width (resolution) fixed by that chosen for the rectangle function in $m(x)$. Hence, any image blurred by linear motion can be separated into two enhanced negative versions of one another. Since the separation L between the two spikes is arbitrarily large, fixed by the extent chosen for $m(x)$, the two restored images can always be made to not overlap, i.e., interfere, with each other. This is a distinct advantage over restoring the image by differentiation, where the two enhanced images are separated by the *fixed* amount of image motion. (After Swindell [5.32]) ·

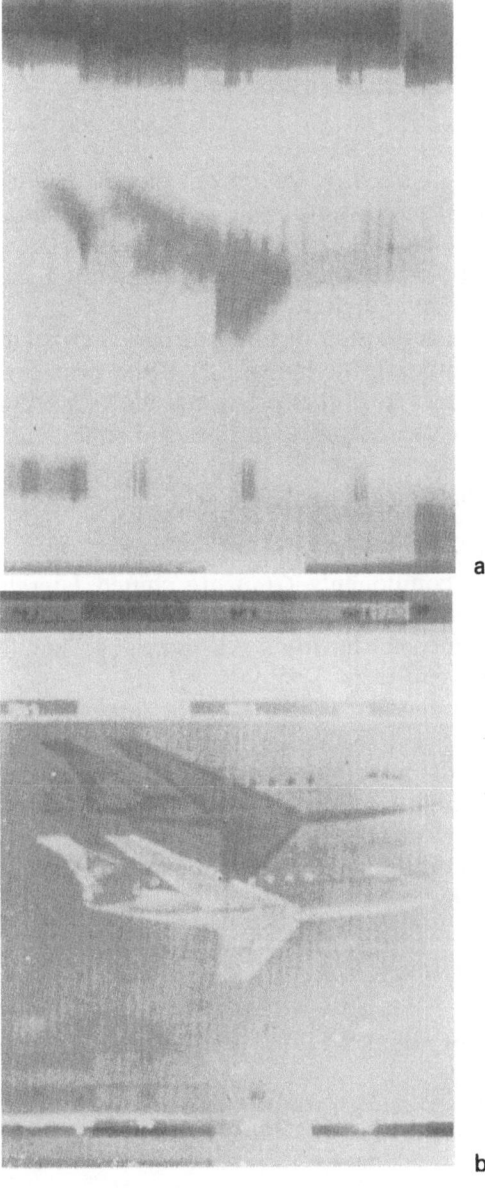

Fig. 5.5a and b. Use of the method illustrated in Fig. 5.4. The image in (a) suffers purely from linear image motion in the vertical direction. After deconvolution with filter $m(x)$, two enhanced versions of the airplane appear in (b). Filter $m(x)$ was made long enough to just separate the output images in (b). By inspection, a great deal of detail that is lost in the blur of (a) is retrieved is (b). (After SWINDELL [5.32])

notation (5.11)]

$$\hat{o}(x) \equiv m \otimes i = m \otimes [o \otimes s] = o \otimes [m \otimes s],$$

$$\equiv o \otimes s_{net}$$
(5.56)

by definition of s_{net}. Hence, the output picture $\hat{o}(x)$ will consist of two displaced, sharpened versions, with as respective spread functions the positive and negative spikes in Fig. 5.4. An example of use of the method is shown in Fig. 5.5. Although this was an analog use of the method, it could have been done digitally as well.

Although image-motion of this type can alternatively be reduced by spatially differentiating the image $i(x)$, it has been found in a separate study by Hawman [5.33] that use of Swindell's mask induces less noise propagation into the output than does the differentiation method.

5.10.3. Pure Phase Compensation

At frequencies for which the imaging transfer function $\tau(\omega)$ is negative, object spectral components $O(\omega)$ are shifted laterally in the image. These, in turn, interfere destructively with the *unshifted* spectral components for frequencies having a positive $\tau(\omega)$. The physiological effect is a blurred image.

It has been suggested, then, that for imaging processes for which $\tau(\omega)$ is *strongly* negative this translation- or phase-type of blur is the dominant cause of image blur, over and above that due to pure attenuation of $\tau(\omega)$ at high ω. Hence, it should be possible to enhance such a degraded image merely by advancing or retarding the phase part of $I(\omega)$ so as to cancel the phase errors due to $\tau(\omega)$. That is, if $\tau(\omega) \equiv |\tau(\omega)|$ · $\exp[j\phi(\omega)]$, enhance by making

$$\hat{O}(\omega) = I(\omega) \exp[-j\phi(\omega)].$$
(5.57)

This operation produces a $\tau_{net}(\omega) = |\tau(\omega)|$, and hence an

$$s_{net}(x) = (2\pi)^{-\frac{1}{2}} \int_{-\Omega_p}^{\Omega_p} d\omega |\tau(\omega)| \exp(+j\omega x).$$
(5.58)

Since, for linear image-motion $\tau(\omega) = \text{F.T.}[\text{Rect}(x/L)] = 2L \text{sinc}(\omega L)$, and since the sinc function has substantial negative regions, an image suffering from image-motion ought to be usefully processed in this manner. This has *not* been found to be decisively true, however. In a computer experiment (see [5.34]) of this type the output was not very

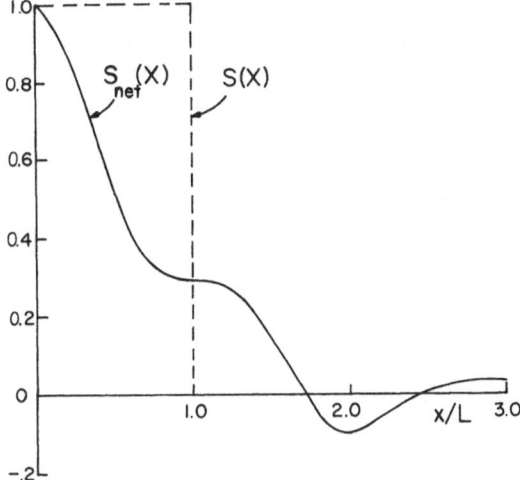

Fig. 5.6. Spread function $s_{net}(x)$ after pure phase compensation of the image spectrum corresponding to linear image motion of extent $2L$. The compensation is truncated at the second zero of the transfer function $\text{sinc}(\omega L)$ of the image motion. Hence, the region between the first and second zeros is merely flipped from negative to positive values. Comparing $s_{net}(x)$ with the original blur (dashed curve), we see that a slight gain in resolution has been obtained. Phase compensation is a very conservative restoring operation

strongly enhanced by use of algorithm (5.57), in comparison with use of pure inverse-filtering.

To check this result, we show in Fig. 5.6 the calculated, net spread function (5.58) for this experiment when the processing cutoff-frequency Ω_p is at the second zero of $\text{sinc}(\omega L)$. Thus, operation (5.57) acted simply to flip positively the first (and largest) negative sidelobe of the sinc function. From Fig. 5.6, $s_{net}(x)$ basically consists of two additive psf's; one with first-zero $x_0 \simeq L$ and the other with $x_0 \simeq 2L$. We conclude that pure phase compensation is worthwhile only when small enhancement (and, consequently, small noise propagation as well) is desired.

5.10.4. Van Cittert's Method of Successive Convolution

We have seen at (5.31) that linear filtering may be carried through completely by operation in direct, or x, space. Function $y(x)$ in (5.31) is often called a "de-convolving" function. Evidentally, if $y(x)$ were everywhere positive the output $\hat{o}(x)$ would necessarily be even smoother, or more degraded, than the input image. Hence, for enhancement purposes $y(x)$ must contain significant negative regions.

Suppose the user wishes to accomplish enhancement by use of a linear filtering in x-space, but is constrained (e.g., by the needs of an analog device) to use a purely *positive* function $y(x)$. According to the previous reasoning, this would appear impossible. However, the above arguments held for use of *one* convolution $i \otimes y$. If, on the other hand, the user can multiply-convolve then a positive kernel $y(x)$ can be used. Interestingly, the choice $s(x)$ for $y(x)$ works, as discovered by VAN CITTERT [5.35] in 1931. JANSSON [5.10] discussed some modern spectroscopic uses of van Cittert's method, which is defined as follows.

Let superscript (k) indicate the iteration number. Start with $k = 0$ and the initial $\hat{o}^{(0)}(x) = i(x)$, and use the following algorithm:

(i) $\hat{i}^{(k)}(x) = \hat{o}^{(k)}(x) \otimes s(x)$,

(ii) $\hat{o}^{(k+1)}(x) = \hat{o}^{(k)}(x) + [i(x) - \hat{i}^{(k)}(x)]$,

(iii) $k \rightarrow k + 1$,   (5.59)

(iv) Go to (i).

It has been empirically found [5.36] that VAN CITTERT's method does not always converge, especially when the image $i(x)$ is strongly blurred or when there is significant noise in $i(x)$. We show next why this is so; that in fact VAN CITTERT's method converges in the limit $k \rightarrow \infty$ to the *inverse-filtering estimate* [5.35] providing $|1 - \tau(\omega)| < 1$ at all ω. Also, at frequencies ω for which $\tau(\omega) = 0$, we will find that the image noise is *linearly magnified* with successive iteration!

The analysis is very simple when (5.59) is transformed into frequency space. These yield the result (arguments ω suppressed)

$$\hat{O}^{(k+1)} = \hat{O}^{(k)} \cdot (1 - \tau) + I$$

when combined. By starting with $\hat{O}^{(0)} = I$ on the right, and successively substituting the left-hand side into the right, a geometric series is developed

$$\hat{O}^{(k)} = I \sum_{j=0}^{k} (1 - \tau)^j$$

which may be analytically summed to yield

$$\hat{O}^{(k)} = (I/\tau) [1 - (1 - \tau)^{k+1}].$$ (5.60)

This result, true at each ω, shows that:

1) In the limit $k \to \infty$, $\hat{O}^{(k)}(\omega)$ approaches the inverse-filter estimate $I(\omega)/\tau(\omega)$, provided $|1 - \tau(\omega)| < 1$. Hence, if, for example $\tau(\omega) < 0$ due to linear image motion, defocus or other blurs, $\hat{O}^{(k)}(\omega)$ will *not* converge; and hence, neither will its F.T. $\hat{O}^{(k)}(x)$, the observable of the algorithm (5.59).

2) For finite k, the square-bracketed function in (5.60) acts as an apodising, or smoothing, window function of the type in Subsection 5.9.2. Hence, for finite k van Cittert's output gives a smoothed version of inverse filtering. As k increases, the window approaches Rect(ω/Ω).

3) Taking the limit $\tau \to 0$ in (5.60), and using (5.29) and l'Hôpital's rule, yields the result that at frequencies for which $\tau(\omega) = 0$, the estimate $\hat{O}^{(k)}(\omega) = (k+1) N(\omega)$, i.e., is merely a linearly enhanced version of the *noise*.

4) Finally, for small τ and moderate k in (5.60) we find the estimate $\hat{O}^{(k)}(\omega)$ to be merely a linear version $(k+1) I(\omega)$ of the input image $I(\omega)$.

These results seem to account for Jansson's observations [5.36] described above on the narrowness of scope of application of the method. The requirement that nowhere should $\tau(\omega) < 0$, and effect 4), are particularly important; the latter especially, since for all optical images $\tau(\omega)$ is small near cutoff at Ω.

In view of the shortcomings of this method, it is surprising and impressive that JANSSON et al. [5.10] found a simple twist that makes the approach highly useful for real, strongly blurred images. See Section 5.16 below.

5.10.5. Methods of Discrete Deconvolution

Despite the speed of the FFT computer algorithm (see Section 5.7), some workers have bound it useful to digitally operate in the *space* domain. This is because, basically, the "de-convolution" operation (5.31) may be cast as a discrete sum containing *few* points. This may be seen as follows.

In order to use (5.31) on the computer, it must be put in a discrete form:

$$\hat{o}(x_n) = \sum_{m=1}^{M} y_m i(x_n - x_m) \tag{5.61}$$

$$n = 1, 2, ..., N,$$

where data points x_m are spaced at, say, the Nyquist interval (see Section 5.2). The de-convolving kernel function y is now regarded as a vector of M input numbers, chosen as follows.

Image i relates back to object o via (5.1). Substituting (5.1) into (5.61) yields

$$\hat{o}(x_n) = \int_{-X}^{X} dx\, o(x)\, s_{\text{net}}(x - x_n) \qquad (5.62a)$$

in terms of the net input-output psf

$$s_{\text{net}}(x) \equiv \sum_{m=1}^{M} y_m\, s(x - x_m). \qquad (5.62b)$$

From (5.62a), we want s_{net} to best approximate Dirac $\delta(x)$. On the other hand, from (5.62b) s_{net} is constrained as a weighted sum of sampled values of the known psf. The aim of selecting weights y_m is, then, to make this weighted sum best approximate, in some sense, a delta function.

As an example, suppose we have diffraction-limited imagery, where

$$s(x) \equiv \text{sinc}^2(x). \qquad (5.63a)$$

The bandwidth Ω, Nyquist interval and first-order zero position x_0 are then values 2, $\pi/2$ and π, respectively. Let us fix the image translations x_m in (5.61) as multiples of the Nyquist interval.

As a target design, let us require $s_{\text{net}}(x)$ in (5.62b) to have its first-zero at value $x_0 = \pi/2$, i.e., half the value for the physical psf (5.63a); and suppose we wish to do this using only $M = 3$ terms in (5.61) and (5.62b). With these constraints, and since $s_{\text{net}}(x)$ ought to by symmetric in x, (5.62b) takes the form

$$s_{\text{net}}(x) = \text{sinc}^2(x) + y_1 \left[\text{sinc}^2(x - \pi/2) + \text{sinc}^2(x + \pi/2)\right]. \qquad (5.63b)$$

Weight y_1 must satisfy the resolution requirement $s_{\text{net}}(\pi/2) = 0$, so that

$$0 = \text{sinc}^2(\pi/2) + y_1 \left[\text{sinc}^2(0) + \text{sinc}^2(\pi)\right]. \qquad (5.63c)$$

This of course can be uniquely solved for y_1,

$$y_1 = -\text{sinc}^2(\pi/2) = -4/\pi^2. \qquad (5.64)$$

In summary, then, for this case ($M = 3$) the image displacements $\{x_m\}$ $= (0, -\pi/2, +\pi/2)$ with corresponding weights $(1, -4/\pi^2, +4/\pi^2)$ will yield a net psf whose first-zero is *half* that of the physical psf.

To judge this result, we may use as a benchmark comparison the result s_{net} for inverse-filtering, which is here $\text{sinc}(2x)$; see (5.38). This

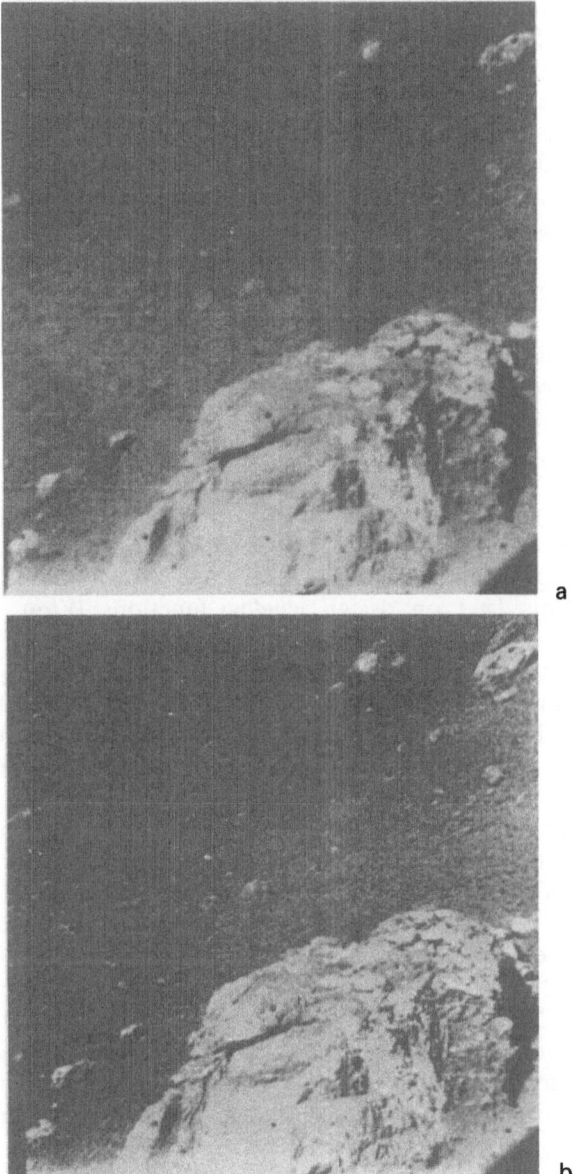

a

b

Fig. 5.7. (a) A view of the Moon, due to Ranger VIII, before transfer function compensation. (b) The compensated picture. The window function $W(\omega)$ for this restoring procedure is unity (full compensation) out to the frequency for which $\tau(\omega)$ falls to 0.2; thereafter $W(\omega) = 5\tau(\omega)$, so that $W(\omega) \rightarrow 0$ eventually. Computer implementation was in the space domain, as in (5.61). (After NATHAN [5.38])

again has a first-zero $x_0 = \pi/2$, putting it on an equal footing with our design (5.63b) and (5.64). However, a plot of (5.63b) shows a largest sidelobe of 0.39 whereas that of $\mathrm{sinc}(2x)$ is about 0.21. Hence, inverse-filtering would be superior to the use of $M = 3$.

The situation is remedied by simply taking larger values of M. For general M, the approach taken at (5.63b) leads to $(M-1)/2$ independent $\{y_m\}$ to fix. Since there is but one constraint equation $s_{\mathrm{net}}(x_0) = 0$ for the $\{y_m\}$ to satisfy, $(M-3)/2$ of them are free parameters. These may be adjusted to minimize the $s_{\mathrm{net}}(x)$ sidelobes in some sense (see [5.37]). For example, $M = 7$ produces about comparable sidelobe characteristics as by inverse-filtering if again resolution $x_0 = \pi/2$. Alternatively, if the user can accept larger sidelobes than 0.21 he can achieve *better* resolution than the inverse-filter value (see [5.37]).

Finally, it was found in [5.37] that first-zero position $x_0 = \pi/2$ can be attained by weights $\{y_m\}$ entirely constrained to the three *integers* $(0, -1, \text{ and } +1)$. When used in approach (5.61), there results a method of *restoring by pure addition and subtraction* of adjacent image values.

The discrete approach (5.61) for processing data was also taken by ARGUELLO et al. [5.28]. However, these authors based their choice of coefficients $\{y_m\}$ on design *window* functions $W(\omega)$, in frequency space. This approach differs philosophically (is contrast or point spread the more basic determinant of image quality?) and, indeed quantitatively from the direct-space design approach [(5.63) and (5.64)] above. For example, it is impossible by inspecting a general curve $W(\omega)$ to infer the largest sidelobe and first-zero position in its Fourier transform $s_{\mathrm{net}}(x)$.

Nevertheless, as in [5.37] ARGUELLO and co-workers concluded that values of M the order of 3 or 5 suffice to produce acceptable restorations. Hence, from two differing points of view it has been concluded that it is as practical to restore data using direct-space computer processing as it is to use frequency space operations.

Indeed, the most widely known examples of picture enhancement, those due to NATHAN [5.38] and co-workers at Jet Propulsion Laboratory on the Ranger moon lander photos and others, were produced using the *space* de-convolution approach (5.61). Figure 5.7 shows a typical "before" and "after" example of picture enhancement using Nathan's approach.

5.11. Linear Methods Based on a Statistical Approach

The arbitrary nature of the window function $W(\omega)$ previously spoken of has led some authors to seek derived ones, based on some promising criterion such as minimum mean-square error *mmse*. Now, the aim of using windows $W(\omega)$ is to reduce spurious oscillation in $\hat{o}(x)$ due to functional ringing and statistical noise in the data. The latter effect, in part suggests we work with *expected* values.

5.11.1. A Sharpness-Constrained Wiener-Filter Approach

These factors led HELSTROM [5.39] to take a Wiener-filter approach, and seek restoring filter $Y(\omega)$ that multiplies the data-transform $I(\omega)$ and obeys a *mmse* quality criterion

$$\left\langle \int_{-\Omega}^{\Omega} d\omega |O(\omega) - \hat{O}(\omega)|^2 \right\rangle = \text{minimum}, \tag{5.65a}$$

where

$$\hat{O}(\omega) \equiv \begin{cases} Y(\omega) \cdot I(\omega) & \text{for } |\omega| \leq \Omega \\ 0 & \text{for } |\omega| > \Omega. \end{cases} \tag{5.65b}$$

(BACKUS and GILBERT [5.42] take a slightly different approach, to be described below.)

The preceding problem is easily solved for $Y(\omega)$, as shown in [5.39]. However, it is of further interest to see what happens when we tack onto criterion (5.65a) a constraint on the total *sharpness S* in the restoration, defined as

$$S \equiv \left\langle \int_{-\infty}^{\infty} dx |d\hat{o}(x)/dx|^2 \right\rangle \tag{5.66a}$$

$$= \left\langle \int_{-\Omega}^{\Omega} d\omega\, \omega^2 |\hat{O}(\omega)|^2 \right\rangle, \tag{5.66b}$$

the last quantity obtained by use of mathematical identities (5.5), (5.10), and (5.12). Sharpness S is seen to measure the total edge-gradient "content" in output $\hat{o}(x)$.

The new, net criterion becomes

$$\int_{-\Omega}^{\Omega} d\omega \langle |O(\omega) - Y(\omega) I(\omega)|^2 \rangle + \lambda \int_{-\Omega}^{\Omega} d\omega\, \omega^2 \langle |Y(\omega) I(\omega)|^2 \rangle \tag{5.67}$$

$$= \text{minimum}.$$

We see from (5.67) that more, or less, emphasis on sharpness control relative to expected error (or, data consistency) is exercised by choice of the parameter λ. This corresponds somewhat to PHILLIPS' use of smoothing factor γ, in algorithm (5.15).

By expanding out the modulus-squared terms in (5.67), using the imaging Eq. (5.29), and assuming the statistical independence

$$\langle N^*(\omega) O(\omega) \rangle = 0 ,$$

we get

$$\int_{-\Omega}^{\Omega} d\omega \, [\phi_o - Y^* \tau^* \phi_o - Y \tau \phi_o + Y Y^* (|\tau|^2 \phi_o + \phi_n)$$

$$+ \lambda \omega^2 Y Y^* (|\tau|^2 \phi_o + \phi_n)] = \text{minimum} ,$$

where ϕ_o and ϕ_n are the power spectra for object and noise, per definition (5.4a). The frequency dependence of all quantities in the square brackets has been suppressed, for brevity.

The problem is now recognizable as of the Euler-Lagrange type, where the square bracketed quantity is Lagrangian $\mathscr{L}$, and $\mathscr{L}$ is not a function of $\partial Y / \partial \omega$ or $\partial Y^* / \partial \omega$. Hence, the solution is simply

$$\partial \mathscr{L} / \partial Y^* = \partial \mathscr{L} / \partial Y = 0 .$$

Taking either of these derivatives produces the same answer,

$$Y(\omega) = \left(\frac{\tau^* \phi_o}{|\tau|^2 \phi_o + \phi_n} \right) \left(\frac{1}{1 + \lambda \omega^2} \right) . \tag{5.68}$$

Discussion

We see that the *mmse-sharpness* filter naturally separates into two sequential filters, that operate in precisely that order. The first is Helstrom's restoring filter as discussed in [5.39]. The second may by regarded as a "sharpness-control" filter—for a choice of $\lambda > 0$ it acts to attenuate high frequencies [and hence to smooth output $\hat{o}(x)$], while for choice $\lambda < 0$ it acts to boost the high frequencies. Actual experimentation with different λ's has borne out its utility for the choice $\lambda > 0$, i.e., in a smoothing mode; see [5.34].

Since $\hat{O}_{\text{sig}} = Y I_{\text{sig}} = (Y\tau) O$, the derived filter (5.68) corresponds to a *derived* window function, $W = \tau \cdot Y$. This was our aim at the outset.

The first bracketed quantity in (5.68) is Helstrom's restoring filter. From its form, we see that it goes toward inverse-filtering, i.e. $1/\tau(\omega)$, as ratio $\phi_n/\phi_o \to 0$; and conversely, it approaches zero as $\phi_n/\phi_o \to \infty$

(noise rejection). These are, of course, the expected properties of a filter which minimizes output error in the presence of input noise.

Although the filter (5.68) has much flexibility, and has a built-in noise rejection propensity, it has drawbacks, as listed in Section 5.9. In addition, in practice the power spectra $\phi_o(\omega)$, $\phi_n(\omega)$ are rarely known; and the basic *mmse* criterion (5.65a) has its shortcomings. Note that (5.65a) aims to produce only small error *on the average*, i.e., over a multitude of restorations corresponding to the known (see preceding) power spectra; whereas the user may need confidence in the quality of but one or two specific restorations. In fact, the very premise that a large number of restorations are to be made using the filter (5.68) is often untrue for the image enhancement problem.

Information Aspect of Optimum Filtering

We next concern ourselves with the case $\lambda = 0$ in (5.68), which, from (5.67) represents the absolute *minimum mse* situation. It is not widely realized, but $Y(\omega)$ in (5.68) relates *directly* to the Shannon information content in spectral image data $I(\omega)$. In a classic paper, FELLGETT and LINFOOT [5.40] established the maximum Shannon information $SI(\omega)$ at a given frequency to be

$$SI(\omega) = \log\left[1 + \frac{|\tau(\omega)|^2 \, \phi_o(\omega)}{\phi_n(\omega)}\right].$$
(5.69)

This is also called the "channel capacity" for the process of image formation. By combining (5.68) and (5.69) we find the suggestive result

$$Y(\omega) = [1/\tau(\omega)] \cdot [1 - e^{-SI(\omega)}].$$
(5.70)

This relation shows directly how information limits the ability to optimally restore in this *linear* fashion. The full compensation factor $1/\tau$ is modulated by a pure function of the information SI. We observe that full compensation is permitted only by the optimum filter Y when the information is infinite; and conversely, for zero information content no compensation is allowed.

The *mmse* is likewise limited by information content, as shown next. With a choice $\lambda = 0$, the use of optimum filter $Y(\omega)$ in (5.68) produces a minimized *mse* of

$$\varepsilon^2 = \int_{-\Omega}^{\Omega} d\omega \, \frac{\phi_n(\omega) \, \phi_o(\omega)}{|\tau(\omega)|^2 \, \phi_o(\omega) + \phi_n(\omega)}.$$
(5.71)

Again using relation (5.69), we find

$$\varepsilon^2 = \int_{-\Omega}^{\Omega} d\omega \, \phi_0(\omega) \, e^{-SI(\omega)}. \tag{5.72}$$

Since $\phi_0(\omega) \geq 0$, this shows that the *mmse* is attenuated exponentially with increasing information content; see [5.41].

5.11.2. The Backus-Gilbert Optimization Approach

With $\lambda = 0$, the criterion (5.67) is a single-parameter approach to optimization, that parameter being *mmse*. This is not, however, the only useful one. For example, a possible drawback to the criterion (5.67) is that the resulting filter Y might attenuate inputs $I(\omega)$ too strongly, on the basis of relatively large values for input function $\phi_n(\omega)$. Hence, it might produce too blurred an output $\hat{o}(x)$ to satisfy the user's needs. Although he can get around this problem by making free parameter λ negative (as discussed), another tack is possible.

BACKUS and GILBERT [5.42] suggest instead to *separately* minimize, in some convenient sense, i) the output noise *and* ii) the output departure of *signal* restoration from true object. For example, as a measure of i) the user might choose the total mean-square output noise, from (5.30b),

$$\eta^2 \equiv \left\langle \int_{-\infty}^{\infty} d\omega |N\,Y|^2 \right\rangle. \tag{5.73a}$$

A convenient measure of ii) is the mean-square departure of the filter output $Y\tau O$ [see (5.30b)] from a required, windowed estimate WO.

$$\mu^2 \equiv \left\langle \int_{-\infty}^{\infty} d\omega |W(\omega)\,O(\omega) - Y(\omega)\,\tau(\omega)\,O(\omega)|^2 \right\rangle. \tag{5.73b}$$

A bit later on, we shall use these specific measures.

Another possible choice for ii) is the actual choice of BACKUS and GILBERT, measure

$$l^2 \equiv \int_{-\infty}^{\infty} dx \, x^2 [s_{net}(x)]^2. \tag{5.74}$$

[The smaller (5.74) can be made the closer $s_{net}(x)$ is to a Dirac $\delta(x)$.] It is shown in [5.42] that the combined criterion

$$\lambda_1 \eta^2 + \lambda_2 l^2 = \text{minimum} \tag{5.75}$$

may be solved for the de-convolving function $y(x)$, for different sets of inputs λ_1, λ_2. These solutions result in useful tradeoff curves of attainable (shortest) resolution "length" l [square root of (5.74)] vs. tolerable noise η level, for η required below a fixed value.

However, this approach has the following drawbacks:

1) Since it is a linear method, its output $\hat{o}(x)$ is unconstrained and, accordingly, can have negative regions.

2) Measure (5.74) for resolution enhancement is quite arbitrary and is, to a good extent, insensitive to the sidelobe structure in $s_{net}(x)$. Hence, the statement of an achieved resolution length l does not impart information on the erroneous oscillation to expect in the output.

3) For an image of $M \times M$ input values, the method requires solving a linear system of equations with dimensions $M^2 \times M^2$. Hence, for even a modestly extended image of, say, 32×32 a formidable numerical problem arises. It does not appear, at this time, that anyone has tried the approach on two-dimensional imagery (one-dimensional imagery has been tested by SALEH [5.43]).

Evidentally, a filtering approach would avoid the last problem. Accordingly, let us deviate from criterion (5.75) by replacing criterion (5.74) with (5.73b). Accordingly, we seek a "BACKUS-GILBERT filter" $Y_{BG}(\omega)$ obeying

$$\lambda_1 \eta^2 + \lambda_2 \mu^2 = \text{minimum} . \tag{5.76}$$

Parameters λ_1, λ_2 are chosen by the user so as to provide emphasis on either noise-suppression (λ_1/λ_2 large) or resolution enhancement (λ_1/λ_2 small).

By substituting (5.73a) and (5.73b) into (5.76), squaring out the moduli and taking the derivative $\partial/\partial Y^*$ [as in derivation of (5.68)], we find a solution

$$Y_{BG}(\omega) = \frac{\tau^*(\omega)\, \phi_0(\omega)\, W(\omega)}{|\tau(\omega)|^2\, \phi_0(\omega) + (\lambda_1/\lambda_2)\, \phi_n(\omega)} . \tag{5.77}$$

The resemblance to the unconstrained ($\lambda = 0$) Wiener restoring filter (5.68) is remarkable. On the other hand, note that (5.77) has the required properties that for λ_1/λ_2 made small, resolution is emphasized (with $Y_{BG} \to W/\tau$, a windowed-inverse filter), while for λ_1/λ_2 large, noise rejection is emphasized.

We see, then, that the BACKUS-GILBERT philosophy of separately weighting resolution and noise capabilities may be cast as a filtering approach, which readily lends itself to the needs of the two-dimensional enhancement problem. In combination with a fast Fourier transform algorithm, filter (5.77) provides a flexible approach to restoring extended images.

However, as with the standard BACKUS-GILBERT approach in [5.42], it suffers the drawbacks of nonphysical (negative) output regions, and a lack of direct control over the sidelobes of $s_{net}(x)$ (F.T. of τY_{BG}).

5.12. The Possibility of Bandwidth Extrapolation

The standard filtering approach defined by (5.30a) and (5.30c) limits itself to spectral operations within the finite bandwidth $|\omega| \leqq \Omega_p$, with $\Omega_p < \Omega$ (optical cutoff). On the other hand, an astute user may note that, on occasion, the restored spectrum $\hat{O}(\omega)$ is a very *smooth* function within the processing bandwidth $|\omega| \leqq \Omega_p$. Making the not unreasonable assumption that $\hat{O}(\omega)$ *remains* smooth for frequencies $|\omega| > \Omega_p$, he might be tempted to extrapolate $\hat{O}(\omega)$ to frequencies beyond the absolute cutoff Ω, so that $0 \leqq |\omega| \leqq \Omega_p$ with *now* $\Omega_p > \Omega$. If his extrapolation were correct, this would gain him resolution in proportion to the bandwidth extension; see (5.40). How correct is this operation? How much extrapolation is needed for a measurable effect?

To approach the problem, let us assume for simplicity that the object consists of two equally bright point sources with separation a,

$$o(x) = \delta(x - a/2) + \delta(x + a/2) . \tag{5.78a}$$

Let $\hat{o}_{ext}(x)$ denote the restored output due to perfect spectral restoration $\hat{O}(\omega) = O(\omega)$, $0 \leqq |\omega| \leqq \Omega_p$, with $\Omega_p > \Omega$. The plan will be to see how much bandwidth extension, measured by the ratio Ω_p/Ω, is needed to "just" resolve the two points in output $\hat{o}_{ext}(x)$. Assuming unaided observation of $\hat{o}_{ext}(x)$, we shall make Rayleigh's assumption and assume the two points just resolved when the dip in intensity between them is roughly 20%.

It is useful to express the separation a as a fraction r of the Rayleigh resolution distance $2\pi/\Omega$,

$$a \equiv r(2\pi/\Omega) . \tag{5.78b}$$

We suspect that the smaller r is, the greater will be the required bandwidth extension ratio Ω_p/Ω for just resolving the 2 points.

The object spectrum is the F.T. of (5.78a).

$$O(\omega) = (2/\pi)^{\frac{1}{2}} \cos(\pi r \omega/\Omega) . \tag{5.78c}$$

This is presumed to be known perfectly over the entire processing bandwidth $O \leq |\omega| \leq \Omega_p$. Accordingly, we substitute (5.78c) into the inverse operation (5.30c) to get the final estimate

$$\hat{o}_{ext}(x)$$
$$= (\pi)^{-1} \Omega_p \{\text{sinc}[\Omega_p(x + \pi r/\Omega)] + \text{sinc}[\Omega_p(x - \pi r/\Omega)]\} . \tag{5.78d}$$

We assume (as discussed above) that in order to just resolve the two points, $\hat{o}_{ext}(x)$ must have at least 20% dip at $x = 0$ from its values at $x = \pm a/2 = \pm r\pi/\Omega$ [latter, by the identity (5.78b)]. From (5.78d) the ratio in question is

$$\hat{o}_{ext}(0)/\hat{o}_{ext}(a/2) = \frac{2\,\text{sinc}(\pi r \Omega_p/\Omega)}{1 + \text{sinc}(2\pi r \Omega_p/\Omega)} . \tag{5.78e}$$

By direct substitution of trial values $r\Omega_p/\Omega$, (5.78e) shows the required 20% dip when approximately

$$\Omega_p/\Omega = 3/(4r) . \tag{5.79}$$

Hence, as interesting examples if $r = \frac{1}{2}$ (object points separated by $\frac{1}{2}$ the Rayleigh resolution distance) then the necessary bandwidth extrapolation ratio $\Omega_p/\Omega = 3/2$ (50% extension needed); whereas if $r = 1/3$, the necessary ratio is 9/4—quite a bit larger.

In fact, all the required extrapolations (for different r) go out to the *same point* on the object-spectral cosine curve (5.78c): Substituting Ω_p from (5.79) for ω in (5.78c), we find this to be value

$$O(\omega) = \cos(3\pi/4) , \tag{5.80}$$

i.e., *just beyond the first zero of the cosine curve.* Hence, of the entire $(|\omega| \leq \infty)$ spectrum describing the two-point object only the tiniest fraction must be incorporated into output $\hat{o}_{ext}(x)$ to resolve the two points. Moreover, this spectrum (a cosine curve) is very smooth (almost linear) out to value (5.80). The situation is sillustrated in Fig. 5.8.

In summary, then, for the two-point object resolution problem the requisite extrapolation need extend only to just beyond the first-zero of the object (cosine) spectrum; such extrapolation sees only a very smooth, almost linear, part of the spectrum; and merely about a 50% extension is required for object separations of about $\frac{1}{2}$ Rayleigh's resolution length.

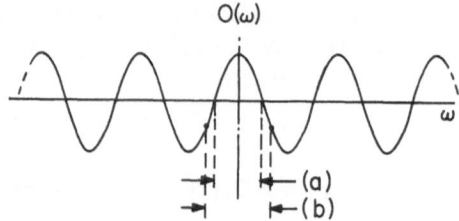

Fig. 5.8. The complete cosine curve represents the spectrum due to a two-point object. What segment of the cosine curve is needed to just resolve the two points, after Fourier transformation of the curve segment? The region indicated by (b) suffices. The region (a) covers the spectral region passed by the optics. Its Fourier transform does not resolve the two points. However, by a smooth and small extension of the curve (a) to (b), enough resolution is added to now resolve the points

It would be exercising considerable hindsight to claim that the proceding proves the feasibility of extrapolation methods. However, what we have shown is that an extrapolation method which tends to produce a *smooth* extension of a given curve $\hat{O}(\omega)$ stands a change of being effective, for objects whose spectra *are* smooth. One class of such objects $o(x)$ is a sequence of closely spaced delta functions, as used above.

As a negative corollary to the preceding, an object consisting of a *large* number of closely spaced impulses (e.g., Dirac "comb" function) would *not* be effectively restored by an extrapolation technique, since its spectrum is itself like a Dirac comb and therefore not at all smooth. In fact, this appears to correlate with experience (discussed below).

5.13. The Central Role Played by a priori Knowledge

There is an old adage that "it is hazardous to extrapolate". The meaning of this statement is well-appreciated by people who have attacked the problem at hand. Nevertheless, the temptations afforded by such dazzling trinkets as doctoral degrees, government contracts, etc., has caused that other adage, "where there's a will there's a way", to predominate. Some extrapolation schemes actually work.

The most effective extrapolation schemes in use incorporate the largest number of *a priori* constraints on output $\hat{o}(x)$. (Subscript *ext* is suppressed, from this point onward.) These constraints describe features of $o(x)$ that exist independent of the given data $i(y_m)$, and which therefore must be *independently injected* into the output.

5.13.1. Knowledge of Limited Spatial Extent

One important constraint of this type is knowledge that $o(x)$ is nonzero only over a finite interval $|x| \leq X$; see (5.1). Incredibly, this constraint *alone* permits a unique extension of a given $\hat{O}(\omega)$, providing image data $\{i(y_m)\}$ is noise-free. This point seems to have been first made to the optical community by WOLTER [5.44]. The reasoning is as follows.

If $o(x) = 0$ for $|x| > X$ then its spectrum obeys

$$O(\omega) = (2\pi)^{-\frac{1}{2}} \int_{-X}^{X} dx\, o(x) \exp(-j\omega x). \tag{5.81}$$

Then the n'th derivative $O^{(n)}(\omega)$ may be expressed as

$$O^{(n)}(\omega) = (2\pi)^{-\frac{1}{2}} (-j)^n \int_{-X}^{X} dx\, x^n o(x) \exp(-j\omega x).$$

For any physically realistic $o(x)$, the right-hand integrals must be unique and finite (the latter, because the limits are finite). Since these derivatives are well-defined, $O(\omega)$ may be expressed as a Taylor series about any one frequency ω_0. And in turn, this means that $O(\omega)$ is an *analytic function*.

An analytic function $O(\omega)$ may be uniquely determined for *all* ω from knowledge of either a) all its derivatives at one frequency, or b) its values $O(\omega)$ over a finite range of frequencies [indeed, knowledge of b) implies knowledge of a)]. Now, in fact by inverse-filtering [see (5.35) and (5.36)] an estimate $\hat{O}(\omega)$ of $O(\omega)$ is known over a finite range of frequencies $|\omega| \leq \Omega_p$. Therefore, inverse-filtering supplies all the inputs needed to extrapolate frequency-space information.

Basically, two restoring schemes have been proposed that incorporate bandwidth-extrapolation based on *solely* the constraint of limited spatial extent. We will briefly describe them here, mainly because of the insights they give into the extrapolation problem. We do not believe they have proved useful in practice. Noise propagation into the output is the major problem [in the context of Section 5.12, the spectral extensions of $O(\omega)$ are not smooth enough]. The basic reason for this is that one *a priori* constraint is simply not enough.

Harris' Method

Equation (5.81) is the statement that $O(\omega)$ is a space band-limited function. Therefore, it obeys the sampling theorem (5.7a) which is here

$$O(\omega) = \sum_{n=-\infty}^{\infty} O(n\pi/X) \operatorname{sinc}(X\omega - n\pi). \tag{5.82a}$$

all ω

Now in principle the left-hand side is known, by inverse-filtering, *inside* the optical bandwidth $|\omega| \leq \Omega$. Then, by substitution of, say, $2N$ values of the known $O(\omega)$ from inside the optical bandwidth into (5.82a) we generate $2N$ equations linear in unknowns $O(n\pi/X)$, some of which lie *outside* the optical bandwidth.

Now by (5.7b) we have

$$o(x) = (\pi/X) \sum_{n=-\infty}^{\infty} O(n\pi/X) \exp(-jn\pi x/X) \tag{5.82b}$$

$$|x| \leq X$$

which shows that solution for the central $2N$ values of $O(n\pi/X)$ yields a $2N$-term Fourier series for the unknown $o(x)$, which will exhibit bandwidth extrapolation if $N > X\Omega/\pi$.

Hence, it suffices to cut off the limits in (5.82a) at $n = \pm N$, N sufficiently large, and solve the $2N$ linear equations [acquired from knowing $O(\omega)$ inside $|\omega| \leq \Omega$] for the $2N$ unknowns $O(n\pi/X)$. This idea was proposed by Harris [5.45].

But solution of the $2N$ linear Eqs. (5.82a) is mathematically equivalent to solving the $2N$ linear Eqs. (5.3). Note the correspondances of experimental data $\{O(\omega_m)\}$ here with data $\{i(y_m)\}$ there, and psf $\mathrm{sinc}(X\omega_m - n\pi)$ here for general psf $s(y_m; x_n)$ there. Hence, this method will have the same fatal pitfalls as the direct matrix inversion approach described in Subsection 5.3.1. Recourse to smoothing techniques (but now in the *spectral* space) such as due to Phillips (Section 5.4) or Twomey (Section 5.5) would be necessary.

Since a smooth spectrum $O(\omega)$ has a sharp F.T. $\hat{o}(x)$, such smoothing might make Harris' approach effective for the two-point resolution problem (where a smooth spectrum is required; see Section 5.12). On the other hand, for a general object this approach appears to hold no advantage over *direct* use of the imaging equation, as in Sections 5.4 and 5.5.

Use of an "Extrapolating Window" Function

Taking a filtering approach, we next ask whether an extrapolating window $W(\omega)$ can exist. This would require product $W(\omega) O(\omega)$ over the *finite* interval $|\omega| \leq \Omega$ to act like $O(\omega)$ alone over the *infinite* interval, or

$$\int_{-\Omega}^{\Omega} d\omega \, W(\omega) O(\omega) \exp(j\omega x) \simeq A \int_{-\infty}^{\infty} d\omega \, O(\omega) \exp(j\omega x). \tag{5.83a}$$

The left-hand integral represents the output $\hat{o}(x)$ of the windowed restoration, of course. Since $o(x)$ is of limited spatial extent, (5.83a) would have to be satisfied only at values

$$|x| \leqq X .$$ (5.83b)

In (5.83a), constant A would hopefully be finite.

The requirement (5.83a) appears impossible to meet. And yet, HARRIS' approach (5.82a) and (5.82b) implies that an overall *linear* operation on $O(\omega)$ over $|\omega| \leqq \Omega$ will produce an extrapolated output. This is encouraging, since the sought filter $W(\omega)$ would perform precisely a linear operation.

In fact, a function $W(\omega)$ satisfying requirements (5.83a) and (5.83b) has been found (see [5.46]). This is providing function $O(\omega)$ is space band-limited (which it is, by hypothesis of limited extent). The solution is

$$W_N(\omega) = \sum_{\substack{n(\text{even})=0 \\ c \equiv 2X\Omega}}^{N} (-1)^{n/2} \lambda_n(c)^{-\frac{3}{4}} \psi_n(c, 0) \, \psi_n(c, 2\omega X/\Omega) .$$ (5.84a)

The larger N is made, the better requirement (5.83a) is satisfied, but the smaller multiplier A becomes. In the limit $N \rightarrow \infty$ (5.83a) is perfectly satisfied but $A \rightarrow 0$ (the restoration $\hat{o}(x)$ becomes uniformly black; such are the ways of nature).

In (5.84a), functions ψ_n and numbers λ_n are the prolate functions and eigenvalues, which were developed by SLEPIAN and co-workers (see, e.g., [5.47]). A summary of their properties and their past applications to optics may be found in [5.48].

Use of the special case $O(\omega) = 1$ for $|\omega| \leqq \infty$ in (5.83a) yields the requirement on $W(\omega)$ that its finite F.T., $s_{\text{net}}(x)$ by (5.53), be a Dirac delta function. In fact, function $W_N(\omega)$ has been found in [5.30] to obey

$$s_{\text{net}}(x) \equiv (2\pi)^{-\frac{1}{2}} \int_{-\Omega}^{\Omega} d\omega \, W_N(\omega) \, e^{j\omega x} = A \, \delta_N(x) ,$$ (5.84b)

$$|x| \leqq 2X ,$$

where $\delta_N(x)$ is an N-th-order approximation to the Dirac delta function,

$$\delta_N(x) \equiv \sum_{n(\text{even})=0}^{N} \lambda_n(c)^{-1} \psi_n(c, 0) \, \psi_n(c, x) ,$$ (5.84c)

$$|x| \leqq 2X .$$

Plots of $W_N(\varphi)$ and $\delta_N(x)$ for the case $N = 40$, $c = 6.25$ are shown in Figs. 5.9 and 5.10.

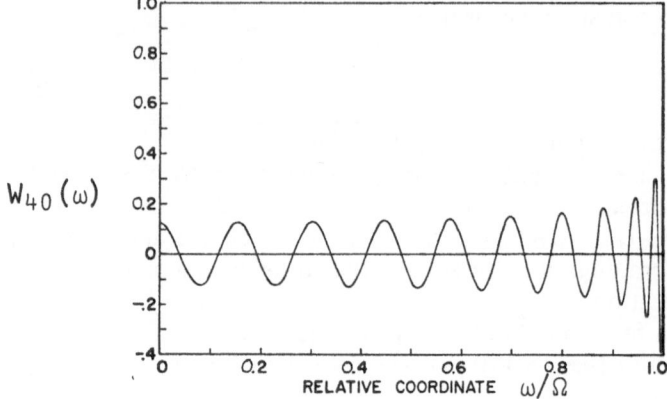

Fig. 5.9. Extrapolating window function $W_{40}(\omega)$, for the case $c = 6.25$. The curve is symmetric about the origin. Over a wide region about the origin, W_{40} is well approximated by a cosine curve. Both the frequency and amplitude of W_{40} rapidly increase as cutoff frequency Ω is approached. (After FRIEDEN [5.47])

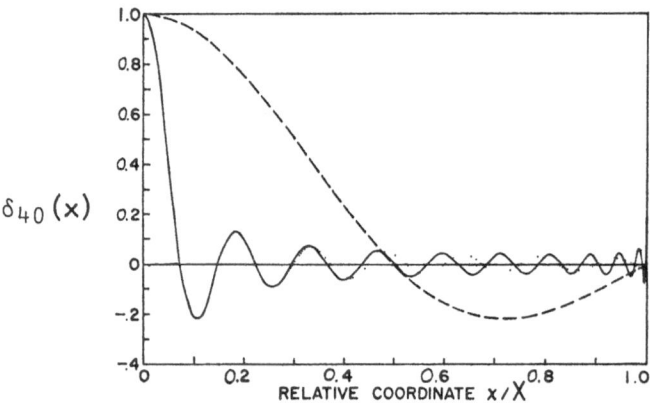

Fig. 5.10. The Fourier transform $\delta_{40}(x)$ (solid curve) of the window function in Fig. 5.9. This is also $s_{net}(x)$. For comparison, the dashed curve is the net spread function for inverse filtering over the same frequency range as for $W_{40}(\omega)$. The dotted curve is the net spread function for inverse-filtering over a frequency range of 6 times that of $W_{40}(\omega)$. Hence, window function W_{40} acts to extrapolate the object spectrum it multiplies in (5.83a) by a factor of 6, for this c-value. (After FRIEDEN [5.47])

The present author was led to the discovery of function $W_N(\omega)$ after reading a paper by BARNES [5.49], which was the first to apply functions $\psi_n(\omega)$ to the restoration problem. Function $W_N(\omega)$ has since found interesting application when used as a real-time pupil coating for an optical system (see [5.30, 48]).

Of course, the problem with actual use of window $W_N(\omega)$ is that it extrapolates indiscriminately both the signal and the noise parts of input image $I(\omega)$. Since the *net* filter operating upon $I(\omega)$ is $W_N(\omega)/\tau(\omega)$, the total filter output is $W_N(\omega)\,O(\omega) + W_N(\omega)[N(\omega)/\tau(\omega)]$. We have previously considered only the first of these two terms in the output integral (left-hand side of (5.83a)).

The second term causes the noise output

$$\int_{-\Omega}^{\Omega} d\omega\; W_N(\omega)[N(\omega)/\tau(\omega)] \exp(j\omega x)\,. \qquad (5.84d)$$

The problem stemming from this term is a compounding of the problem of noise-propagation due to inverse-filtering alone. Since noise term $N(\omega)$ tends, on the average, to remain flat as ω increases whereas $\tau(\omega) \to 0$ the quotient N/τ in (5.83c) tends to increase as cutoff Ω is approached. But then, function W_N acts to *extrapolate* this *ascending* noise function term (assuming for simplicity that N/τ is also a space band-limited function). This extrapolated part must therefore ascend higher yet, over a finite band of frequencies ω beyond cutoff $\pm\,\Omega$, before it falls back toward zero again. Clearly this results in a significant amplification of noise, over and above the inverse-filtering noise effect.

Although attempts have been made to build some noise-rejection into this basic approach (see [5.17, 50, and 51]), the prognostications on practical use of the method have not been good.

Aside from the serious problem of *noise* extrapolation, the overall approach (5.83a) suffers the drawbacks of significant spurious oscillation, negative outputs, and the purely computational problem of generating the required functions $\psi_n(\omega)$. Evidentally, knowledge of finite extension alone does not exert a sufficiently strong effect on the estimate $\hat{o}(x)$ to allow significant extrapolation in the presence of noise. Something in addition is needed.

5.13.2. Knowledge of Positivity: Effects on Spurious Oscillation and Resolution

The simple constraint of forcing estimates $\hat{o}(x)$ to be positive exerts a powerful and beneficial effect on the estimate. This can be shown as follows, extending the argument of Subsection 5.3.1 (last paragraph).

Suppose, for simplicity, that the object $o(x)$ consists of one delta function $\delta(x)$. Then its image is sampled values of the spread function $s(y_m)$. These are practically zero beyond a central-core region $|y_m| \leq y_0$.

Now let use consider object points that are conjugate to these very small image values. Assuming the restored object to be constrained to *positive* values, if it oscillates spuriously at these points it must oscillate about some *positive* local mean. Therefore, the net contributions to the image integral (5.1) from these spurious oscillations would likewise be positive, finite image contributions. But by hypothesis the input image is negligible at these values, $|y_m| > y_0$. Hence such object oscillations would be inconsistent with the given image. For this reason, a *positive* constrained object that is made *consistent* with the image data cannot have spurious oscillation over conjugate regions to where the true object is negligible.

This would be a prevailing effect for an object consisting of mainly the zero value, i.e., a finite number of delta functions against zero background. Now, it is *empirically* found that the lack of spurious oscillation is often accompanied by increased *resolution* in the output $\hat{o}(x)$. This can be explained as follows, using similar reasoning to that of the preceding paragraph.

If the restored object is allowed to go negative, i.e., is not constrained by positivity, a typical restoration of $o(x) = \delta(x)$ is $\mathrm{sinc}(\Omega x)$. This has spurious oscillations beyond its first-zero position $x = \pi/\Omega$, beginning with a major negative one. Evidently, this negative oscillation is cancelled in its image contribution by neighboring points in the positive central core in $\mathrm{sinc}(\Omega x)$. Hence, if this negative oscillation did *not* exist (due to a positivity constraint on the restoration) the neighboring central core values would not have to be finite and positive. They would remain zero, thereby reducing the central core region and adding resolution. This is observed, for positive-constrained restorations.

5.13.3. Knowledge of an Upper Bound

By viewing an image, the observer can always guess at some upper bound B to the unknown object. (Notice that B does not have to be the *least* upper bound, i.e., a number actually attained by $\hat{o}$ at some x.) Hence, the user could now constrain $\hat{o}(x)$ to obey

$$0 \leqq o(x) \leqq B . \tag{5.85}$$

Jansson [5.10] has established empirically just how important is this added information. The "objects" in question were absorption spectra, which must be bounded by levels 0 and 100%. We describe, below, Jansson's restoring method in some detail.

An explanation of *why* knowledge of the upper bound is important can be made using arguments that are completely analogous to those in Subsection 5.13.2 regarding the lower bound. The two situations are entirely complementary. Hence, we may conclude that upper bound information is mainly useful over spatial regions corresponding to level B in the true object. This means that, to be especially useful, the upper bound B presumed by the user should be the *least* upper bound, i.e. a level actually attained at various points (the more the better) in the object.

Regarding the *combined* constraint of lower bound and upper bound information, from the foregoing reasoning this would be optimally used for restoring objects which consist *entirely* of a known upper- and a known lower bound. In this binary situation, the restoring algorithm then degenerates to one which merely estimates the *positions* of *boundaries* in the object. Some objects of this type are portions of absorption spectra and alphanumeric text.

5.14. On Positive-Constrained Restoring Methods in General

If the reader is by now convinced of the importance of forcing positive restorations $\hat{o}(x)$, he might well ask how this can be accomplished. More precisely, how can a positive $\hat{o}(x)$ be formed *consistent with* the given image data $i(y_m)$ via convolution of (5.1)? This is the key question. For example if a positive $\hat{o}(x)$ is formed merely by taking the absolute value of a *non*-constrained estimate $\hat{o}(x)$, this positive $\hat{o}(x)$ will no longer be consistent with the image data. Neither will it exhibit any improvement in resolution or smoothness, since the arguments of Subsection 5.13.2 presume such consistency.

We may first note that any linear, convolution method cannot be guaranteed to yield a positive output at all x. This follows because, in order to enhance resolution, the convolution kernel must have some negative oscillations (otherwise it merely smooths, or blurs, the image further). And, these negative oscillations will show up in the output in regions corresponding to zero background in the true object.

For this reason, recourse must be made to *non*-linear operations on image data $\{i(y_m)\}$. A scheme such as $\hat{o}(x) \equiv [i(x)]^2$ would fall into this category—however, this $\hat{o}(x)$ would obviously not be consistent with the image data. We note in passing, however, that if $i(x)$ is bandlimited with frequency extent 2Ω, this $\hat{o}(x)$ has a bandwidth 4Ω, or double that of the input. This is a tendency to extend bandwidth, which is common to all non-linear methods.

The problem, then, is to find a non-linear function of $i(y)$ which guarantees a) positivity and b) consistency with the image. A few algorithms have been developed for accomplishing this double requirement. These will be described in the following sections. In order to accomplish b), it will be seen that all but one of these methods *iteratively* seek a solution; hence, these tend to be more time-consuming than the linear methods described in preceding sections.

5.15. Method of Schell and Biraud

One obvious way to force a positive output $o(x)^+$ is to set

$$o(x)^+ \equiv a(x)^2 \tag{5.85a}$$

and then try to find $a(x)$. In frequency space,

$$O(\omega)^+ = A(\omega) \otimes A(\omega) \tag{5.85b}$$

by identity (5.9).

The requirement that $A(\omega)$ be consistent with the image leads to its determination. Consistency demands that

$$I(\omega) = \tau(\omega)[A(\omega) \otimes A(\omega)] \tag{5.86a}$$

over the data bandwidth $|\omega| \leq \Omega_p$. Another way of stating this is

$$I(\omega)/\tau(\omega) = A(\omega) \otimes A(\omega), \quad |\omega| \leq \Omega_p. \tag{5.86b}$$

But the left-hand side of (5.86b) is the inverse-filter estimate of Section 5.9. We see then that sampled values of the inverse-filter solution $\hat{O}(\omega_m)$ may be used as inputs to a positive-constrained algorithm.

Consistency now means (5.86b). However, due to noise in the measured image the inputs $\hat{O}(\omega_m)$ will be in error. Hence, rather than demand strict equality in (5.86b) we can use the weaker requirement that

$$E \equiv \sum_{\omega_m = -\Omega_p}^{\Omega_p} |\hat{O}(\omega_m) - A(\omega_m) \otimes A(\omega_m)|^2 = \text{minimum} \tag{5.87a}$$

through choice of $A(\omega)$.

On the other hand, at frequency $\omega = 0$, where $\tau(\omega) = 1$, the error due to inverse-filtering is minimal [see (5.34)]. Hence, here we preserve the stronger requirement of equality:

$$\hat{O}(0) = A(\omega) \otimes A(\omega)|_{\omega = 0}. \tag{5.87b}$$

Equations (5.87a) and (5.87b) comprise Biraud's definition of the problem.

Something a bit further should be said about requirement (5.87b). The two sides of (5.87b) represent the areas under space curves $\hat{o}(x)$ and $o(x)^+$. The requirement of equality means that our final output

$o(x)^+$ must be purely a re-distribution (without amplification) of the energy under the band-limited input curve $\hat{o}(x)$. Further, since $\hat{o}(x)$ is linearly linked to the true $o(x)$ the equality (5.87b) actually fixes the total energy in output $o(x)^+$ at that of the *true* object (plus some noise). Since this tends to be a reliable estimate, it could be expected to aid in producing a convergent solution to (5.87a). In fact, this is empirically found to be true.

It is important to note at this point that in order to approximate $\hat{O}(\omega)$ over $|\omega| \leq \Omega_p$ in (5.87a) and (5.87b), $A(\omega)$ need only be defined over the *half*-interval $|\omega| \leq \Omega_p/2$. This is a property of the convolution operation in (5.87a). However, if $A(\omega) \neq 0$ for frequencies $|\omega_m| > \Omega_p/2$, these also contribute to (5.87a) and (5.87b); they are mixed into interval $|\omega| \leq \Omega_p$ by the convolution operation. This is a way of packing more degrees of freedom of A into data interval $|\omega| < \Omega_p$ so as to further reduce the minimum attained in (5.87a). We return to this important point below.

Suppose it is known that certain components $\hat{O}(\omega_m)$ are more likely to be in error than others (by a priori knowledge of the power spectra for signal and noise). Then it is useful to modify (5.87a) by injecting a multiplicative weight at each ω_m so as to reduce the *expected value* of the minimum; see [5.8]. For example, a small weight at ω_k would allow $A \otimes A(\omega_k)$ to depart widely from $\hat{O}(\omega_k)$ without seriously affecting the contribution to the minimum. This is logical if it is known that at ω_k the signal to noise is low. However, without knowledge of signal and noise statistics, this approach is not feasible.

SCHELL and BIRAUD have solved the problem (5.87a) and (5.87b) in different ways, but each with an iterative algorithm. SCHELL uses the algorithm

$$A^{(k+1)}(\omega_m) = \text{constant} \times \sum_{n=-(M+m)/2}^{(M-m)/2} [\hat{O}(\omega_n) - A^{(k)} \otimes A^{(k)}(\omega_n)] \tag{5.88}$$

$$\cdot A^{(k)}(\omega_{n+m}).$$

$$k = 0, 1, 2, \dots .$$

Superscript (k) denotes iterate number of the estimate $A(\omega_m)$, where $m = 0, \pm 1, \pm 2, \dots, \pm P$ and $P \geq M/2$. $2M$ is the total number of input data values $\hat{O}(\omega_n)$. An initial estimate $\{A^{(0)}(\omega_n)\}$ is substituted into the right-hand side, thereby yielding $\{A^{(1)}(\omega_n)\}$ for the left-hand side. $\{A^{(1)}(\omega_n)\}$ is substituted into the right-hand side, etc., until some iterate $k = K$ is reached for which $\{A^{(K+1)}(\omega_n)\} = \{A^{(K)}(\omega_n)\} + $ tolerable error. This numerical procedure is called the method of "successive substitutions" (see, e.g., [5.52]).

Algorithm (5.88) may be derived by injecting (5.87b) into (5.87a) as an additive Lagrange constraint, and setting the derivative $\partial/\partial A(\omega_k)$ of the total expression to zero. The left-hand term of (5.88) is essentially the derivative of constraint term (5.87b) in that expression. The delegation of iterate numbers $(k+1)$, (k) as in (5.88) is intrinsic to the method of successive substitutions.

Biraud's method of solving (5.87a) and (5.87b) is outlined in [5.9]. It is based on successive perturbation of the unknowns $A(\omega_n)$ defined as

$$A(\omega_n) = A(n\varDelta\omega), \qquad n = 0, \pm 1, \pm 2, ..., \pm P.$$ (5.89a)

The inputs are

$$\hat{O}(\omega_m) = \hat{O}(m\varDelta\omega), \qquad m = 0, \pm 1, \pm 2, ..., \pm M,$$ (5.89b)

where

$$M\varDelta\omega = \Omega$$

is the data interval in frequency space. Spacing $\varDelta\omega$ is fixed as the sampling interval [note (5.7b)] corresponding to knowledge of finite extension $|x| \leq X$ for the spatial object,

$$\varDelta\omega = \pi/X.$$ (5.89c)

For the first series of perturbations, described next, $P = M/2$. With this situation, there are just enough values $A(\omega_n)$ to contribute to (5.87a) over the entire data interval $|m| \leq M$.

First, the three quantities $A(\omega_0)$, $A(\omega_1)$, $A(\omega_{-1})$ are regarded as unknowns in (5.87a) and (5.87b), with all other A set to zero. These three unknowns are determined as the solution to three cubic equations obtained in the following way. Attach constraint (5.87b) to (5.87a) as an additive Lagrange term. Then set derivatives $\partial/\partial A(\omega_0)$, $\partial/\partial A(\omega_1)$ and $\partial/\partial A(\omega_{-1})$ of the equation equal to zero. The resulting three cubic equations are solvable by an efficient algorithm.

Having obtained a first-approximation to $A(\omega_0)$, $A(\omega_1)$ and $\mathrm{A}(\omega_{-1})$, attention is shifted to $A(\omega_0)$, $A(\omega_2)$ and $A(\omega_{-2})$. The previous solution for $A(\omega_{\pm1})$ fixes those two numbers, and the remaining A are made zero. Again three cubic equations in these three unknowns are obtained and solved, as above.

Attention is now shifted to finding $A(\omega_0)$, $A(\omega_{\pm3})$. Previously obtained values for $A(\omega_{\pm1})$ and $A(\omega_{\pm2})$ are used to fix those numbers, and the remaining A are made zero. The three new unknowns $A(\omega_0)$, $A(\omega_{\pm3})$

are found as outlined above. This procedure is continued until the trio of unknowns $A(\omega_0)$, $A(\omega_{\pm P})$ is solved for.

These perturbations of the unknowns, taken three at a time in turn, out to $A(\omega_{\pm P})$ comprise *one cycle* of the perturbation procedure.

A new cycle is now initiated, once more regarding the central trio $A(\omega_0)$, $A(\omega_{\pm 1})$ as unknowns, all the other A fixed at their previous values; etc. for $\omega_{\pm 2}, \ldots$. After the trio of unknowns $A(\omega_0)$, $A(\omega_{\pm P})$ are once again found, this completes the second cycle.

The effect of each new cycle k is to further reduce the obtained minimum $E(k)$ in (5.87a). It is found, however, that after a finite number of cycles, say K, the minimum does not further reduce significantly. What should be done now? First, should a further minimization be sought?

We note that the inputs $\hat{O}(\omega_m)$ are inverse-filtered values, and hence suffer noise $N(\omega_m)/\tau(\omega_m)$; see (5.34). Substituting $O + N/\tau$ for $\hat{O}$ in (5.87a) shows, in fact, that if $A \otimes A = O$ (a perfectly accurate estimate A) there is still a residual, non-zero minimum of

$$\sum_{m=-M}^{M} |N(\omega_m)/\tau(\omega_m)|^2 = E_{\text{res}}. \tag{5.90}$$

The expected value $\langle E_{\text{res}} \rangle$ could be known to the user from knowledge of the noise power spectrum and systems transfer function.

It is logical, therefore, to seek to further reduce the minimum $E^{(k)}$ in (5.87a) only if it exceeds value $\langle E_{\text{res}} \rangle$. Let us assume this is true. But, from above, $E^{(k)}$ has saturated at an irreducible number. The only recourse left to the user is the knowledge that *permitting values* $A(\omega_{P+1})$, $A(\omega_{-P-1})$, *etc., to be non-zero will contribute new degrees of freedom inside the sum in* (5.87a), *and hence will permit E to be further reduced.*

At this point, then, the trio $A(\omega_0)$, $A(\omega_{P+1})$, $A(\omega_{-P-1})$ are perturbed, with all the other A left constant at their previous values. A dramatic decrease in E usually occurs now. The next perturbation returns to the central-most trio, and the process is recycled as before, now using frequencies ω_{P+1}, ω_{P-1} as the outermost.

If a new saturation for E occurs, and if $E > \langle E_{\text{res}} \rangle$, the next outermost frequencies ω_{P+2}, ω_{-P-2} are permitted within the range of perturbation, etc. Finally, E falls below $\langle E_{\text{res}} \rangle$, and the spatial solution $o(x)^+$ is obtained

$$o(x)^+ = \sum_{n=-Q}^{Q} O(\omega_n)^+ \, e^{j\omega_n x} \tag{5.91}$$

with $O(\omega_n)^+$ given by (5.85b). ω_Q is the outermost frequency used in the perturbation scheme. In general, $\omega_Q > \Omega$ the data frequency band. Hence, there is usually bandwidth extrapolation.

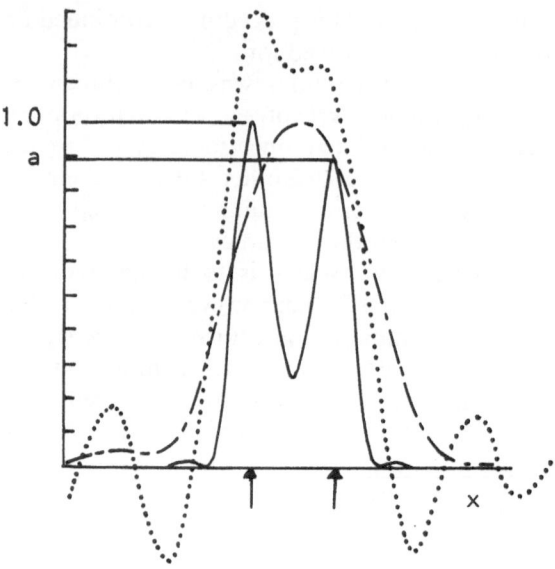

Fig. 5.11. Various versions of the Cygnus A brightness distribution obtained on the Nançay Radiotelescope. — – — Recorded measurement, ⋯⋯⋯ Result of inverse filtering the preceding, ——— Biraud positive restoration. The separation and relative brightness a of the two peaks agrees quite well with direct measurements made with larger-aperture telescopes. (After Biraud [5.9])

An effective demonstration of the use of Biraud's method was its application to finding the brightness profile of a double-star, Cygnus A. The Nancay Radiotelescope provided the experimental inputs, with an overall signal-to-noise ratio of about 50. The relevant curves are shown in Fig. 5.11.

An extensive investigation of Biraud's algorithm was the subject of a thesis by Wong [5.53].

5.15.1 Discussion

The essential ingredient in Biraud's and Schell's approaches is the *necessity* for exceeding the data bandwidth if a positive output *consistent with the inputs* is to be obtained. This is providing noise-propagation residual E_{res} [in (5.90)] is small enough. In practice, the method is tolerant to quite significant levels of noise in the inputs; and extrapolation by factors of two or more is the rule rather than the exception.

However, these benfits accrue only for objects that consist of a general array of impulses against known (hence, zero) background. For example, an object consisting of random steps does not restore by this method much better than by plain inverse filtering. The edge gradients are enhanced better than by inverse filtering, but there is as much objectionable spurious oscillation in the plateau regions. This drawback is the tendency of all extant, positive-constrained restoring schemes. However, see Subsection 5.19.2.

5.16. The Jansson-van Cittert Method

We examined in Subsection 5.10.4 van Cittert's method of successive convolution; and saw why it is usually incapable of strong resolution enhancement of noisy data. Recently, JANSSON [5.10] found a modification of the technique which has transformed it into a useful tool for enhancing spectral data.

JANSSON modified van Cittert's basic algorithm (5.59) to read: Initial conditions $k = 0$, $\hat{o}^{(0)}(x) = i(x)$. Then iterate

i) $\hat{i}^{(k)}(x) = \hat{o}^{(k)}(x) \otimes s(x)$,

ii) $\hat{o}^{(k+1)}(x) = \hat{o}^{(k)}(x) + r(x)[i(x) - \hat{i}^{(k)}(x)]$,

where

$$r(x) = C[1 - 2|\hat{o}^{(k)}(x) - \tfrac{1}{2}|], \quad C = \text{constant}, \tag{5.92}$$

iii) $k \to k + 1$,

iv) go to (i)

(actually, $\hat{o}^{(k)}(x)$ on the right-hand side of step i) incorporates updated values $o^{(k-1)}(x)$ from step ii) as they are formed for preceding x-values; for simplicity, we suppress this from the notation).

Comparing this approach with VAN CITTERT's, we observe the insertion of a relaxation factor $r(x)$ at step ii). This is the key modification of the method. The dependence of $r(x)$ upon $\hat{o}^{(k)}(x)$ is such that $r(x) \to 0$ as either $\hat{o}^{(k)}(x) \to 0$ or 1. Hence, if at any x an iterate $\hat{o}^{(k)}$ is close to either 0 or 1, by step ii) the next iterate $\hat{o}^{(k+1)}$ and all subsequent ones will will remain at that value. This constrains the output $\hat{o}(x)$ to obey $0 \leq \hat{o}(x) \leq 1$. Precisely this physical need occurs with absorption spectra, for which the algorithm was designed.

Regarding step ii) above, we see that initially r is linear in $\hat{o}^{(0)}$, which in turn is linear in image i. But then $\hat{o}^{(1)}$ formed in ii) becomes quadratic in image i, $\hat{o}^{(2)}$ is cubic in i, etc. In general, then, the output $\hat{o}(x)$ is highly non-linear in the image data. The factor $r(x)$, then, transforms the linear, unconstrained van Cittert technique into a *nonlinear, constrained* method.

Regarding possible use of the method upon images other than absorption spectra, it is possible to tailor the choice of $r(x)$ to other constraint needs. If, for example, it is desired that $A \leqq \hat{o}(x) \leqq B$, with A, B known constants, one can use a new relaxation factor

$$r(x) = C[1 - 2(B-A)^{-1}|\hat{o}^{(k)}(x) - 2^{-1}(A+B)|].$$

Or, if simply a positive output $\hat{o}(x) \geqq 0$ is required, we may set $A = 0$ in the preceding and use any B which bounds from above the anticipated $o(x)$ pattern.

It is rather interesting that the procedure can even be generalized to the case where the user knows bounding *curves* $A(x)$, $B(x)$ between which the object lies. Now $A(x)$, $B(x)$ replace the constants A, B in the preceding formula.

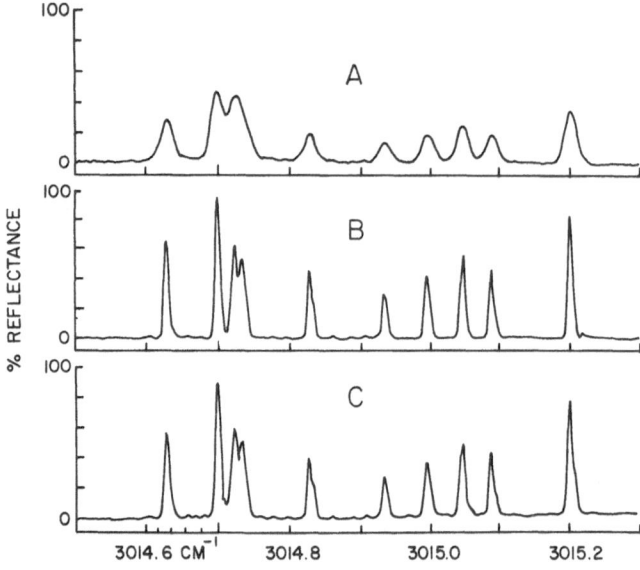

Fig. 5.12. Jansson's restoring technique applied to spectral data, curve A (a portion of the Q branch of the v_3 band of CH_4). Curve B is the output of the restoring algorithm, using the measured psf $= s(x)$. Curve C is the output of the restoring algorithm based on a psf modeled as a gaussian function, whose half-width is the same as that of the measured psf. This approximation results in only a slight degradation of quality in the restoration. (After JANSSON [5.10])

Figure 5.12 shows a typical use of algorithm (5.92) upon spectroscopic image data. The resolution in curves B and C greatly exceeds that of the data in A, to the extent of splitting one apparent line into two. Curve C shows that the method is not overly sensitive to accuracy in the spread function $s(x)$; here, a gaussian-approximation to the experimentally obtained one was used in the algorithm (5.92).

5.17. Maximum Entropy Restorations: Two Philosophies, Two Approaches

The constrained restoring algorithms of Sections 5.15 and 5.16 were *ad hoc* in nature. That is, each of the two algorithms reflected a pragmatic course of operations based on intuition. A logical question to ask, is whether a positive formalism exists which logically follows from some prior, meaningful principle (an example is the way Wiener-Helstrom filtering follows the *mmse* criterion in Subsection 5.11.1). The principle of maximum entropy falls into this category.

The rationale for using this principle is as follows. A very similar problem to inversion of (5.1) is inversion of

$$q_m = \int\limits_{-\infty}^{\infty} dx\, x^m \hat{p}(x), \quad m = 1, 2, \dots, M \tag{5.93a}$$

for the unknown probability density $\hat{p}(x)$ given M of its moments $\{q_m\}$. The latter correspond to our input image data $\{i(y_m)\}$. The similarity goes deeper, in that $\hat{p}(x)$ *must be positive* to represent a real probability density. For this probability-estimation problem, JAYNES [5.11] showed that the "least-biased" estimate $\hat{p}(x)$ is the unique function which has a maximum entropy

$$H \equiv - \int\limits_{-\infty}^{\infty} dx\, \hat{p}(x) \ln \hat{p}(x), \tag{5.93b}$$

subject to the preceding data equations. "Least biased" means as equiprobable as the moment inputs will allow; the maximum entropy condition thereby exerts a *smoothing influence* on the estimate. We have seen, from Subsection 5.3.1 onward, how important an effect this is. But there is a further benefit.

The solution $\hat{p}(x)$ to Jaynes' maximum entropy approach is

$$\hat{p}(x) = \exp\left[- \sum_{m=1}^{M} \lambda_m x^m \right], \tag{5.93c}$$

where unknowns λ_m must be adjusted to satisfy the moment equations. This form for $\hat{p}(x)$ *cannot be negative*. Hence, the maximum entropy criterion automatically leads to an all-positive output.

Convinced of the utility of this approach, how do we apply it to estimating a space (and not probability) function? Evidentally, o (at any x) must now be modelled as a random variable so that $p(o)$, its probability, may be formed. With $p(o)$ known entropy H is determined from (5.93b).

At this point, there is a divergence of approaches between models due to BURG [5.5], and due to FRIEDEN [5.18].

In the former case, o is regarded as the square of another variable, call it a (as in Section 5.15). This guarantees that o is positive. Next, the first two moments of a are presumed fixed; this is an arbitrary assumption for the optical problem, at least. With this assumption, (5.93c) shows that the least-biased representation for $p(a)$ is normal (or, gaussian, assuming zero mean).

Under these circumstances, the entropy for o is known. It is

$$H = -\alpha \ln \langle o \rangle + \beta, \quad \alpha, \beta \text{ pure numbers}$$

for the one point x. Regarding all x, it is least-biased to assume the $o(x)$ statistically independent. Then the total entropy is simply the sum of the individual entropies,

$$H = -\alpha \int dx \ln \langle o(x) \rangle + \beta.$$

The last step is to presume that $\langle o(x) \rangle$ represents the estimate $\hat{o}(x)$ we are seeking. Hence, we end up with the criterion

$$H = -\int dx \ln \hat{o}(x) = \text{maximum}. \tag{5.94}$$

[Factors α, β are unimportant in determining the $\hat{o}(x)$ satisfying (5.94).] This is Burg's maximum entropy approach, as it adapts to the optical problem.

An alternative use of the maximum entropy principle was independently developed by FRIEDEN [5.18]. A statistical model for the object is assumed which differs fundamentally from Burg's choice. Here, the object is imagined to be composed of discrete, mathematical "grains" of small intensity $\Delta(o)$ which are distributed over the object scene. The scene is subdivided into cells centered on a subdivision of points $\{x_n\}$, and the unknown object is assumed to have O_n grains in cell n. Thus, $o(x_n) = O_n \Delta o$. Let p_n represent the probability of a grain locating in cell n. Then if a large number of grains are distributed over the object, by the law of large numbers

$$p_n = O_n/O_T \, ,$$

where O_T is the total number of grains in the object. O_T is assumed known by conservation of energy from the image data.

The entropy, then, is

$$H \equiv - \sum_{n=1}^{N} p_n \ln p_n = - \sum_{n} (O_n/O_T) \ln (O_n/O_T)$$
$$= - \sum_{n} (o(x_n)/\Delta o\, O_T) \ln [o(x_n)/\Delta o\, O_T]$$

from the preceding. By the definition of O_T, we finally get

$$H = -\alpha \sum_{n} o(x_n) \ln o(x_n) + \beta \qquad (a, \beta \text{ constant}) .$$

The principle of maximum entropy then becomes

$$H = - \sum_{n} \hat{o}(x_n) \ln \hat{o}(x_n) = \text{maximum} . \tag{5.95}$$

Comparison with the alternative criterion (5.94) reveals that the two approaches differ by a factor $o(x)$. This factor gives criterion (5.95) a mathematical form that is closer to that of classical entropy (5.93b) than is criterion (5.94). However, a price paid for this resemblance is the lack of a closed form solution (shown below) for $\hat{o}(x)$. We may observe, finally, that (5.95) has followed from perhaps a weaker set of assumptions than did (5.94).

The essential point, however, is that both criteria exert a smoothing influence on estimate $\hat{o}(x)$: If any two object values $o(x_1)$, $o(x_2)$ are made to differentially change to a pair of more equal values (a smoothing tendency) either H *increases*. Therefore, a *maximum* H by either criterion fosters a maximally smooth estimate. The absolute maximum H, in fact, results from the *perfectly* smooth estimate $\hat{o}(x) = $ constant. The input data, however, act as constraints which force fluctuations in $\hat{o}(x)$ and hence departure from the absolute maximum H situation.

5.18. Burg's Maximum Entropy Solution

As with SCHELL and BIRAUD, assume the data inputs are the inverse-filter estimates $\{\hat{O}(\omega_m)\}$. These are used to constrain the maximum

232   B. R. FRIEDEN

entropy criterion (5.94) in the conventional way, delegating a Lagrange multiplier λ_m to each

$$-\int_{-X}^{X} dx \ln \hat{o}(x) - \sum_{m=1}^{m} \lambda_m \left[\int_{-X}^{X} dx\, \hat{o}(x)\, e^{-j\omega_m x} - \hat{O}(\omega_m) \right] \tag{5.96a}$$
$$= \text{maximum}.$$

The derivation was obtained (but not published) by BURG, and has recently appeared in the open literature; see [5.54]. As it is rather lengthy, we shall not repeat it here. The result is

$$\hat{o}(x) = \frac{(2X)^{-1} \gamma_{M+1}}{\left| 1 + \sum\limits_{m=1}^{M} \gamma_m \exp[-jm\pi(2X)^{-1}(x+X)] \right|^2} \tag{5.96b}$$

$$-X \leqq x \leqq X$$

with the $M+1$ unknowns $\{\gamma_m\}$ determined as the solution to the $M+1$ linear equations

$$
\begin{bmatrix}
\hat{\mathscr{O}}(0) & \hat{\mathscr{O}}(\omega_1) & \hat{\mathscr{O}}(\omega_2) & \hat{\mathscr{O}}(\omega_3)...\hat{\mathscr{O}}(\omega_M) \\
\hat{\mathscr{O}}(\omega_1) & \hat{\mathscr{O}}(0) & \hat{\mathscr{O}}(\omega_1) & \hat{\mathscr{O}}(\omega_2)...\hat{\mathscr{O}}(\omega_{M-1}) \\
\hat{\mathscr{O}}(\omega_2) & \hat{\mathscr{O}}(\omega_1) & \hat{\mathscr{O}}(0) & \hat{\mathscr{O}}(\omega_1)...\hat{\mathscr{O}}(\omega_{M-2}) \\
\cdot & \cdot & & \cdot \\
\cdot & \cdot & & \cdot \\
\cdot & \cdot & & \cdot \\
\hat{\mathscr{O}}(\omega_M) & \hat{\mathscr{O}}(\omega_{M-1}) & \cdot & \hat{\mathscr{O}}(0)
\end{bmatrix}
\begin{pmatrix} 1 \\ \gamma_1 \\ \gamma_2 \\ \cdot \\ \cdot \\ \cdot \\ \gamma_M \end{pmatrix}
=
\begin{pmatrix} \gamma_{M+1} \\ 0 \\ 0 \\ \cdot \\ \cdot \\ \cdot \\ 0 \end{pmatrix},
\tag{5.96c}
$$

where $\omega_m = m\pi(2X)^{-1}$, $m = 0, 1, ..., M$, and

$$\hat{\mathscr{O}}(\omega) \equiv 2\,\text{Re}\,\{\hat{O}(\omega)\exp(-j\omega X)\}. \tag{5.96d}$$

The latter relates intermediate quantities $\{\hat{\mathscr{O}}(\omega_m)\}$ to input data $\{\hat{O}(\omega_m)\}$.

5.18.1. Discussion

The elegant and useful aspect of the solution (5.96b)–(5.96d) is its closed and simple form: one matrix solution, in (5.96c), followed by substitution into (5.96b). Hence, the requirement of a positive estimate consistent with the data does not here require an iterative algorithm (as it did in Sections 5.15 and 5.16). In fact, these simple computer operations are basically the same as those we found for linear, *non*-positive restoring

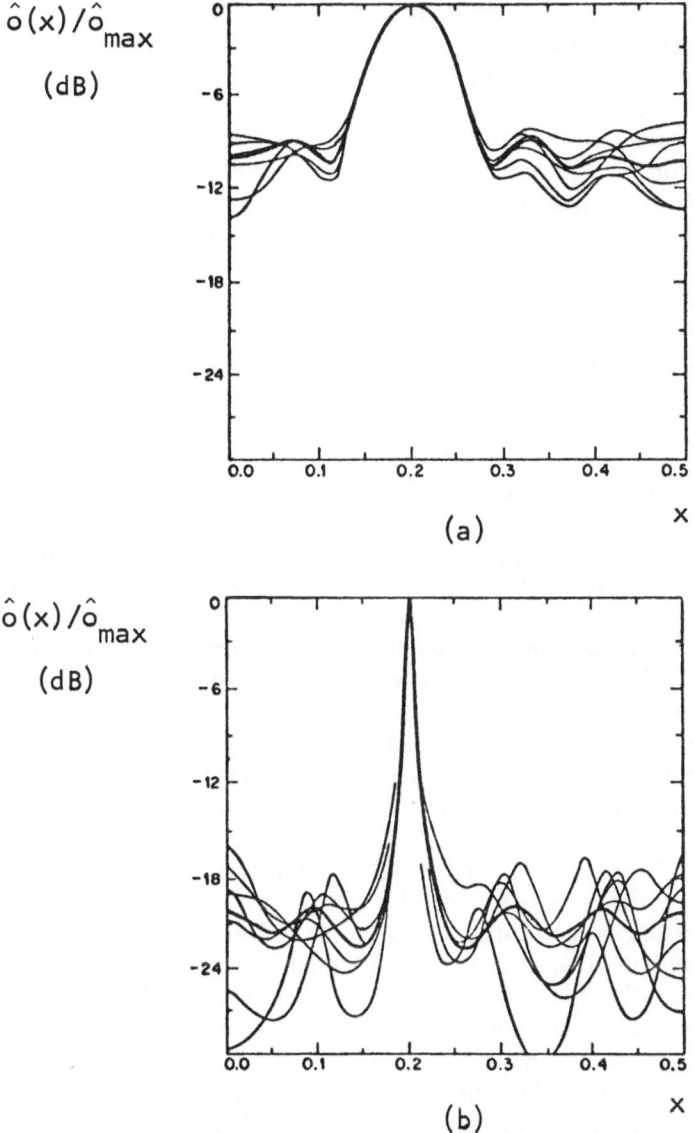

$\hat{o}(x)/\hat{o}_{max}$

(dB)

(a)

x

$\hat{o}(x)/\hat{o}_{max}$

(dB)

(b)

x

Fig. 5.13a and b. Comparison of restoring by (a) conventional triangular window function $W(\omega)$, and (b) Burg's maximum entropy algorithm (5.96b)-(5.96d). Seven sets of noisy input data were restored by each of (a) and (b), and are displayed. The bold curves correspond to the noiseless-input case. Over the x-domain shown, the true object was a δ-function at $x = 0.2$. Burg's method shows much higher resolution than does the standard approach. Also, the sidelobes are quite small (note the logarithmic scale used). However, the estimated peak values are significantly off from the true peak value (see text). (After Lacoss [5.6])

algorithms [see, e.g., (5.26)]. The potential saving in computer time this represents is a basic advantage over all other presently known, positive restoring algorithms (which *are* iterative).

The main drawback to this method is its relative sensitivity to error in the data $\{\hat{O}(\omega)\}$. This stems from an ignoring of such error in the derivation of the algorithm. For example, by its derivation the matrix in (5.96c) is *semipositive definite*. The latter would indeed be true if the $\{\hat{O}(\omega_m)\}$ were the *true* values $\{O(\omega_m)\}$; but this is not necessarily true in the presence of error in these inputs.

This problem has led to spotty results during actual use of the method. Lacoss [5.6] has extensively tested the method on noisy data. For example, Fig. 5.13b shows the estimated object (5.96b) based on seven sets of noisy data $\{\hat{O}(\omega_m)\}$ arising from the one object $o(x) = 1 + \delta(x + 0.2) + \delta(x - 0.2)$. The data were corrupted by uniformly random noise of zero mean and standard deviation 3.3% of maximum signal. Seven different sets of noise were used. In addition, the bold curve is the result of using perfect data (zero noise). We see that the sidelobe structure, representing false detail, is quite small, whereas the "core" structure corresponding to the true object spike is quite narrow. These appear to be good restorations, especially as compared with Fig. 5.13a, which shows linear restorations, using the same data sets, based on use of a triangular window function (see Subsection 5.9.2). The comparison is a bit undemocratic, however.

The maximum entropy restorations in Fig. 5.13b have all been renormalized (after direct use of the algorithm) to unity as a maximum level. Prior to renormalization, these maximum levels depart fairly strongly from the true level, with excursions of $\pm 20\%$ rather typical, see [Ref. 5.6, Table 3]. And, with an increase in noise level from 3.3% to 5% the errors rise dramatically.

As a partial solution to this problem, Lacoss found that the *area* under a core region better approximates the true peak value than does the estimated peak itself. But, this is only true for a very narrow object spike. At what width does the tendency break down, and how does it depend on shape of the spike? Further, what if the user is confronted with an unknown array of spikes and broad plateaus. These basic questions are currently being pursued by members of the geophysics and astronomy communities.

5.19. Frieden's Maximum Entropy Solution

This solution is based on the use of (5.95) as the basic entropy measure. An additional departure from Burg's approach is to regard the image noise as another unknown array, which is to be estimated also. An unbiased, or maximum entropy, *estimate of noise* is therefore now desired as well.

However, the noise in the image data can be both positive and negative; and $\ln n_m$ for n_m negative is undefined. A way around this problem is to define a new set of noise values

$$N(y_m) = n(y_m) + B, \quad B \geqq 0. \tag{5.97a}$$

B is a positive constant of sufficient size that it cancels the most negative-going values $n(y_m)$, and thereby makes all $N(y_m) \geqq 0$. We wish to find a maximum-entropy estimate of the $N(y_m)$ and the $o(x_n)$.

An unbiased situation for $N(y_m)$ permits *a priori* all N values from 0 to ∞. Therefore, from (5.97a) B should ideally be the value

$$B = - \text{most negative } n(y_m). \tag{5.97b}$$

But obviously, the user cannot know this number precisely. However, he *can* input a reasonable value for B, such as -2σ noise (say), if such *a priori* information is at hand. We have found empirically that the quality of solution for the unknown $\{o(x_n)\}$ and $\{N(y_m)\}$ does not critically depend on use of a value of B precisely satisfying (5.97b); however, the solutions are best when (5.97b) is indeed satisfied.

Since the object and the noise are independent arrays of numbers, the total entropy from both is the sum of the two entropies. The input data which is to constrain the maximum entropy solutions away from perfectly flat estimates (see Section 5.17, last paragraph) is the image data and its total energy I_0. The latter value is equated to the total object flux, by conservation of energy. The restoring algorithm is then [compare with Burg's equation (5.96a)]

$$- \sum_{n=1}^{N} \hat{o}(x_n) \ln \hat{o}(x_n) - \varrho \sum_{m=1}^{M} \hat{N}(y_m) \ln \hat{N}(y_m)$$

$$- \sum_{m=1}^{M} \lambda_m \left[\sum_{n=1}^{N} \hat{o}(x_n) s(y_m - x_n) + \hat{N}(y_m) - B - i(y_m) \right] \tag{5.97c}$$

$$- \lambda_{M+1} \left(\sum_{n=1}^{N} \hat{o}(x_n) - I_0 \right) = \text{maximum}.$$

The new input parameter ϱ permits the user to emphasize smoothness in one or the other of $\{\hat{o}(x_n)\}$ or $\{\hat{N}(y_m)\}$. The larger ϱ is made, the more emphasis is placed upon maximizing the entropy of the $\{\hat{N}(y_m)\}$, and hence upon making it smooth; and, vice versa. Test cases, to be described below, imply a value $\varrho = 20$ as about optimum for a wide range of object and noise situations.

The solution to (5.97c) is gained by differentiating $\partial/\partial\hat{o}(x_n)$ and $\partial/\partial\hat{N}_m$ over all m, n and equating the results to zero. This results in a separate solution for each of the unknown arrays:

$$\hat{o}(x_n) = \exp\left[-1 - \lambda_{M+1} - \sum_{m=1}^{M} \lambda_m s(y_m - x_n)\right], \tag{5.98a}$$

$$\hat{N}(y_m) = \exp(-1 - \lambda_m/\varrho). \tag{5.98b}$$

The unknowns $\{\lambda_m\}$ which define these solutions are found by making these solutions obey the input constraint equations

$$\sum_{n=1}^{N} \hat{o}(x_n) s(y_m - x_n) + \hat{N}(y_n) - B = i(y_m), \tag{5.98c}$$

$$m = 1, 2, ..., M$$

$$\sum_{n=1}^{N} \hat{o}(x_n) = I_o.$$

The right-hand sides of (5.98c) are the input data. By substitution of (5.98a) and (5.98b) into (5.98c), we have a system of $M + 1$ equations that are nonlinear in $M + 1$ unknowns $\{\lambda_m\}$.

Although there may be other ways of effecting a solution, we have found a Newton-Raphson, or relaxation, algorithm (see, e.g., [5.55]) works. An initial set $\{\lambda_m\}$ is chosen [we use $\lambda_m = 0$ for all $m < (M + 1)$; with λ_{M+1} chosen so as to satisfy the $(M + 1)$st constraint equation (5.98c)]. The $\{\lambda_m\}$ are then changed in the direction of satisfying all the constraint equations, by assuming all higher-order derivatives $\partial^{(k)} i(y_m)/\partial\lambda_p^{(k)}$ are zero, $k = 2, \dots$. After anywhere from 8 to 40 iterations of this algorithm, a set of $\{\lambda_m\}$ satisfying (5.98c) is found.

5.19.1. Discussion

Although we have started the derivation from an *a priori assumption* of maximum entropy, (5.97c), it is of further interest that this prior assumption may itself be *derived* from a *maximum-likelihood* approach to the problem; see [5.18]. Hence, the solution (5.98a) and (5.98b) is both a maximum-entropy and maximum-likelihood estimate.

Regarding positivity, we see that the functional form of representation (5.98a) for the object guarantees a positive estimate.

Regarding spurious oscillations in output $\hat{o}(x)$, we see from (5.98a) that

$$d^{(k)}\,\hat{o}(x)/dx^{(k)} \propto \hat{o}(x), \quad k = 1, 2, \dots .$$

This means that $\hat{o}(x)$ cannot oscillate where it is at zero background and can oscillate little where $\hat{o}(x)$ is small. In practice, typically $\hat{o}(x)$ has values like 10^{-8} *throughout* regions where the true object $o(x)$ is

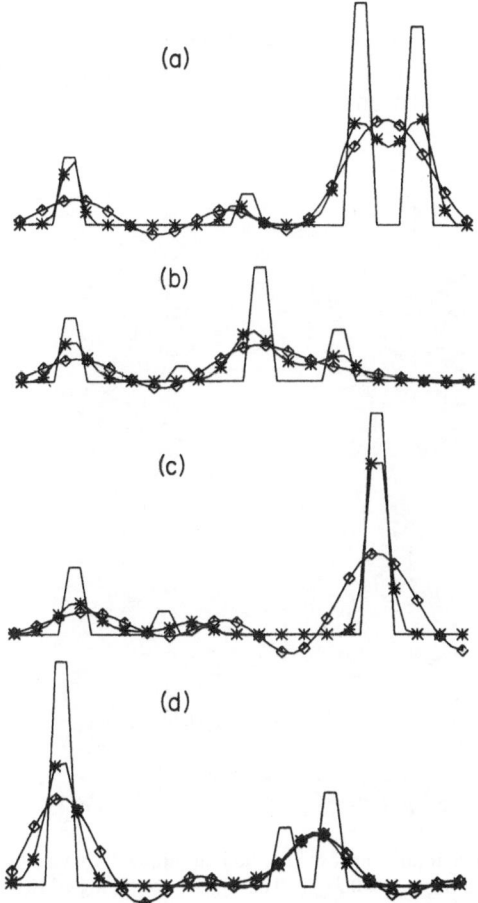

(a)

(b)

(c)

(d)

Fig. 5.14a–d. Restorations by Frieden's maximum entropy method (starred curves), of some impulsive objects (solid curves). A value $\varrho = 1$ was used. The input image data $\{i(y_m)\}$, which are not displayed, were generated in the computer and degraded by pure diffraction blur and uniformly random noise of maximum relative error 5%. To show the advantages of the method over linear restoring methods, the inverse-filter restorations based on noise*less* data are also shown (boxed curves)

zero. Hence, it is very smooth over these regions, completely lacking spurious oscillation. It is for this very reason that the method works quite well on objects consisting of multiple impulses, for here all but a finite number of object values *are* zero.

5.19.2. Numerical Tests

Figure 5.14 illustrates application of the method to the restoration of diffraction-blurred impulse arrays. These cases were constructed on the computer.

Figure 5.15 shows results when applied to a real photograph, that of a diffraction-blurred double slit. The slit separation is half the Rayleigh resolution length, and non-restorable by linear methods. As is evident

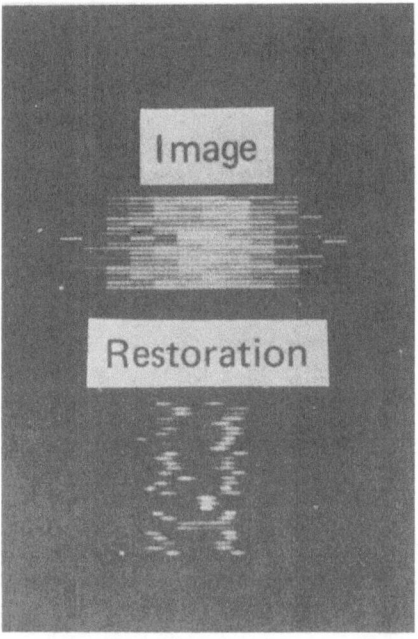

Fig. 5.15. Frieden's maximum entropy method as applied to very noisy experimental image data. The noise has a $\sigma \simeq 0.2 i_{max}$. The object, consisting of two slits separated by 1/2 the Rayleigh resolution distance, was blurred across the slits by pure diffraction (due to a slit-optical aperture). The resulting image was photographed, and then scanned and restored linewise. A value $\varrho = 20$ was used. The first 30 lines of image data and restoration are shown. Although nearly each line of data is restored to two impulses—the correct tendency— the severe noise in the data gives rise to random shifts in the positions of these impulses. These shifts, however, are small; the order of 1/10 of the Rayleigh resolution distance. (After FRIEDEN [5.63])

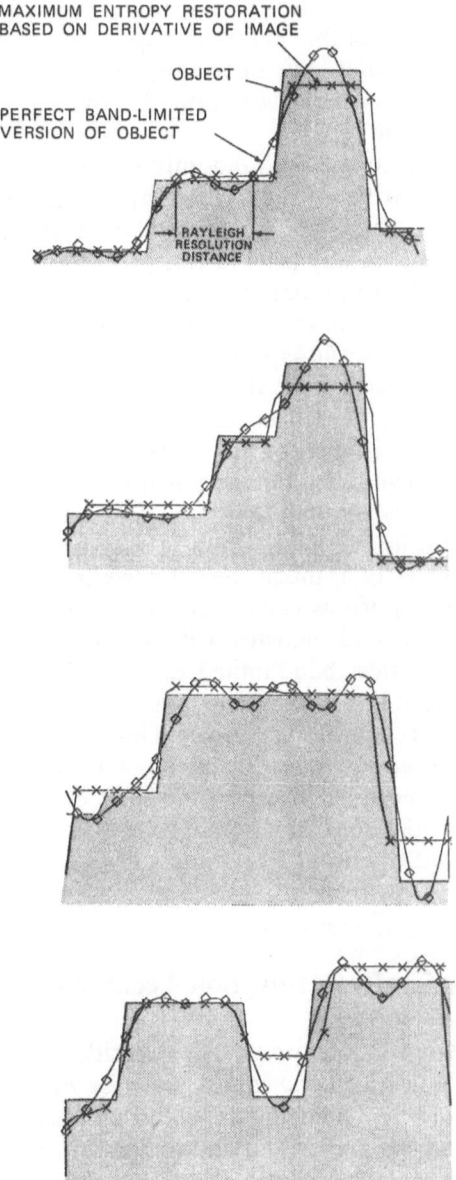

Fig. 5.16. Application of Frieden's maximum entropy approach to randomly-stepped objects. The image data, generated on the computer, were blurred by a $sinc^2(x)$ kernel and sampled at 1/2 the Nyquist interval. This finer-than-usual sampling permits a fairly good first-difference approximation to the first derivative. Use of this method is constrained to cases where the user knows a priori that the object consists of random steps

in the photo, the image suffers considerable noise, with an estimated $\sigma_{noise} \simeq 21\%$. The effect of this noise upon each line of the restoration is, remarkably, but a spatial shift of the two impulses that are generally restored. This gives to the restoration the overall visual effect of a shimmering double slit.

When applied to images of randomly stepped objects, the method gives only slightly better results than by optimum linear filtering. The main drawback is the same Gibbs oscillations that appear familiarly in linear restorations. A way around this problem was found, however. If an object consists of random steps, its *derivative* consists of random impulses. The latter are well-restored by the method (see above). Hence, if user knows his object scene is randomly stepped he may differentiate the given image data and input these numbers into the algorithm. The output will be $d\hat{o}/dx$. Integration across the field retrieves $\hat{o}(x)$.

We have tested this method, with the results shown in Fig. 5.16. The images (not shown) were differentiated by finite difference between adjacent values spaced at half the Nyquist interval. The comparison with a perfect band-limited restoration of each object [defined at (5.8)] shows the advantages of a) much steeper edge gradients; and b) a complete lack of Gibbs spurious oscillation. As far as its ability to identify the presence of edges and plateaus, without false detail, the algorithm is superior to the perfect bandlimited estimate. However, on a mean-squared error basis, it is not. This is because the *heights* of the maximum entropy plateaus are usually in error. This results from cumulative error due to integration across the field of restored delta functions, whose areas are randomly in error. We may conclude, then, that in applications where the determination of geometrical boundaries is more important than that of absolute brightness levels, the proposed method should be of use.

5.20. A Monte Carlo Restoration Technique

All of the restoring methods previously described in this chapter have been based on the use of a *functional form* for $\hat{o}(x)$, which ultimately links back to the inputs. These functional forms either contain an explicit solution for $\hat{o}(x)$, as e.g. in (5.61), or an implicit solution as in, e.g., (5.88).

The probable reason for the prevalence of such functional approaches is that human nature is biased toward analytical answers to problems. The use of statistics is most often regarded as a last resort, after analytical efforts fail. A further factor is simply that training in statistical methods is lacking in the education of many workers in the field (excluding those with E.E. degrees, but including physicists!).

One widely used tool of statistics is the Monte Carlo calculation. Its premise is that a representative sample (say, the object can be built up in the computer by a large number of trials selected randomly from the probabilistic and deterministic laws that define all possible outcomes (objects).

Applying this idea to the restoring problem, we imagine the object space to be initially composed of empty resolution cells. Then, mathematical "grains" of fixed intensity increment do are randomly allocated to the cells by some decision rule. The restored object is thereby built up by successive grain allocations, until all grains are placed. (The total number of grains is known, by conservation of energy, from the image data.)

The possible benefits of this kind of approach are many. First, by divesting ourselves of an analytic form for $\hat{o}(x)$, the output has infinite flexibility regarding fluctuations: edge gradients and linear resolution distances are unconstrained by bandwidth, or other, considerations. In a sense, this is the "ultimate" in nonlinear approaches.

Second, all desired constraints, such as positivity, an upper bound, etc., can be easily effected, by merely ruling our certain "events" from the realm of possibilities. We have seen the importance of enforcing these constraints.

Third, the method is potentially very fast; permitting two-dimensional restorations, for the first time, to enjoy the extra resolution previously afforded only by one-dimensional entities. The speed of the algorithm must ultimately depend on the decision-rule that is chosen for acceptance of a grain in a given cell.

One decision rule that appears to have merit is the following [5.19]: Accept grain k in the cell at x_n if the cumulative image $\hat{i}^{(k)}(y_m)$ formed as

$$\hat{i}^{(k)}(y_m) \equiv \hat{i}^{(k-1)}(y_m) + do \cdot s(y_m - x_n) \tag{5.99a}$$

obeys

$$\hat{i}^{(k)}(y_m) \leqq r \cdot i(y_m) \quad \text{at all} \quad m = 1, \ldots, M \tag{5.99b}$$

with r a minimum. By a minimum in r, we mean that a smaller r would not permit the k-th grain to be placed in *any* cell x_n by the rule (5.99b). As usual $\{y_m\}$ are the data positions, and the $\{i(y_m)\}$ are the image data. When a grain k is accepted in a cell x_p, the object is incremented to

$$o^{(k)}(x_n) = o^{(k-1)}(x_n) + do \cdot \delta_{np}. \tag{5.99c}$$

The algorithm (5.99a)–(5.99c) is very efficient for computer use. It consists operationally of multiple trial additions and comparisons

242   B. R. FRIEDEN

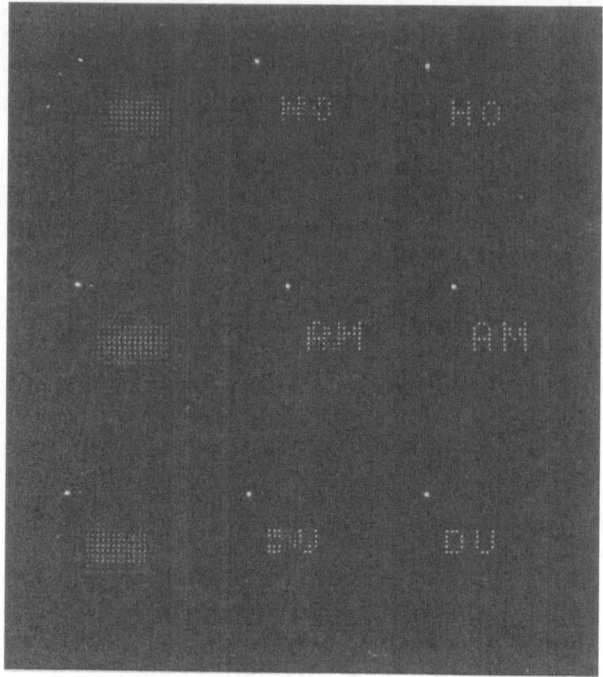

Fig. 5.17. Sample restorations using the Monte Carlo grain-allocation approach. Each row shows the image data, restoration and true object. The image data is noiseless. A linear filtering of the data cannot restore the details of the letters. By comparison, the Monte Carlo approach permits recognition, or near-recognition, of the letters. (After FRIEDEN [5.19])

(5.99a) and (5.99b) followed by one addition (5.99c) when a grain is placed. Note that the sampled values $s(y_m - x_n)$ of the point spread function called for in the algorithm may be computed and stored at the onset of the program, for later recall by the algorithm.

In Fig. 5.17, we show application of the method to the restoration of severely blurred printed text. The point spread function was $\text{sinc}^2(x)$ $\cdot \text{sinc}^2(y)$. The blur is such that opposite extremes of each letter, and the spaces between letters, are separated by half a Rayleigh's resolution distance. With this amount of blur, inverse-filtering cannot resolve the letters even if the image data are noiseless. There is not sufficient band-width, and extrapolation is necessary.

Each input image was a 21×21 matrix of sampled values spaced at 3/4 of the Nyquist interval. With these binary objects at levels 0 and 1, we used a value do of 0.2. Thus, 5 grains fill a given cell. In the cases shown, no noise was added to the image data prior to use of the algorithm.

Subsequent runs with the same data were made, now using the value $do = 0.1$. Although the required computer time increased from about 1 min per two-letter restoration to about 2 min, the accuracy of restoration increased markedly. With the addition of noise characterized by a uniformly random error of maximum value 5% of each image input, the restorations deteriorated about to the level shown in Fig. 5.17.

This algorithm is still in the development stage.

5.21. Other Approaches

In recent years, a great many restoring methods have been proposed in the open literature. The ones we have described in previous sections are those which we have sufficient experience with, on either a theoretical or empirical level, to permit intelligible discussion.

Some methods which we have not yet had the opportunity to digest are due to: RICHARDSON [5.56], MACADAM [5.57], and SAWCHUK [5.58]. Given the scope of the problem, others probably exist as well.

In all of the preceding, we have presumed psf $s(x)$ to be a deterministic quantity known to some accuracy. In the case of atmospheric turbulence, however, it is proper to regard $s(x)$ as a random variable. For this case, SLEPIAN [5.59] has found an *mmse* restoring filter which is a generalization of that derived at (5.68) for a deterministic $s(x)$.

Most images are in the physical form of photographs; and most restorations are acquired for specifically visual observation. All restorations of photographs would profit by reduction of granularity in the image data, and by optimization of the restoration for viewing by the human eye. These problems have not been considered in this chapter. See, e.g., work by HUANG [5.60] and STOCKHAM [5.61].

5.22. Synopsis of Results

Since student days, we have come to believe in condensing all results onto one or two finite pages. This kind of presentation blurs the trees, for an overall view of the forest. Table 5.1 offers a summary of the methods we have previously discussed. The order follows that of the main text. Each restoring method is described as to its "type": linear, or not; bandlimited, or not; constrained to physical values, or not; its resolution properties; and its possible application to extended, two-dimensional images. These are the usual benchmark comparisons of interest to the user. In addition, some comments of a random nature are given in the last column.

Table 5.1. Summary of methods

Method	Type[a]	See Section/Subsection	Remarks on resolution (R = Rayleigh length)	On application to 2-D extended images (matrix $M \times M$)	General remarks
Inverse of matrix image equation	1 NB NI	5.3.1	Perfect in theory; but resolution length $x_0 \gtrsim R/2$ in practice	Solution of $M^2 \times M^2$ matrix required	Hypersensitive to noise in data
Phillips	1 NB NI	5.4	Perfect in theory; in practice, tradeoff with spurious detail	As above	Uses 2nd-difference smoothing parameter γ
Twomey	1 NB NI	5.5	As above	As above. Input matrix may be rectangle	General quadratic smoothing approach
Inverse filtering	1 B NI	5.9	Resolution length $x_0 = R/2$	Demonstrated use	Augmented with FFT-algorithm, for increased speed
With optimum bandwidth $2\Omega_p$	1 B NI	5.9.1	$x_0 = (R/2)(1 - \phi_n/\phi_o)^{-\frac{1}{4}}$ (particular case)	No problem anticipated	General solution Ω obeys $\lvert \tau(\Omega_p)\rvert^2 = \phi_n(\Omega_p)/\phi_o(\Omega_p)$
Windowed filtering	1 B NI	5.9.2	$x_0 \simeq R/2$ in practice	As above	Permits tradeoff between resolution and spurious detail
DC-suppression of bandwidth $2\omega_0$	1 B NI	5.10.1	$x_0 \to R/4$ as $\omega_0 \to \Omega$ (optical cutoff)	As above	Much spurious detail for resolution $x_0 \lesssim R/2$
Pure phase compensation	1 B NI	5.10.3	$x_0 \simeq R$	As above	Relatively insensitive to noise in data
van Cittert	1 B I NI	5.10.4	$x_0 = R/2$	As above	At finite iteration k, like windowed filtering
Discrete convolution	1 B NI	5.10.5	$x_0 \gtrsim R/2$ in practice	Demonstrated use	Space-equivalent to frequency-plane filtering

Method	Type[a]	Eq.			
Wiener-Helstrom	1 B, NI	5.11.1	$x_0 \geqq R/2$	No problem anticipated	Standard *mmse*-filter approach
With sharpness constraint	1 B, NI	5.11.1	As above	No problem anticipated	Allows constraint on sharpness of output
Backus-Gilbert direct approach	1 NB, NI	5.11.2	Flexible, by hypothesis $x_0 \gtrsim R/2$ in practice	Solution of $M^2 \times M^2$ matrix required	Optimum tradeoff between noise propagation and resolution
Backus-Gilbert filter approach	1 B, NI	5.11.2	As above	No problem anticipated	As above
Linear methods that attempt extrapolation (Harris, Barnes, Frieden)	1 NB, NI	5.13.1, 5.13.2	Perfect in theory, but $x_0 \gtrsim R/2$ in practice	Possible, but probably not worth the effort	Extreme problem of noise sensitivity
Schell and Biraud	2 NB, I	5.15	$x_0 \simeq R/4$ accomplished in practice	Possible use	Best use with impulsive objects
Jansson-van Cittert	2 NB, I	5.16	As above	Promising use	As above
Burg	2 NB, NI	5.18	As above	Solution of $M^2 \times M^2$ toeplitz matrix required	As above
Frieden (max. entropy)	2 NB, I	5.19	As above	Iterative solution of $M^2 \times M^2$ matrix required	As above
Frieden (Monte Carlo)	2 NB, I	5.20	As above	Demonstrated ability	As above

[a] Type 1: output linear with input data; unconstrained output.
Type 2: output non-linear with input data; positivity constraint.
NB ≐ non-bandlimited.
B ≐ bandlimited.
NI ≐ non-iterative.
I ≐ iterative.

5.23. Possible Directions for Future Research

Perusal of the table shows that resolution the order of $R/2$ is fairly easy to come by. This is about the practical limit of all linear, non-constrained restoring methods; as explained in Subsection 5.13.2. It is only the newer, *non*-linear methods that permit appreciably lower resolution limits, approximately $R/4$ in practice.

On the other hand, the major domain of use for the nonlinear methods is with impulsive objects such as stars, line spectra, etc. The nonlinear methods fail to work very effectively on randomly stepped, or edge-type, objects. The expedient of differentiating edge-image data works (see Subsection 5.19.2) fairly well, but is undesirable because of extra noise-propagation error, and because the user must know that the object *is* of the edge type. Also, all but Burg's have the further drawback of being iterative.

It appears, then, that major effort needs to be directed toward discovering algorithms which intrinsically work well on edge-type scenes, as well as on impulses; and which are preferably of a closedform type as opposed to being iterative. Perhaps the coming years will see progress on these problems. However, it is by now abundantly clear that to keep abreast of such progress will require the abandonment of parochial viewpoints; the scientific literature of many, seemingly unrelated, fields will have to be followed.

References

5.1. D.L.PHILLIPS: J. ACM **9**, 84 (1962).
5.2. S.TWOMEY: J. Franklin Inst. **179**, 95 (1965).
5.3. G.YAMAMOTO, M.TANAKA: Appl. Opt. **8**, 447 (1969).
5.4. G.A.VANASSE, H.SAKAI: *Progress in Optics VI*, ed. by E.WOLF (North-Holland Publishing Co., Amsterdam, 1967).
5.5. J.P.BURG: *"Maximum Entropy Spectral Analysis"* (Stanford University, Geophysics Department) (paper presented at 37th Annual Society of Exploration Geophysicists Meeting, Oklahoma City, 1967).
5.6. R.T.LACOSS: Geophysics **36**, 661 (1971).
5.7. R.GORDON, G.T.HERMAN: Comm. ACM-Graphics and Image Processing **14**, 759 (1971).
5.8. A.C.SCHELL: Radio Electronic Eng. **29**, 21 (1965).
5.9. Y.BIRAUD: Astron. Astrophys. **1**, 124 (1969).
5.10. P.A.JANSSON, R.H.HUNT, E.K.PLYLER: J. Opt. Soc. Am. **60**, 596 (1970).
5.11. E.T.JAYNES: IEEE Trans. Syst. Sci. Cybern. SSC-**4**, 227 (1968).
5.12. A.WRAGG, D.C.DOWSON: IEEE Trans. Inf. Theory IT-**16**, 226 (1970).
5.13. E.L.O'NEILL: *Introduction to Statistical Optics* (Addison-Wesley Publishing Company, Reading, Mass. 1963).
5.14. S.TWOMEY: J. ACM **10**, 97 (1963).
5.15. O.N.STRAND, E.R.WESTWATER: SIAM J. Numer. Anal. **5**, 287 (1968).

5.16. W. K. PRATT, Ed.: *"Bibliography on Digital Image Processing and Related Topics"* (Univ. of Southern California-USCEE Report 453, Los Angeles, 1973).

5.17. T. S. HUANG, W. F. SCHREIBER, O. J. TRETIAK: Proc. IEEE **59**, 1568 (1971).

5.18. B. R. FRIEDEN: J. Opt. Soc. Am. **62**, 511 (1972).

5.19. B. R. FRIEDEN: *"Image Restoration by Decision-Rule Allocations of Pseudo-Grains"* (paper presented at Spring Meeting of the Optical Society of America, Washington, D.C., 1974).

5.20. I. J. GOOD: J. Roy. Stat. Soc. B **20**, 361 (1958); B **22**, 372 (1960).

5.21. J. W. COOLEY, J. W. TUKEY: Math. Computation **19**, 297 (1965).

5.22. R. C. SINGLETON: IEEE Trans. Audio Electroacoustics AU-**15**, 91 (1967).

5.23. B. J. MCGLAMERY: J. Opt. Soc. Am. **57**, 293 (1967).

5.24. A. PAPOULIS: *Probability, Random Variables, and Stochastic Processes* (McGraw-Hill Book Co., New York, 1965).

5.25. W. B. DAVENPORT, W. L. ROOT: *Random Signals and Noise* (McGraw-Hill Book Co., New York, 1958).

5.26. M. M. SONDHI: Proc. IEEE **60**, 842 (1972).

5.27. J. L. HARRIS: J. Opt. Soc. Am. **56**, 569 (1966).

5.28. R. J. ARGUELLO, H. R. SELLNER, J. A. STULLER: IEEE Trans. Computers C-**21**, 812 (1972).

5.29. P. JACQUINOT, B. ROIZEN-DOSSIER: In *Progress in Optics III*, ed. by E. WOLF (North-Holland Publishing Company, Amsterdam, 1964).

5.30. B. R. FRIEDEN: Optica Acta **16**, 795 (1969).

5.31. E. L. O'NEILL: IRE Trans. Inf. Theory IT-**2**, 56 (1956).

5.32. W. SWINDELL: Appl. Opt. **9**, 2459 (1970).

5.33. E. W. HAWMAN: Internal report, Optical Sciences Center, University of Arizona, 1973.

5.34. R. BAKER, J. BURKE, R. FRIEDEN: *"Progress in Digital Image Processing, 1969"* (Techn. Report 50, Optical Sciences Center, University of Arizona, 1970).

5.35. P. H. VAN CITTERT: Z. Physik **69**, 298 (1931).

5.36. P. A. JANSSON, R. H. HUNT, E. K. PLYLER: J. Opt. Soc. Am. **58**, 1665 (1968).

5.37. B. R. FRIEDEN: J. Opt. Soc. Am. **64**, 682 (1974).

5.38. R. NATHAN: In *Pictorial Pattern Recognition* (G. C. CHENG, THOMPSON, Washington, D.C., 1968).

5.39. C. W. HELSTROM: J. Opt. Soc. Am. **57**, 297 (1967).

5.40. P. B. FELLGETT, E. H. LINFOOT: Phil. Trans. Roy. Soc. (London) A **247**, 369 (1955).

5.41. B. R. FRIEDEN: J. Opt. Soc. Am. **60**, 575 (1970).

5.42. G. BACKUS, F. GILBERT: Phil. Trans. Roy. Soc. (London) A **266**, 123 (1970).

5.43. B. E. A. SALEH: Appl. Opt. **13**, 1833 (1974).

5.44. H. WOLTER: In *Progress in Optics I*, ed by E. WOLF (North-Holland Publishing Co., Amsterdam, 1961).

5.45. J. L. HARRIS: J. Opt. Soc. Am. **54**, 931 (1964).

5.46. B. R. FRIEDEN: Appl. Opt. **9**, 2489 (1970).

5.47. D. SLEPIAN, H. O. POLLAK: Bell Syst. Tech. J. **40**, 43 (1961).

5.48. B. R. FRIEDEN: In *Progress in Optics IX*, ed by E. WOLF (North-Holland Publishing Co., Amsterdam, 1971).

5.49. C. W. BARNES: J. Opt. Soc. Am. **56**, 575 (1966).

5.50. C. K. RUSHFORTH, R. W. HARRIS: J. Opt. Soc. Am. **58**, 539 (1968).

5.51. G. J. BUCK, J. J. GUSTINCIC: IEEE Trans. Antennas Propagation AP-**15**, 376 (1967).

5.52. F. B. HILDEBRAND: *Introduction to Numerical Analysis* (McGraw-Hill Book Co., New York, 1956), p. 443.

5.53. H. F. WONG: *"A Study and Evaluation of Biraud's Deconvolution Algorithm"* (Masters Thesis, Electrical Engineering Dept., Queen's University, Kingston, Ontario, Canada, 1971).

5.54. J. A. EDWARD, M. M. FITELSON: IEEE Trans. Inf. Theory IT-**19**, 232 (1973).

5.55. L. A. PIPES: *Mathematics for Engineers and Physicists* (McGraw-Hill, Book Co., New York, 1958), p. 116.

5.56. W. H. RICHARDSON: J. Opt. Soc. Am. **62**, 55 (1972).

5.57. D. P. MACADAM: J. Opt. Soc. Am. **60**, 1617 (1970).

5.58. A. A. SAWCHUK: Proc. IEEE **60**, 854 (1972).

5.59. D. SLEPIAN: J. Opt. Soc. Am. **57**, 918 (1967).

5.60. T. S. HUANG: In *Woods Hole Summer Study on Restoration of Atmospherically Degraded Images*, Vol. 2 (Defense Documentation Center, Alexandria, Virginia, 1966).

5.61. T. G. STOCKHAM: Proc. IEEE **60**, 828 (1972).

5.62. D. A. O'HANDLEY, W. B. GREEN: Proc. IEEE **60**, 821 (1972).

5.63. B. R. FRIEDEN: J. Opt. Soc. Am. **62**, 1202 (1972).

Further References with Titles

A. O. ABOUTALIB, L. M. SILVERMAN: Restoration of images degraded by curvi-linear motion; in *Second International Joint Conference on Pattern Recognition* (Copenhagen, 1974; IEEE Cat. No. 74CH0885-4C).

J. J. BURKE: Estimating objects from their blurred and grainy images; in *Proceedings of the Technical Program. Electro-Optical Systems Design Conference* (West International Laser Exposition, San Francisco, 1974).

B. R. FRIEDEN: Restoration of Pictures by Monte Carlo allocation of pseudograins; in *Second International Joint Conference on Pattern Recognition* (Copenhagen, 1974; IEEE Catalog No. 74CH0885-4C), p. 141.

6. Noise Considerations in Digital Image Processing Hardware

F. C. BILLINGSLEY

With 16 Figures

The limiting factor in image processing is noise. Therefore, in the design of any imaging system, one must consider closely the noise performance of the system, especially in the manner in which the noise interacts with the visual appearance of the picture and in which it perturbs the subsequent digital processing.

A number of areas in which this is important will be discussed briefly, However, since this paper is concerned with the noise sources *per se* rather than the specific effects of the noise on subsequent processing, these examples will not be pursued in detail.

The effect of noise on digital reconstruction and enhancement are determined from the statistics of the amount of perturbation caused by the noise. Accordingly, the effects of noise on quantization accuracy are derived, and the resulting curves used to estimate the performance of the system in terms of the probability of assigning the correct digital number (DN) in the analog-digital conversion process.

This derivation leads to the concept of an invariant (β: quantizing step size/rms noise) which may be used in estimating the noise performance of a system. The system chosen for analysis is a flying spot scanner in which the detection is by photomultiplier. Analysis of several "normal" systems indicates their strengths and weaknesses, and leads to a proposed new configuration which combines some of the best features of each.

6.1. Effects of Noise on Viewing and Analysis

Since these two considerations (viewing of an image and the analysis of it) lead to somewhat different noise requirements, they must be discussed separately. The examples given are not meant to be exhaustive, but will serve to set the stage for discussion. It is hoped that these examples will lead the reader to consider other areas in which the prediction of noise performance may be beneficial.

6.1.1. Viewing for Esthetics

The effect of two-dimensional noise upon the human observer has attracted the attention of a large number of workers, see, for example

[6.1–6]. The visual effects of noise are tied up intimately with the desired resolution, appearance, contrast, surroundings, brightness, and other aspects of the picture as well as the magnitude, spectral content and other aspects of the noise itself. As a very general rule of thumb (and recognizing the likelihood of oversimplification) a 40 dB (100/1) signal-to-noise ratio will give a reasonably clean picture. More noise will degrade the ability to differentiate by eye two areas of different brightness (or in the case of film, of different transmittances) [6.7, 8].

Noise may have a beneficial effect under certain conditions. ROBERTS, for example, achieved data compression by breaking up the annoying contours caused by coarse quantization by the addition and subtraction of an identical pseudorandom noise at the transmitter and receiver [6.9]. This technique, which takes advantage of the physiological properties of the human viewer, is not an information preserving technique, so that the effects on possible subsequent processing are detrimental.

Image processing for human viewing will be quite limited by the noise. For example, the pictures received from the ERTS Satellite may have quite low contrast and may require a contrast stretch of 5 × or more. This amount of stretch will cause an apparently clean picture to look quite noisy. Similarly, it has been found that the high spatial frequency enhancement used to sharpen the visual appearance of pictures must be limited to some relatively low value, say 5 (that is, the high spatial frequencies have been amplified by 5 × relative to the amplification of low frequencies), to prevent the enhanced noise from becoming objectionable. This limits the amount of image sharpening that may be accomplished.

Although the visual effects are of importance, they are secondary to data considerations when image processing is to be attempted, and will therefore not be discussed in further detail here.

6.1.2. Effects on Data Processing

Two-dimensional noise in a picture will degrade the ability to measure the modulation transfer function [6.10–13]. If the measurement is done visually, the degradation is related to the human observer's ability to determine known patterns (square-wave or sinusoidal bars) of small size and low contrast when embedded in noise. Mechanization of this measurement by using a microdensitometer having a narrow slit will reduce the noise pickup by averaging the picture elements along the length of the bar. Analysis of the result of scanning may be, by estimation or by computer analysis, by extraction of the simusoidal fundamental component at each spatial frequency. Extraction of the fundamental sinusoid from a noisy signal by digital computation is sufficiently efficacious to allow the use of a scanning spot instead of requiring the long slit, although of course, even here the averaging of several lines together will produce cleaner results.

BLACKMAN points out the effect that noise will have on distorting the measured shape of the MTF curve [6.14]. Where the image processing is dependent on the shape of this curve (as in some cases of high frequency enhancement), the distortion in the curve shape will cause inaccuracies in the resulting processing.

Since the noise power spectrum and the image power spectrum overlap, the noise and signal cannot be separated. Therefore, the spectrum restoration used to restore the image spectrum will cause the noise to be amplified also. To minimize this effect, the shape of the correction function may be modified provided the shapes of the noise power spectrum and the image power spectrum can be obtained. Specifically, letting FT(*I*) and FT(*O*) be the two-dimensional Fourier transform of the image and object, and OTF the optical transfer function (complex, of which the MTF is the modulus), for the noiseless case the image formation can be represented by

$$FT(I) = OTF \cdot FT(O).$$

Provided that OTF exists for all spatial frequencies, recovery of the true object from the image may be made by

$$FT(O) = FT(I)/OFT.$$

In the presence of noise, however, the image is modified by an unconvolved signal, so that the restoration convolves this noise and mixes it with the restored signal. Restoration in this case may be optimized by Weiner filtering, in which the restoration function is [6.15]

$$FT(O) = [FT(I)/OTF] PS(S)/[PS(S) + PS(N)],$$

where PS(*S*) and PS(*N*) are the power spectra of the signal and noise (assumed to be uncorrelated). Since usually neither PS(*S*) nor PS(*N*) are known, the noise has introduced a complexity in requiring that these be determined or estimated, generally from PS(*S* + *N*).

Extensive work has been done on coding and the effects of noise [6.16]. Since there is no way to review the field here with any justice, a single example will suffice to illustrate the situation. The extremely large amount of data produced by modern multispectral scanning systems, such as that on board the ERTS-1 Satellite[1], has led to extensive effort in data

---

[1] The ERTS Scanner can produce an image of 3240 × 3240 picture elements in each of 4 spectral bands approximately every 20 sec. Each image covers approximately 185 × 185 KM on the ground, and is timed to get repeat coverage every 18 days.

compression to alleviate transmission bandwidth, data processing time, and data storage problems. Since the satellite data is to be used by many experimenters having many criteria for usefulness, the compression must be data-preserving. That is, the compression must not destroy data past a defined limit. The extent to which multispectral data can be compressed is proportional to the redundancy which exists between data points. Since the data points are related spatially (in each spectral band there is correlation in the brightness of neighboring samples) and spectrally (at each sample point there is correlation between the four spectral measurements), compression may be made either spatially or spectrally or both. Noise will degrade both types of correlation and thus reduce the possibilities for data compression.

Related to this correlation is the basic usefulness of the data as it is affected by noise, for example, in multispectral classification. In this procedure, the desire is to be able to determine the class (e.g., material, such as corn, water, sand, etc.) to which each sample belongs. This is done by comparing the sample measurement vector to the statistics of the set of known material vectors representing all possible classes, and by using one of several decision methods, determining which of the knowns it

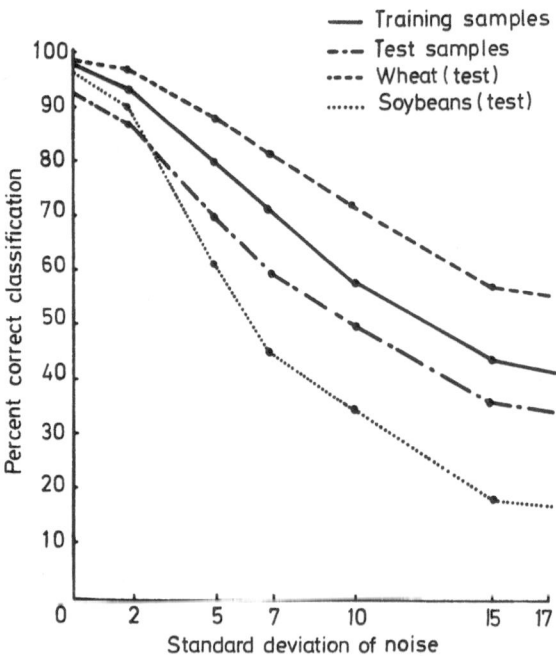

Fig. 6.1. Effect of random noise on the accuracy of multispectral classification. (From READY et al. [6.17])

most nearly matches. Accuracy of this procedure is greatest when the known classes have tight statistics and are relatively widely separated in vector space. Addition of noise will cause the known class statistics to spread, so that the separation/spread ratio decreases, and also causes uncertainty in the vector position of the unknown being classified. READY et al. [6.17] has investigated this effect, and shows in his Fig. 12 (reproduced here as Fig. 6.1) the decrease in classification accuracy with increasing noise.

6.1.3. Some Correlative Considerations

Since film systems and most television cameras are not linear, the effect of noise will be different depending upon the brightness of the part of the picture on which the noise is superimposed [6.18]. Since the effective amount of granularity or noise is a function of the picture element size, the question of granularity is tightly coupled to that of resolution [6.8, 19–24].

In some examples of image processing, such as that shown by OPPEN-HEIM et al. [6.25] and that discussed by SHELTON [6.26] the quantizing step spacing may not be linear in transmittance, but rather may be linear in density or have some other distribution. However, the quantity that is seen by the measuring device is light, which is linearly related to film transmittance. Transmittance, therefore, is the film parameter of importance, although the user may choose to think in other terms. Therefore, the noise considerations are in terms of transmittance noise rather than the density noise which is of importance to considerations of visual effect [6.8].

6.2. Effect of Noise on Quantization

Three different measures of signal-to-noise must be recognized and appropriately used. These will be designated as follows:

S/σ is the ratio of full scale (i.e., black-white) signal to rms noise.
SNR is the ratio of the actual (partial scale) signal to rms noise.
β is the ratio of a quantizing step size to the rms noise.

"Signal" as used here is the measure of amount of light measured from zero = black, or in the case of film scanning, the film transmittance measured from zero = opaque. It is, thus, a one-sided signal referenced to black. In general, the system must handle a signal anywhere within a fixed and predetermined range (e.g., $0 - 1$ when measuring film transmittance), so that arbitrary gain and offset changes to deal with partial-range excursions are not permitted. The dynamic range to be digitized is fixed and bounded.

6.2.1. Quantizing a Noisy Signal

Briefly, a quantizer transforms the magnitude of the signal into a discrete number of steps. In a noiseless system, there is no ambiguity in the designation of a particular signal level as a certain digital number (DN). However, in the presence of noise (assumed to be random) it is the signal + noise which is quantized, and the level of the signal alone is somewhat uncertain from inspection of the digital number. Specifically, there is a finite probability, measured by the relative areas bounded by the quantized step boundaries, of assigning the wrong DN to a given measurement. This is illustrated in Fig. 6.2. A derivation and discussion of the effect of noise on quantization was given by FRIEDMAN [6.27] and the resulting curve is the ±0 DN (digital number) curve of Fig. 6.3. This ±0 curve is derived by assuming a signal which can be anywhere

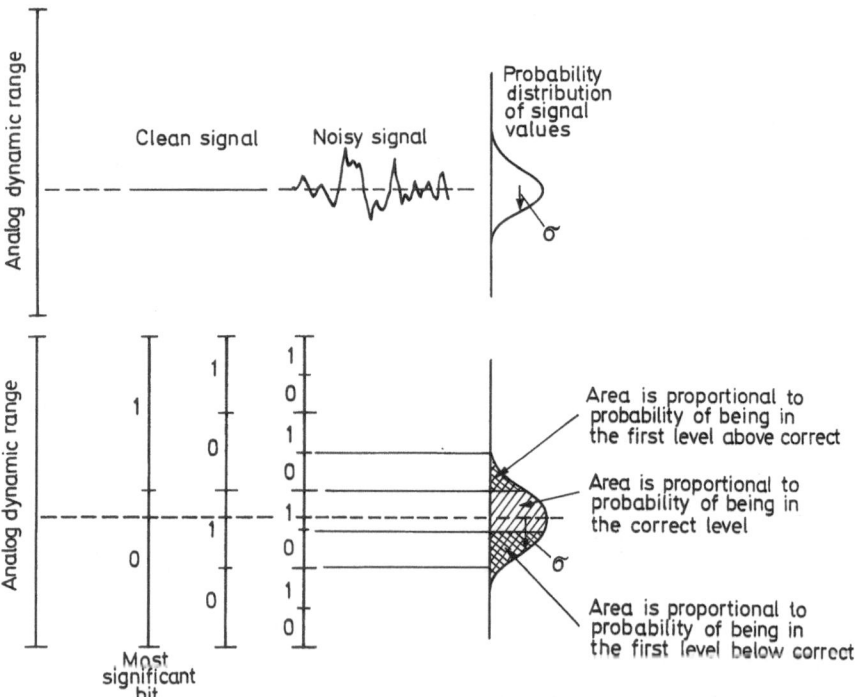

Fig. 6.2. A clean signal contaminated with noise will have a probability of assignment of an incorrect digital level in the analog-digital process

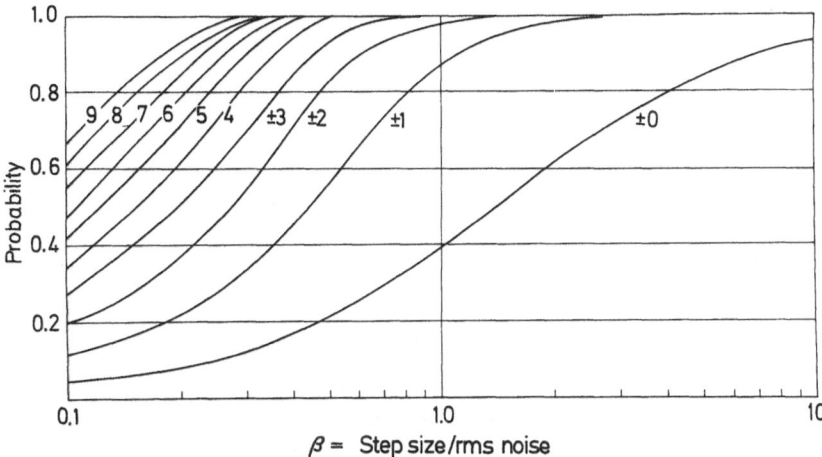

Fig. 6.3. Given a signal uniformly distributed over the quantization intervals. Given a Gaussian noise of value $= \sigma$. The curves show probability of correctly assigning a digital value corresponding to the noise-free signal within ± 0, ± 1, ... ± 9 DN (inclusive) as a function of the ratio $\beta =$ step size/σ

withing the quantizing range with uniform probability, added to which is a Gaussian noise with a distribution equal to σ. The probability distribution of the signal + noise is found by convolving the probability distribution of the signal with that of the noise. The probability of correct digitizing is then found by integrating the probability distribution between appropriate limits representative of the quantization interval boundaries.

There is an interesting corollary to this concerning the usefulness and effects of repeat sampling. Consider first the situation in which there is zero noise, so that the signal level is precisely the same each time it is sampled. In this case, the DN assigned to the signal will always be precisely the same and it will be impossible to determine the precise value of the signal from inspecting the DN, other than to say that it is somewhere within the range occupied by the step indicated. On the other hand, if there is even a slight amount of noise, there is a finite probability that the instantaneous value of the signal plus noise will be somewhat, and perhaps widely, different than the signal alone. There will therefore be a finite probability that the indicated DN will be different than that of the clean signal. Therefore, by taking a large number of samples, and averaging the DN values obtained, an average DN will result which, in its fractional part, will indicate the position of the true signal within the DN range. Does this indicate that it may be better to have a noisy signal than one which is without noise?

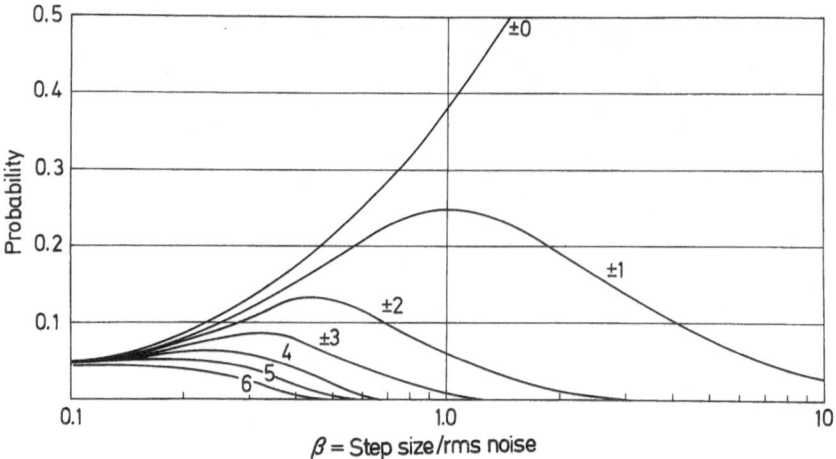

Fig. 6.4. Given a signal uniformly distributed over the quantization intervals. Given a Gaussian noise of value $= \sigma$. The curves show probability of assigning the quantized value to the correct level (± 0), the level above ($+1$), the level below (-1), ... $+6$, -6 DN as a function of the ratio $\beta = $ step size$/\sigma$

FULTZ carried this analysis one step further [6.28] and widened the boundaries to determine the probability of classification to within $\pm L$ digital levels of the correct value. This probability is given by the expression

$$PC_L = (L+1)\,\mathrm{erf}\,(L+1)\alpha - L\,\mathrm{erf}\,L\alpha$$
$$- \{\exp(-L^2\alpha^2) - \exp[-(L+1)^2\alpha^2]\}/\alpha\sqrt{\pi}, \tag{6.1}$$

where α denotes the step size$/\sigma\sqrt{2}$ and L is the digital level from correct level. For $L=0$ this reduces to the probability of correct classification. A plot of this probability for various L up to 9 is given in Fig. 6.3, plotted against $\beta = \sqrt{2}\alpha = $ step size$/\alpha$ for convenience.

The probability of being in a particular level may be found by successively subtracting the cumulative values for the next narrower level as given in Fig. 6.3 from the cumulative value for that level. The results are given in Fig. 6.4.

6.2.2. Detectability of Two Areas of Different Brightness

Let us now determine the detectability by analytical means of two areas of different brightness or, in the case of film, of different transmittances. This will be done first for the continuum case without digitizing in which

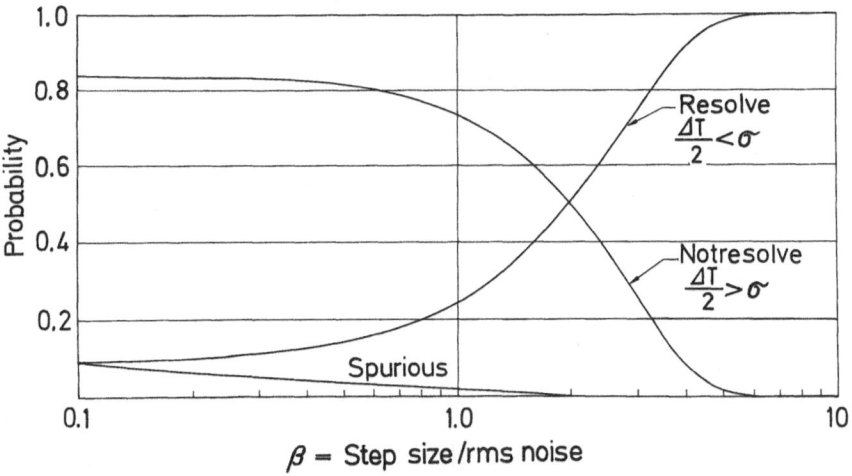

Fig. 6.5. Given two areas of film with an average difference in transmittance between them equal to ΔT. Each area has a Gaussian distribution equal to σ. The curves give the probability of detecting the difference between them (resolving), not resolving, and spurious resolving as a function of the ratio ($\beta' = \overline{\Delta T}/\sigma$)

analog measurements are made, and then for the case in which digital samples are taken and compared.

For the analog case, assume an average difference in transmittance equal to $\overline{\Delta T}$, and a Gaussian noise on each individual level equal to σ. This is equivalent to the situation of measuring two "uniform" but somewhat noisy areas with a transmittance meter and attempting to tell from the meter readings whether the areas are different (resolved) or not. We can define that the two areas are "resolved" if $\overline{\Delta T}/2 > \sigma$, "not resolved" if $-\sigma < \overline{\Delta T}/2 < +\sigma$, and "spuriously resolved" (i.e., indicating they are different when they really are not) when $-\infty < \overline{\Delta T}/2 < -\sigma$. The probabilities for these three cases are given by

$$P \text{ (resolve)} = [1 + \operatorname{erf}(\alpha'/\sqrt{2} - 1)]/2, \qquad (6.2a)$$

$$P \text{ (not resolve)} = [\operatorname{erf}(\alpha'/\sqrt{2} + 1) - \operatorname{erf}(\alpha'/\sqrt{2} - 1)]/2, \qquad (6.2b)$$

$$P \text{ (spurious)} = [1 - \operatorname{erf}(\alpha'/\sqrt{2} + 1)]/2, \qquad (6.2c)$$

where $\alpha' = \overline{\Delta T}/\sqrt{2}\sigma$.

Curves for these are given in Fig. 6.5, plotted against $\beta' = \sqrt{2}\alpha' = \overline{\Delta T}/\sigma$. Note that these are for the continuum case and do not involve quantization.

In the equivalent quantized case in which each of the noisy levels is independently digitized, the probability distributions of Fig. 6.4 are used.

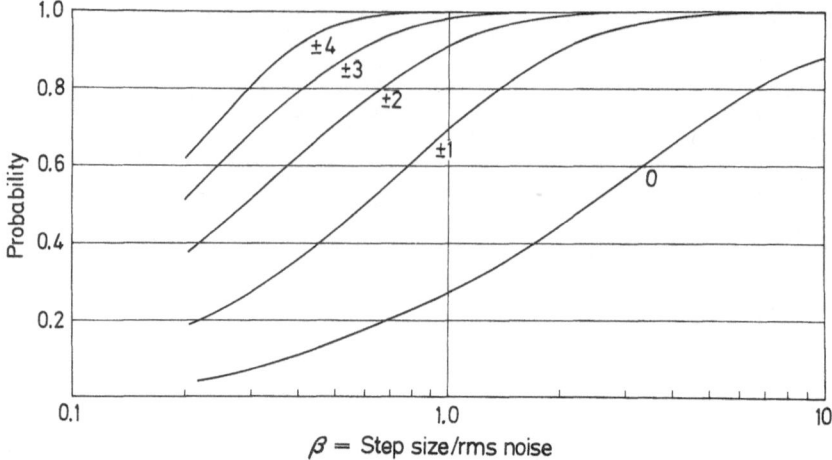

Fig. 6.6. Given two signals which have been perturbed by Gaussian noise of value equal to σ. Each is quantized to the same number of bits. The curves give the probability of correctly determining the true difference in the two levels within $\pm 0, \pm 1, \ldots \pm 4$ (inclusive) DN as function of the ratio $\beta =$ step size$/\sigma$

Figure 6.6 is the probability of correctly indicating the difference in average transmittance or brightness of two levels to within digitally indicated error limits up to ± 4 DN.

With these tools, the analysis of the digital performance of a noisy system can be begun.

We will use as a framework discussion an "8-bit" system, using various definitions of what "an 8-bit system" might mean.

6.2.3. Derivation of a Performance Measure

Let us start by defining the statement "digitizing to M useful bits" to mean that as a result of noise, the $M + 1$ bit is correct with a probability $= 0.5$. From Fig. 6.3, it is seen that a value of $\beta = 1.4$ is required for the $M + 1$ bit, where $\beta =$ step size/rms noise. This gives a $\beta = 2.8$ for the M-th bit.

Thus for an 8 bit system the 9-th bit must be $1.4 \times$ the rms noise, or

$$S/\sigma = 2^{(M+1)} \cdot \beta = 2^9 \times 1.4 = 717 .$$

With this S/σ the 8-th bit (for which $\beta = 2.8$) will have a $P = 0.72$ of being correct. Essentially all measurements will be within ± 2 9-bit DN of correct, and with $P = 0.96$ of being within ± 1 9-bit DN. The probability that a difference in level between two areas will be correctly measured equals 0.34, $P(\pm 1) = 0.8$, and $P(\pm 2) = 0.96$.

If the required criterion were that two levels having a difference in brightness of 1 DN be resolved to 50% probability, the required $\beta = 2.4$, requiring a $S/\sigma = 1215$.

Alternatively, system performance may be approached in terms of noise introduced by the digitizer. The value of the noise power introduced by the act of digitizing is given by [6.29]

$$\sigma_Q^2 = \text{step size power}/12$$
$$= (\text{step size})^2/12 .$$

We can define a balanced system as one in which the quantizing noise signal is equal to the rms sum of all prior noise sources in the system. Thus

$$\sigma_{\text{prior to digitizing}} = \text{step size}/\sqrt{12} \tag{6.3}$$

or

$$\beta = \sqrt{12} = 3.56$$

giving a $S/\sigma = 256 \times 3.56 = 915$

The three criteria discussed lead to the required β into the digitizer of 2.8, 2.4, and 3.5, respectively. We will use a general rule of thumb that the rms noise into the digitizer should be $\leq 1/3$ step size ($\beta = 3$) as representing a reasonable consensus from these criteria.

6.2.4. System Noise

If we consider the requirements in terms of digitizing a film, the required bk/w signal-to noise ratio (S/σ) may be derived as follows:

Define $\sigma = 1/3 \times$ size of one digital step $\triangleq 1/3 \times \Delta T$ ($\beta = 3$),
$\Delta T = $ size of one digital step
$= 1/256$ for 8-bit, $0 < T < 1$,
$S = $ black-white signal $\triangleq 1$.

The density and transmittance at the first step away from $T = 0$ are

$T_1 = \Delta T = 1/256 = 1/N$ where $N = $ number of steps,
$D_1 = \log(1/T_1)$,
$D_1 = $ film density at first digital step away from $T = 0$.

Therefore,

$S/\sigma = 1/(1/3 \times 1/N) = 3N$,
$S/\sigma = bk/w$ signal to rms noise ratio.

S/σ in dB is expressed by

$$S/\sigma|_{\text{dB}} = 20\log(S/\sigma) = 20(\log N + \log 3).$$ (6.4a)

But $T_1 = 1/N$ and $D_1 = \log(1/T_1) = \log N$

$$S/\sigma|_{\text{dB}} = 20(D_1 + 0.477).$$ (6.4b)

For an 8-bit system $T_1 = 0.039$ and $D_1 = 3.4$. Therefore, $S/\sigma = 77.6$ dB.

6.2.5. Film Grain Noise

The image on any normal silver halide film is produced from the super-position of a number of opaque silver granules. The macroscopic effect is an average density produced as a result of the local exposure, but as the aperture is reduced in an effort to get higher and higher resolution, the silver grains, which appear as black patches on a clear background, now begin to become visible. This is the film granularity, which is usually stated in terms of the rms value of the statistical fluctuation in density (σ_D), when measured with an aperture of 48 μm, and at a density $D \approx 1$ (see Fig. 6.7).

From a consideration of probability of overlap of the silver grains it can be shown that the noise (σ_T) at any value of transmittance (T) may

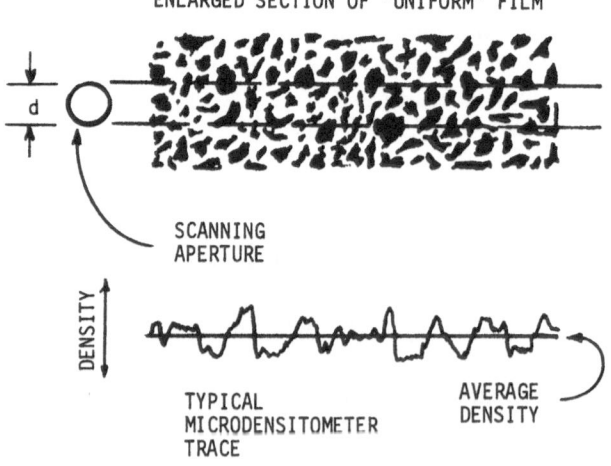

Fig. 6.7. The microdensitometer trace measures the film granularity, which is usually stated in terms of the rms value of the statistical fluctuation in density (σ_D), when measured with an aperture of 48 μm, and at a density $D \approx 1$

be derived from a known value of noise (σ_{T_1}) at a transmittance T_1 [6.30]

$$\sigma_T = \sigma_{T_1} \sqrt{\frac{T}{T_1} \frac{(1-T)}{(1-T_1)}} . \tag{6.5}$$

The maximum σ_T from this expression occurs at $T = 0.5$, and therefore represents the worst-case noise. The maximum value of this worst-case noise is chosen to give $\beta = 3$ as before

$$\sigma_{T_{\max}} = 1/3 \times \text{size of one digital step} .$$

Then, for 8-bit digitizing over the range $0 < T < 1$

$$\sigma_1 = (1/3 \times 256) \sqrt{2T \frac{(1-T)}{(1-1/2)}} .$$
$$= 0.0026 \sqrt{T(1-T)}$$

"Standard Conditions" at which granularities are quoted are approximately at $T = 0.1$ $(D = 1)$. This substitution gives

$$\sigma_{T=0.1} \approx 0.00078 .$$

This is the worst-case σ_T, permissible at $T = 0.1$. It must be converted to σ_D, (i.e., σ measured in density units) since normal specifications quote σ_D. For a density well above fog level [6.30]

$$\sigma_T/T \approx \sigma_D/0.4343 \quad \sigma_D = \text{rms granularity in density units}, \tag{6.6}$$

$$\sigma_D = 0.4343\,\sigma_T/T$$

$ = 0.0034.$ This is the allowable maximum at $T = 0.1$, measured with the aperture as determined by resolution requirements.

But it was found [6.30] that for the practical case in which the noise spectrum is essentially flat within the spatial frequencies of interest:

$$\sigma_D \sqrt{A} = \text{constant} \quad (\text{apertures} \geqslant 10 \times \text{grain diameter}), \tag{6.7}$$

where A is the area of aperture. Conversion to various diameters is then done by

$$\sigma_{D_1}/\sigma_{D_2} = SA_2/SA_1 , \tag{6.8}$$

Table 6.1. Typical granularity of some Kodak films. Granularity numbers obtained from various Kodak data sheets

Description	Film no.	RMS granularity[a]
Extended red (700)	2457, 2479	32
Extended red (700)	2496	22
Panchromatic high speed	2484	30
Panchromatic	2498	28 (Neg.)
		17 (Rev.)
High speed IR (900)	2481, 2424	38
Kodachrome II movie		9
Ektachrome EF	2241	13
Ektachrome MS aerographic	2448	12
Aerocolor neg.	2445	13
Ektachrome IR		17
Aerochrome IR	2443	17

[a] 48 μm aperture, f 2.0 lens, density ≈ 1.0, $\times 1000$

where SA is the diameter of the scanning aperture in μm. Therefore

$$\sigma_{D_{48\,\mu m}} = \sigma_{D_{SA}}(SA/48) = 0.0034 \times SA/48 \approx 0.00007SA .$$

Thus, with this criterion the required granularity is

$$\sigma_D \leqq 0.00007 \times SA , \tag{6.9}$$

where σ_D represents the required rms density granularity at $D = 1$ and aperture of 48 μm to give $\beta = 3$ for 8-bit digitizing.

Granularity as used by the Eastman Kodak Co. is $1000 \times$ this figure, taken at $SA = 48$ μm. Thus for a typical scanning aperture of 100 μm diameter, the required granularity is $\sigma_D = 0.07 \times 100 = 7$. Perusal of film specifications will show that most normal film has a granularity considerably higher than this value and hence the system will be film-limited. For this reason it is desirable where possible to go to direct camera pickup of the light image (provided that camera performance is adequate) rather than use film as an intermediary step.

The variation in β as T varies may be found as follows

$$\sigma_T = 1/\beta_T N$$
$$\sigma_T/\sigma_{T_1} = \sqrt{\frac{T}{T_1}\frac{(1-T)}{(1-T_1)}} = \beta_1/\beta_T . \tag{6.10}$$

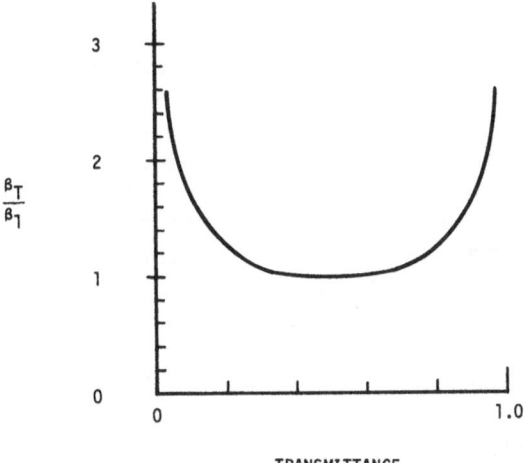

TRANSMITTANCE

Fig. 6.8. Variation of β caused by film grain as a function of film transmittance

The worst-case (lowest) β occurs at $T=0.5$. Defining this as β_1,

$$\beta_T = \tfrac{1}{2}\beta_1\sqrt{T(1-T)}. \tag{6.11}$$

The curve is sketched in Fig. 6.8.

6.3. System Bandwidth and Sampling Methods

The noise in the electronic system (including the pickup device) may be minimized by reducing the bandwidth of the system, with the eventual necessity of reducing the sampling rate. The allowed maximum bandwidth of the system may be derived on either the basis of constant SNR at all transmittances, or on the basis of constant signal/noise (S/σ) at all transmittances. The choice between these will depend on the uses to which the system is put, and the analyst's evaluation of the different results (primarily at low light levels).

While it is realized that the various noise powers introduced into the system will add together, these have different proportions at different parts of the brightness (or transmittance) range. For this reason, the addition is not systematic and each will therefore be separately allowed the full noise allowance.

6.3.1. Detector Noise

For determination of the system bandwidth in terms of the photodetector noise, we will consider a system (Fig. 6.9) in which the measuring light, after attenuation by film being measured, falls on a photomultiplier having a photocathode efficiency $\varepsilon < 1$. The electrons released will be counted, and the count will exhibit a statistical fluctuation. Associated with the photocathode will be an electron multiplier having multiplication noise. Since load resistor thermal noise is generally small compared to the other noise sources, it will be ignored.

The number of photoelectrons released in a given sampling period may be assumed to have Gaussian distribution [6.31], having a standard deviation σ equal to the square root of the average value.

$$\sigma_Q = \sqrt{Q}. \tag{6.12}$$

We will designate those parameters occuring at open aperture $(T = 1)$ by the subscript letter "o". Thus

$$\sigma_{Q_o} = \sqrt{\bar{\bar{Q}}_o}.$$

Since the photocathode efficiency is < 1, the number of photons is greater than the number of photoelectrons, and the photon noise is given by

$$\sigma_\lambda = \sqrt{\bar{\lambda}} = \sqrt{\bar{\bar{Q}}}/\varepsilon = \sigma_Q/\sqrt{\varepsilon}, \tag{6.13}$$

where λ denotes the number of photons during sample period.

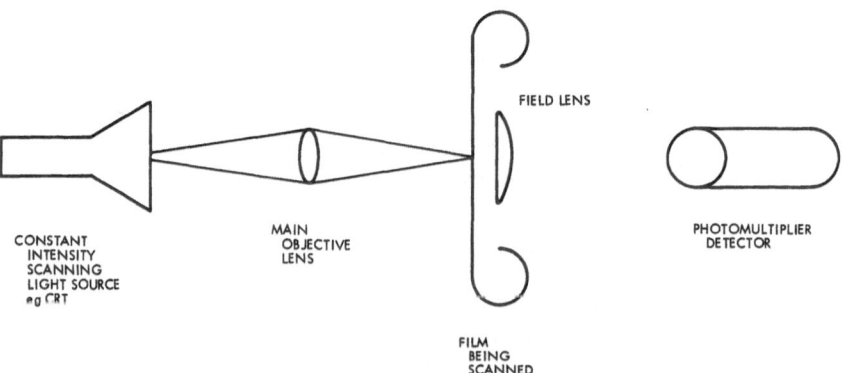

FIELD LENS

CONSTANT INTENSITY SCANNING LIGHT SOURCE e.g CRT

MAIN OBJECTIVE LENS

FILM BEING SCANNED

PHOTOMULTIPLIER DETECTOR

Fig. 6.9. Flying-spot-scanner system

The photon noise will add to the photoemission noise to give the total emitted noise from the photocathode

$$
\begin{aligned}
\sigma_e &= \sqrt{(\sigma_\lambda \cdot \varepsilon)^2 + \sigma_Q^2} \\
&= \sqrt{(\sigma_Q \sqrt{\varepsilon})^2 + \sigma_Q^2} = \sigma_Q \sqrt{(1 + \varepsilon)}.
\end{aligned}
\tag{6.14}
$$

Note that this derivation assumes that all of the noise is generated by fluctuations of the useful signal current. However, any stray light reaching the photodetector will add to the fluctuation noise without a corresponding useful increase in signal, if the total photocathode current is the measured output signal. This would be the case, for instance, in a CRT scanning system, as shown in Fig. 6.9. EBERHARDT [6.31] pointed out that for an image dissector detector, the extraneous light surrounding the desired pixel is eliminated by the dissection process. Furthermore, for the case of detecting very small objects such as star images, the aperture size should be reduced so as to just accept the image to reduce the non-signal noise.

Accordingly, the derivations assume that the noise generated is all caused by the desired signal without additional components caused by a brightness bias. An interesting corollary is implied: a small signal (i.e., low modulation) should be less masked by noise if it is located in an area of low brightness rather than in an area of high brightness.

We will collect photoelectrons for a sampling period τ, i.e.

$$
Q = I\tau/q_e,
\tag{6.15}
$$

where I represents the photocathode current, τ the sampling time, q_e is the charge on electron $= 1.6 \times 10^{-19}$ Coulomb, and Q denotes the number of electrons per sample.

The photomultiplier associated with the photocathode will have a gain G^n and a multiplicative noise contribution $\sqrt{G(G-1)}$ [6.31]. Thus

$$
I_s = q_e Q G^n/T,
$$

$$
\sigma_s = (q_e G^n/\tau)\sqrt{QG(1 + \varepsilon)/(G-1)},
$$

or

$$
Q = [G(G+1)/(G-1)]I_s^2/\sigma_s^2,
$$

$$
= [G(1 + \varepsilon)/(G-1)\,\mathrm{SNR}^2,
$$

$$
= K(\mathrm{SNR})^2,
$$

where I_s is the signal current out of photomultiplier, G the stage gain, n the number of stages, σ_s the noise current out of photomultiplier, and $K = (1 + \varepsilon) G/(G - 1)$. This gives

$$I_s = q_e(1 + \varepsilon) I_s^2 G^n G/\tau \sigma_s^2 (G - 1) \tag{6.16a}$$

with $I_s/\sigma_s = \mathrm{SNR}$ at output, and

$$\tau = q_e(1 + \varepsilon)(\mathrm{SNR})^2 G^{n+1}/I_s(G - 1). \tag{6.16b}$$

The factor $K = (1 + \varepsilon) G/(G - 1)$ represents the extra charge required to increase the output SNR to just compensate for the noise introduced by the photons and the dynode multiplication.

Each element (film, CRT, amplifier, etc.) in the system will cause a loss in high frequency information, or roll-off. Generally, the combination of the various roll-offs will limit the total system bandwidth such that the high-frequency attenuation curve is more or less Gaussian. Assuming that the band-limiting roll-off is Gaussian, the maximum SNR is transmitted through the channel when [6.32]

$$B'\tau' = 1/3, \tag{6.17}$$

where B' denotes the cutoff bandwidth in MHz, and τ' is the pulse width in μsec. However, at the termination of the pulse, the output pulse top is still rising, so that pulse width errors cause amplitude errors. A compromise which does not sacrifice much SNR and yet is less critical to τ variations is

$$B'\tau' \approx 1/2.$$

This relation will be used in subsequent estimations of the required bandwidth as it is determined by pulse width, and gives a bandwidth [Hz]

$$B = I_s(G - 1)/2q_e(1 + \varepsilon)(\mathrm{SNR})^2 G^{n+1}. \tag{6.18}$$

6.3.2. Measurement Accuracy

Following and extending the method of SHELTON [6.26] we have: Let a transmittance range of T_{min} to T_{max} be divided into N equal levels (Fig. 6.10). Each level is then a transmittance size $= \Delta T/N$.

The width of each level in terms of Q is

$$W_Q = Q_o \Delta T/N, \tag{6.19}$$

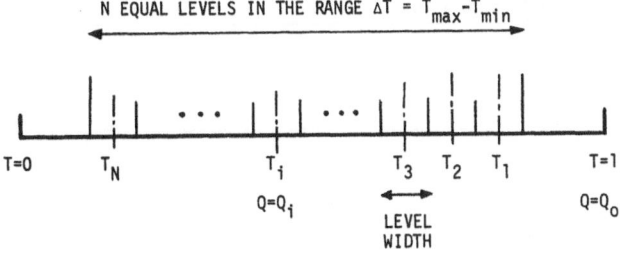

Fig. 6.10. Division of the transmittance range from $T_{\min}$ to $T_{\max}$ into N equal levels. (After SHELTON [6.26])

where $\Delta T = T_{\max} - T_{\min}$, and Q_{o} denotes the open aperture Q. Collect sufficient electrons to produce the required $\beta_{\max}$ at $T_{\max}$, where $\beta =$ charge step size$/\sigma_Q = Q_{\text{o}}\Delta T/N\sigma_Q$. The measured quantity at the step i down from $Q_{\max}$ will be the collected electrons, Q_i. Q_{o} at open aperture to give $\beta_{\max}$ at $T_{\max}$ is

$$Q_{\text{o}} = K\beta^2_{\max}N^2 T_{\max}/(\Delta T)^2 .$$

The average T at the center of the i-th level down from $T_{\max}$ is

$$\bar{T}_i = T_{\max} - \Delta T(2i-1)/2N .$$

The average $\bar{Q}$ in the i-th level down from $T_{\max}$ is given by

$$\bar{Q}_i = Q_{\text{o}}T_{\max} - W(2i-1)/2 .$$

β for the i-th level is

$$\beta = W_Q/\sqrt{\bar{Q}_i \cdot K}$$

$$= (\Delta T \cdot \sqrt{Q_{\text{o}}}/N\sqrt{K})[T_{\max} - \Delta T(2i-1)/2N]^{-\frac{1}{2}} . \qquad (6.20)$$

The value of β_i within the range of T compared to the defined $\beta_{\max}$ at $T_{\max}$ is

$$\beta_i/\beta_{\max} = \sqrt{T_{\max}}/[T_{\max} - \Delta T(2i-1)/2N]^{\frac{1}{2}} . \qquad (6.21)$$

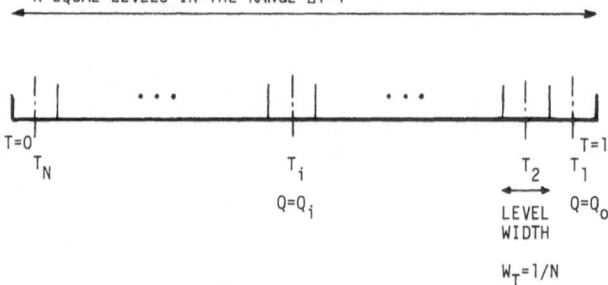

Fig. 6.11. The special practical case of $T_{\min} = 0$ and $T_{\max} = 1$

The attainable β_o at open aperture and $0 \leq T \leq 1$ will be

$$\beta_o = \beta_{\max} \sqrt{T_{\max}/\Delta T}. \tag{6.22}$$

The SNR at any given level i is

$$\begin{aligned} \mathrm{SNR}_i &= \sqrt{Q_i}/\sqrt{K} \\ &= [Q_o T_{\max} - w(2i-1)/2]^{\frac{1}{2}}/\sqrt{K} \\ &= \{Q_o[T_{\max} - \Delta T(2i-1)/2N]\}^{\frac{1}{2}}\sqrt{K}. \end{aligned} \tag{6.23}$$

The SNR_i compared to the $\mathrm{SNR}_{\max}$ at $T_{\max}$ is therefore

$$\mathrm{SNR}_i/\mathrm{SNR}_{\max} = \{[T_{\max} - \Delta T(2i-1)/2N]/(T_{\max} - \Delta T/2N)\}^{\frac{1}{2}}. \tag{6.24}$$

The SNR_i compared to the worst case SNR at $i = N$ is

$$\begin{aligned} \mathrm{SNR}_i/\mathrm{SNR}_{i=N} = \{&[T_{\max} - \Delta T(2i-1)/2N] \\ &\cdot [T_{\max} - \Delta T(2N-1)/2N]\}^{\frac{1}{2}}. \end{aligned} \tag{6.25}$$

It is thus seen that the lowest β is $\beta_{\max}$, occurring at $T_{\max}$, whereas the worst case SNR occurs at the last step, just above $T = 0(i = N)$.

For the special case of $0 < T < 1$, $\beta_{\max} = \beta_o$, $T_{\max} = 1$, $\Delta T = 1$, $\mathrm{SNR}_{\max} = S/\sigma$ (Fig. 6.11).

Case 1A: increase of β from defined β_o with decreasing T.

$$\beta_i/\beta_o|_{0 < T < 1} = [1 - (2i-1)/N]^{-\frac{1}{2}}. \tag{6.26}$$

Case 1B: decrease of SNR from SNR at $T = 1$ with decreasing T

$$\text{SNR}_i/(S/\sigma)|_{0 < T < 1} = \{[1 - (2i - 1)/2N]/(1 - 1/2N)\}^{\frac{1}{2}}. \qquad (6.27)$$

Case 1C: increase of SNR from defined SNR at $i = N$ with increasing T

$$\text{SNR}_i/\text{SNR}_{i=N}|_{0 < T < 1} = \{2N[1 - (2i - 1)/2N]\}^{\frac{1}{2}}. \qquad (6.28)$$

The Q required to obtain a given SNR at the middle of the first step away from zero is obtained when $i = N$. This is also equal to the β for this step, and for $0 < T < 1$, the resultant open aperture Q is found to be:

$$Q_o = 2KN\beta^2 \qquad \text{to satisfy } \beta \text{ at lowest step}. \qquad (6.29)$$

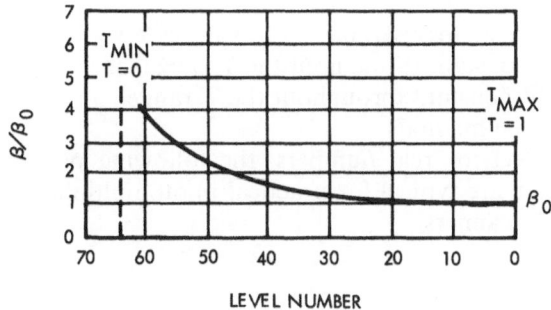

Fig. 6.12. Growth to β from β_o as film transmittance decreases (After SHELTON [6.26])

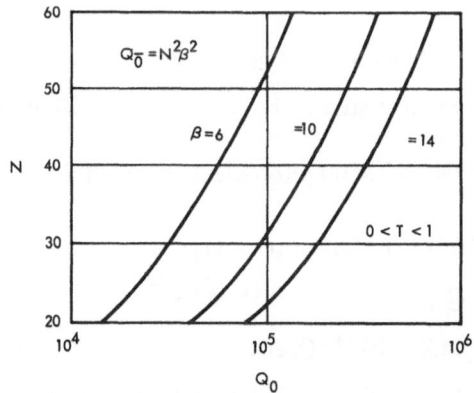

Fig. 6.13. Number of equal transmittance levels available for various β. (After SHELTON [6.26])

This can be compared to the open aperture Q previously derived

$$Q_o = K\beta^2 N^2 \quad \text{to satisfy } \beta \text{ at brightest step} . \tag{6.30}$$

Plots of β_i/β_0 and of Q_{max} are given in Figs. 6.12 and 6.13 for the case of $0 < T < 1$.

6.4. Practical Cases

We can define the required system performance in several ways:
1) Hold system parameters constant (vs T) and require that:
    A) β be at least a certain value at its worst-case end of the T range,
    or
    B) SNR be at least a certain value at its worst-case end of the T range.
2) Allow the system parameters to vary vs T so that:
    A) β is constant throughout the T range, or
    B) SNR constant throughout the T range.
    C) Special methods.

To get a feel for real numbers, the following parameters will be assumed. These are typical for practical photomultipliers when used in fine spot CRT scanners.

$$G = 3$$

$$n = 10$$

$$I_s = 1 \text{ mA at open aperture}$$

$$K \approx 2 .$$

6.4.1. Case 1: Constant System Parameters

Use $\beta = 3$ at the highest step as before, and 6-bit digiting ($N = 64$) over the range $0 < T < 1$:
$$Q_o = 2 \times 3^2 \times 64^2 = 73\,800 \text{ photoelectrons/sample to satisfy } \beta \text{ at the}$$
highest step

$$\tau = (q_e Q_o/I_s) [G^{n+1}/(G-1)]$$

$$= Q_o(1.6 \times 10^{-19} \times 3^{11})/(10^{-3} \times 2)$$

$$- 14.2 \times 10^{-12} Q_o \text{ scc}$$

$$= 14.2 \times 10^{-12} \times 73.8 \times 10^3 = 1 \times 10^{-6} \text{ sec} .$$

$$SNR_{i=N} = \sqrt{(Q_o/K)}\sqrt{(1/2N)} = 3 \times 64/\sqrt{(2 \times 64)} = 24\sqrt{2} .$$

The required system S/σ is

$$S/\sigma = \beta N = 3 \times 64 = 192 \,.$$

The SNR at the lowest step will be less than the S/σ at the highest step by a factor

$$\text{SNR}_{i=N}/(S/\sigma) = (2N - 1)^{-\frac{1}{2}} \approx 1/8\sqrt{2} \quad 0 < T < 1 \,.$$

The β at the lowest step will be greater than β at the highest step by a factor

$$B_{i=N}/\beta_{i=1} = \sqrt{2N} = 8\sqrt{2} \quad 0 < T < 1 \,.$$

Allowing a dead time between pulses of 20 μsec for spot motion, phosphor decay, and digital readout (a practical value), the resultant maximum digitizing rate is about 50000 samples per second, determined primarily by the dead time.

This sampling rate, together with the size of the picture, now determines the total sampling time required for a complete picture. For example, approximately 20 sec is required to completely digitize a 1024×1024 picture sampled at 50000 samples per second.

A corresponding 8-bit system will require Q to be $4^2 = 16$ times as high, and thus require $\tau = 16\,\mu\text{sec}$ and $S/\sigma = 768 = 57.7\,\text{dB}$. Again note the particular assumed parameters for which these numbers apply.

6.4.2. Case 2: Variable System Parameters

There will be situations in which it may not be desirable or possible to maintain the system parameters constant. This may occur for instance where there is not sufficient light to obtain the required β and it is desirable to compensate with an increase in digitizing time as when digitizing to high density values (implying small transmittance steps), when it is desirable to get the maximum number of digital steps into the transmittance range, or when it is desirable to reduce the sampling time to an absolute minimum. In these cases advantage can be taken of the normal variability of β and SNR by varying some of the system parameters as a function of brightness or transmittance to hold β or SNR constant. The normal candidates for varying are the step size (i.e., a variable step size can be made to produce a constant β or constant SNR), sampling time (e.g., "constant confidence" scanning by integrating at each sample point to a constant Q), or scanning illumination (i.e., increasing the scanning brightness at the very dense part of the film).

Variable Step Size to Get Constant β

As in the previous derivation, the distribution of photoelectrons from a photomultiplier cathode is assumed to be Poisson. which for large numbers approaches Gaussian. The standard deviation σ of this distribution is equal to the square root of the average value. If, therefore, the distribution of quantizing levels is made a square root function of T, the β for all levels will be equal. SHELTON [6.26] has considered this case and has derived expression for the required Q_o and N. If we modify his equations to take into account the photon and dynode noise multiplication we arrive at the following

$$\bar{T}_i = (\sqrt{T_{max}} - \beta_i/2)^2 , \tag{6.31a}$$

$$Q_o \approx KN^2\beta^2/4(\sqrt{T_{max}} - \sqrt{T_{min}})^2 , \tag{6.31b}$$

$$N = 2\sqrt{Q_o}(\sqrt{T_{max}} - \sqrt{T_{min}})/\beta\sqrt{K} , \tag{6.31c}$$

for constant β, and variable step size. This scheme makes it possible to achieve more levels within a given range than either the equal transmittance steps or equal density step schemes, but has the disadvantage of introducing nonlinearity into the digital computation. Whether or not this is tolerable depends on the accuracy to which the computation is desired and on the specific picture content. No general rule can be given.

Variable Step Size to Get Constant SNR

Since we have derived a way to digitize with constant β by varying the step size in a certain manner, we may now be tempted to inquire how to digitize to achive constant SNR. As before, the noise σ of the Gaussian distribution of photoelectrons is equal to the square root of the average value, giving an SNR at any transmittance T

$$SNR_T = \sqrt{Q_T}/\sqrt{K} = const\sqrt{T} \tag{6.32}$$

Note that the continuum SNR is completely determined before any reference to step size is made. We can define "digitize to constant SNR" to mean adjusting the step size progression to produce the probability of correct digitizing (i.e., β) proportional to the transmittance T, so that T/β = constant. Since the noise is going up as $\sqrt{T}$, the step size distribution must go up as $T^{\frac{3}{4}}$ to cause β to go up linearly with T. This seems a rather arbitrary condition, and since it produces no apparent advantages, it will not be pursued further. The normal "**constant** confidence" approach will be discussed later under the variable parameter case.

Constant β With Uniform Step Size

β may be held constant by varying the sampling time (and along with it the system bandwidth) as a function of T (or the digital step, i). Beginning with the previous equation for β_i and using the relation $T_i = T_{max} - \Delta T(2i-1)/2N$,

$$\beta_i = \frac{\sqrt{Q_o \Delta T}}{\sqrt{K} N \sqrt{T}} . \tag{6.33}$$

Again, this is derived for a system in which Q_i is the measured quantity. It can be seen that β_i is constant if Q_o/T is constant, or that

$$Q_o \triangleq K \beta^2 N^2 T/(\Delta T)^2 , \tag{6.34a}$$

where $\triangleq$ means "must be made to be equal to...".
  For the case $0 < T < 1$, this reduces to

$$Q_o|_{0<T<1} \triangleq K \beta^2 N^2 T . \tag{6.34b}$$

This variation in Q_o may be obtained either by varying the sampling time or by varying the scanning light intensity. Since

$$Q_o = \tau dQ_o/dt$$

either the sampling time τ or the rate of collection of charge, dQ_o/dt, may be varied. Considering the variation of τ, we have

$$\tau = T(K \beta^2 N^2)/(dQ_o/dt) = \text{scale factor} \cdot T , \tag{6.35}$$

where $K \beta^2 N^2/(dQ_o/dt)$ is the scale factor for the entire system.
  Similarly, if τ is held constant, the rate of flow of photoelectrons, as affected by the scanning light brightness, may be modulated

$$dQ_o/dt \triangleq T(K \beta^2 N^2)/\tau . \tag{6.36}$$

  Dsign of a system under this approach starts at the determination of Q_i at the minimum T step. Using the same typical numbers as before, for $0 < T < 1$ and $N = 64$:

$$Q_o = 2 \times 3^2 \times 64^2 \times 1/64 = 1152 \text{ photoelectrons/sample}$$

$$\tau = 14.2 \times 10^{-12} Q_o = 16.3 \times 10^{-9} \text{ sec/sample} .$$

  τ increases linearly with T to the open-aperture $\tau_o = 1$ µsec. Advantage is thus taken of the normal growth of β at low T; this becomes of more importance as N is increased. For example, if it is desired to digitize to a density of 3.0, the equivalent full-range linear step $N = 1024$, requiring an open aperture $\tau_o = 256\,\mu\text{sec}$ to give $\beta_o = 3$ at open aperture. This in turn, however, gives an unnecessarily high β at the lower steps which may be profitably sacrificed to reduce the digitizing time.

Conversely, if the τ_o obtained (e.g., the 1 µsec in the above example) is satisfactory, the scanning light may be reduced at lower T. Under normal conditions the light used for scanning the film would be set at the maximum brightness available under all conditions, and the situation outlined under constant system parameters would be obtained. If the scanning light were produced from a flying spot CRT, however, this will result in maximum aging of the CRT and hence it is desirable where possible to minimize the brightness of the scanning light. Under these conditions modulation of the scanning light in some way such as outlined above would be considered.

This method, however, poses some practical implementation problems. Since the required τ or scan light is proportional to T, this is basically a positive feedback or an open-loop feed forward system. As such, it can tend to get unstable, resulting in erratic measurements. Also, it requires that τ or the light be zero at $T = 0$, so that if scanning goes across an opaque spot or out of the frame, the light goes off and everything stops. Special efforts are therefore needed to avoid this. It may be more desirable to look to other methods of taking advantage of the normal growth of β to reduce the sampling time.

Constant SNR with Uniform Step Size

Although under normal conditions it is desirable to digitize to constant β to produce a constant probability of error independent of T, under some conditions it may be desirable to produce a constant SNR independent of T. This may occur, for instance, for nonlinear processing as used by Oppenheim [6.25], or where the scanning is for human viewing [6.26]. There will also be times in which it is desirable to measure the absolute value of brightness to a given accuracy. For these cases, we hold the SNR constant (that is, the actual measured signal to σ ratio) as T varies. This results in S/σ variation as the T varies, and produces increased absolute accuracy as $T \rightarrow 0$.

Using the previous equation for SNR_i and the relation $T_i = T_{\max} - \Delta T(2i-1)/2N$,

$$\text{SNR}_T = (1/\sqrt{K})\sqrt{\int_0^{th} Q_o' T\, dt}\,, \tag{6.37}$$

Q_o' being the rate of flow of Q_o.

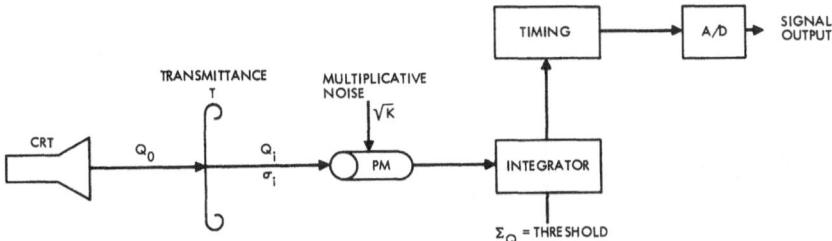

Fig. 6.14. Integration of the sample output to a predetermined threshold allows the time of integration to be the measured parameter

Thus constant SNR is obtained when Q'_o is integrated to a quantity

$$\Sigma_{th} \triangleq K(SNR)^2$$

which represents the total Q_o collected.
This will produce a variable β_T

$$\beta_T = (SNR \cdot \Delta T/N)/T . \tag{6.38}$$

The modulation of Q_o is usually done by varying τ.

Modulation of the sampling time according to $1/T$ is conveniently accomplished by allowing the scanning aperture to dwell on a given picture element while integrating the output signal until it reaches a predetermined threshold value. Figure 6.14 illustrates the system.

The length of time required for this integration is measured. This time is linearly proportional to the reciprocal of the film transmittance or the image light value. This system implicitly produces uneven time intervals between sampling pulses and side effects such as nonsynchronous data recording must be designed for. This is the approach usually termed "constant confidence" scanning. The system bandwidth should be adjusted in accordance with the sampling time to avoid introduction of extra noise due to any excess bandwidth, but it usually is not so modulated.

Three new sources of error (i.e., noise) may be introduced by the integration and time measuring scheme:
1. Any variation in the threshold detection process will be reflected as an apparent variation in the signal.
2. The transmittance being measured must be calculated by the reciprocal of the measured integrating time. Care must be taken to measure the time to sufficient precision to avoid adding truncation and similar digital noise in the division process.

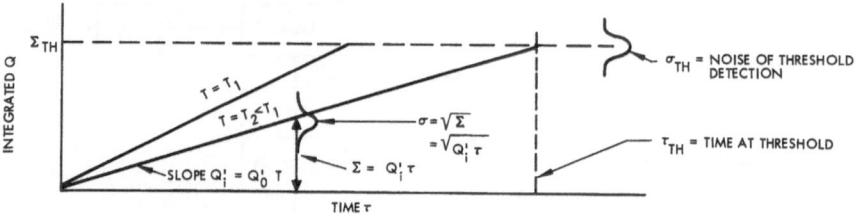

Fig. 6.15. Integrating "constant confidence" detector analysis

3. The success of the integrating depends on stability of the scanning
   light during the integrating interval. Therefore this light must be
   "absolutely" stabilized or a separate integrator used to measure it
   for subsequent use in normalization.

   The integrator analysis is as follows (see Fig. 6.15):

   Output photoelectrons Q_i are collected until a pre-determined
quantity Σ_{th} is collected.

$$\text{Rate of collection} = Q'_o \cdot T \quad Q' = \text{rate of flow of } Q.$$

$$\text{Total collected at any time } \tau: \quad \Sigma = Q'_i \tau = Q_o T \tau.$$

Time required to collect $\Sigma_{th} : \tau_{th} = \Sigma_{th}/Q'_o T = (\Sigma_{th}/Q'_o)/T.$
Thus the time is inversely proportional to T.
Noise at any time τ is $\sigma_\Sigma = \sqrt{\Sigma} = (K Q_i \tau)^{\frac{1}{2}}$.
Noise in determining the threshold is $\triangleq \sigma_{th}$.
The noise involved in the detection of equality of Σ and Σ_{th} is

$$\sigma_\Delta = \sqrt{(\sigma_{th}^2 + \sigma_\Sigma^2)} \quad \text{at threshold.}$$

This is converted into an equivalent uncertainty in measuring τ as a
function of the slope (i.e., rate of approach of Σ to Σ_{th})

$$\sigma_\tau = \sigma_\Delta/Q'_i = \sqrt{(\sigma_{th}^2 + K\Sigma_{th})}/Q'_i.$$

Since $Q'_i = Q'_o T$

$$\sigma_\tau = \sqrt{(\sigma_{th}^2 + K\Sigma_{th})}/Q'_o T. \tag{6.39a}$$

If the integrator threshold uncertainty is very small, this reduces to

$$\sigma_\tau = \sqrt{(K\Sigma_{th})}/Q'_o T, \tag{6.39b}$$

and the SNR is constant

$$\text{SNR} = \tau/\sigma_\tau = \Sigma_{th} Q_o' T / Q_o' T \sqrt{(K\Sigma_{th})} = \sqrt{(\Sigma_{th}/K)}. \tag{6.40}$$

Modulation of Scanning Light

Modulation of Q_o may also be by modulation of the scanning light intensity. The often-suggested feedback scheme in which the input light is increased as a function of T to cause the output light to be constant requires that the resulting input light be the measured variable. This scheme, however, produces neither linear steps of Q_o vs T, constant β, nor constant SNR. Its characteristics are

$$Q_i = Q_o T.$$

For constant Q_i (i.e., clamped at some threshold value), and without taking into account any noise fed back into Q_o by noise in Q_i

$$Q_o = Q_i/T$$

$$\left.\begin{array}{l} NQ_o < Q_o < Q_o \\ 1/N < T < 1 \end{array}\right\},$$

$$\beta = \text{step size}/\sigma_{Q_o}\sqrt{K} = NQ_o/N\sqrt{K}\sqrt{Q_o} = \sqrt{(Qi/K)}/\sqrt{T}, \tag{6.41a}$$

$$\text{SNR}_{Q_o} = \sqrt{(Q_o/K)} = \sqrt{(Qi/K\sqrt{T})}. \tag{6.41b}$$

Modulation of the incoming light inversely to T has an attendent danger: an extremely dark area of the film being scanned (including opaque dirt spots) will cause a scanning light to be boosted to very high values and may easily burn the CRT phosphor at that point. A limiting circuit must therefore be included to limit the CRT light to a safe value, and suitable overload alarms provided.

Modulation of the light as above does not accomplish much alone, as the sampling time is not altered. A combination of the "constant confidence" integrator and the light feedback results in an output measured variable (τ) linear with T, reduced sampling time at lower values of T and hence shorter average frame digitizing time, and noise characteristics between the constant parameter and the constant SNR cases.

The circuit is shown in Fig. 6.16. Feedback is used to stabilize Q_i to a constant value. This raises Q_o to

$$Q_o = Q_i/T.$$

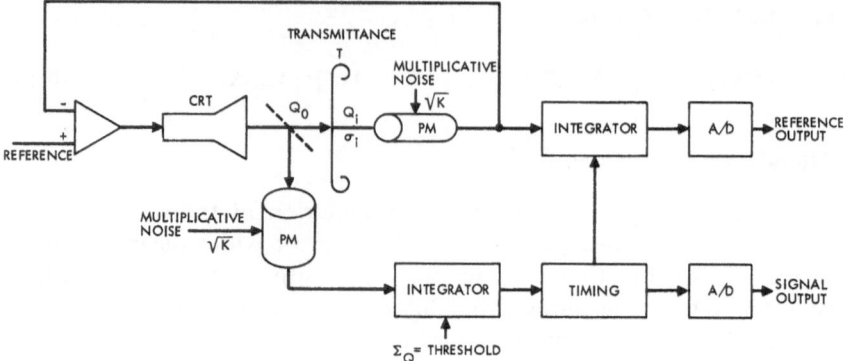

Fig. 6.16. Combination system utilizing integration and feedback

The integrator accumulates Q_o until a threshold Σ_{th} is reached

$$\Sigma_Q = Q'_o \cdot \tau$$
$$\tau_{th} = \Sigma_Q/Q'_o = \Sigma_{th} T/Q'_i .$$

(6.42)

The digitizing (accumulation) time is thus linear with T, thus taking maximum advantage of low values of T.

The feedback causes Q_i to become constant, and

$$Q_o \approx (R - \sigma_i)/T$$
$$\sigma_{Q_o} = \sqrt{Q_o} = \text{scale factor} \cdot 1/\sqrt{T} \quad = f(R - \sigma_i)$$
$$= \sqrt{(M/T)},$$

(6.43)

where M is the modulation scale factor.

The total noise out of the integrator is (by extension of the previous integrator analysis)

$$\sigma_{tot} = T(\sigma_{th}^2 + \sigma_{th} Qi/T + M/T)^{\frac{1}{2}}/Q_i$$
$$= T(\sigma_{th}^2 + \Sigma_{th} + M/T)^{\frac{1}{2}}/Q_i ,$$

(6.44)

where σ_{th} denotes the noise in threshold detection, Σ_{th} is the noise associated with the collected Q_{th}, and M/T the modulation noise introduced into Q_o due to Q_i feedback.

It is thus seen that σ_{tot} is a function of T or of $\sqrt{T}$ or in between depending on the relative contribution of the three noise components. This will be dependent on the actual hardware design.

Since

$$\left.\begin{array}{l} \tau = T\Sigma_{\mathrm{th}}/Q_i' \\[4pt] \mathrm{SNR} = \tau/\sigma_\tau = [\text{scale factor}] \end{array}\right\} \quad \text{if } \sigma_\tau \approx [SF]\,T \qquad (6.45\text{a})$$

or

$$\left.\begin{array}{l} \mathrm{SNR} = [SF]\sqrt{T} \\[4pt] \beta = \text{step size}/\sigma_\tau = [SF]/\sqrt{T} \end{array}\right\} \quad \text{if } \sigma_\tau \approx [SF]\sqrt{T}. \qquad (6.45\text{b})$$

Since the feedback can cause overbright conditions on opaque areas, a limiting circuit is necessary. The reference output is used to measure the actual total Q_i to normalize against overbright limiting and variations in Q_i during integration.

System design starts at the lowest β end (maximum T). σ and Q_o are calculated as before for this point and enough integrator A/D bits allocated to still have a measurement at the low-T end of the range. The reference output is linear with Q_i and must be scaled to contain the maximum Q_i which occurs at the low-T end of the range.

Although this approach does not achieve constant β, it has the following advantages:

Over normal constant confidence:
1) output τ is linear in T, avoiding the division process,
2) faster for $T < 0.5 (D > 0.3)$ (the usual situation),
3) can approach more constant β.

Over normal feedback:
1) takes advantage of the increased light to reduce τ,
2) output is linear in T, avoiding the division process.

Over constant parameter:
1) much faster since constant parameter τ is scaled to worst case,
2) reduces CRT aging by boosting light only in low-T areas.

6.5. Summary

Consideration of noise statistics upon quantization accuracy (in a digitally quantized system) leads to the concept that a useful system measure is the ratio of the quantizing step size to the rms noise (β). Furthermore, it is shown that a reasonable value of this parameter is 3. Several noise sources (film grain, electronic noise, shot noise) are investigated to determine their characteristics relative to the brightness of the signal being quantized. From this, one is led to investigate the characteristics of several typical flying spot scanner systems. Finally, a new hardware configuration is proposed sharing many of the components of the more standard systems, but having superior characteristics with regard to noise performance and quantizing time. Further analysis needs to be done to determine whether this same system concept would be optimum if detectors such as silicon diodes were used instead of photomultiplier tubes, and if illumination sources such as lasers were used instead of the cathode ray tube.

References

6.1. O. H. SCHADE, Sr.: J. Soc. Mot. Pict. Telev. Engrs. **56**, 137 (1951); **58**, 181 (1952); **61**, 97 (1953); **64**, 593 (1955).
6.2. O. H. SCHADE, Sr.: J. Opt. Soc. Am. **46**, 721 (1956).
6.3. K. F. STULTZ, H. J. ZWEIG: Opt. Soc. Am. **49**, 693 (1959).
6.4. K. F. STULTZ, H. J. ZWEIG: J. Opt. Soc. Am. **52**, 45 (1962).
6.5. K. HACKING: J. Brit. IRE **23**, 307 (1962).
6.6. T. S. HUANG: IEEE Trans. Info. Theory IT-**11**, 43–53 (1965).
6.7. J. M. HEYNING: "The Human Observer", Proc. Seminar on Human in the Photo-Opt. Syst., SPIE, New York City (1966).
6.8. J. H. ALTMAN: J. Soc. Mot. Pict. Telev. Engrs. **76**, 629 (1967).
6.9. L. E. ROBERTS: IRE Trans. Info. Theory IT-**8**, 145 (1962).
6.10. G. C. HIGGINS: "Information Capacity of Photographic Materials", in *Photographic Systems for Enginers*, (Soc. Photo. Sci. Engrg., Washington, D.C., 1966), pp. 167–207.
6.11. E. C. DOERNER: J. Opt. Soc. Am. **52**, 669 (1962).
6.12. G. A. FRY: J. Opt. Soc. Am. **53**, 361 (1963).
6.13. O. H. SCHADE, Sr.: J. Soc. Mot. Pict. Telev. Engrs. **73**, 81 (1964).
6.14. E. S. BLACKMAN: Phot. Sci. Eng. **12**, 244 (1968).
6.15. W. BRAULT: Astron. Astrophys. **13**, 169 (1971).
6.16. L. C. WILKINS, P. A. WINTZ: "Bibliography on Data Compression, Picture Properties and Picture Coding"; Purdue University Technical Report TR-EE69-10.
6.17. P. J. READY, P. A. WINTZ, S. J. WHITSITT, D. A. LANDGREBE: "Effects of Compression and Random Noise on Multispectral Data", Proc. 7th Symposium on Remote Sensing of the Environment, Univ. Michigan (1971), pp. 1321–1343.
6.18. T. C. RINDFLEISCH, J. A. DUNNE, H. J. FRIEDEN, W. D. STROMBERG, R. M. RUIZ: J. Geophys. **76**, 394 (1971).
6.19. L. LEVI: J. Opt. Soc. Am. **48**, 9 (1958).
6.20. H. J. ZWEIG, G. C. HIGGINS, D. L. MACADAM: J. Opt. Soc. Am. **48**, 926 (1958).
6.21. R. C. JONES: J. Opt. Soc. Am. **51**, 1159 (1961).
6.22. J. A. EYER: Phot. Sci. Eng. **6**, 71 (1962).
6.23. L. LEVI: Phot. Sci. Eng. **7**, 26 (1963).
6.24. K. HACKING: J. Soc. Mot. Pict. Telev. Engrs. **73**, 1015 (1964).
6.25. A. V. OPPENHEIM, R. W. SCHAFER, T. G. STOCKHAM: Proc. IEEE **56**, 1264 (1968).
6.26. C. F. SHELTON, H. H. HERD, J. J. LEYBOURNE: "Grey-Level Resolution of Flying Spot Scanner Systems", SPIE Photo Optical Systems Seminar, Rochester, N.Y. (1967).
6.27. H. D. FRIEDMAN: Proc. IEEE **53**, 658 (1965).

6.28. G.L.FULTZ: "The Effect of Source Noise on Qunatization Accuracy and on PE Statistics", JPL Tech. Memo 3341-65-5 (1965).
6.29. C.R.WALLI: "Quantizing and Sampling Errors in Hybrid Computation", Proc. Fall Joint Computer Conference (1964), pp. 545–558.
6.30. E.L.O'NEILL: *Introduction to Statistical Optics* (Addison-Wesley, Reading, Mass. Palo Alto, London, 1963), pp. 105–121.
6.31. E.H.EBERHARDT: "Signal-to-Noise Ratio of Image Dissector", Technical Note No. 101, ITT Industrial Laboratories, Fort Wayne, Indiana (1966).
6.32. M.SCHWARTZ: *Information Transmission, Modulation, and Noise* (McGraw Hill, New York, London, 1959).
6.33. T.G.STOCKHAM: Proc. IEEE **60**, 828 (1972).

Further References with Titles

R.T.CHIEN, W.E.SNYDER: Hardware for visual image processing. IEEE Trans. Circuits and Systems CAS-**22**, No. 6, 541–551 (1975).
J.L.MANNOS, D.J.SAKRISON: The effect of visual criterion on the encoding of images. IEEE Trans. Information Theory IT-**20**, No. 4, 525–536 (1974).
W.F.SCHREIBER: The effect of scanning speed on the S/N ratio of camera tubes. Proc. IEEE **52**, No. 2, 217 (1964).

7. Recent Advances in Picture Processing and Digital Filtering

T. S. HUANG

It is the purpose of this brief chapter to review the advances since 1975 in selected areas of digital image processing and two-dimensional filtering. Specifically, the following topics will be discussed: Two-dimensional transforms, two-dimensional recursive and nonrecursive filters, and image restoration and enhancement.

7.1. General References

For progress in picture processing in general, we recommend highly the annual reviews of ROSENFELD, which started around 1970, and most of which appeared in the journal *Computer Graphics and Image Processing* [7.1].

Since 1975, three books [7.2–4] and several volumes of collected papers [7.5–7] on digital image processing have appeared, as well as a book on image restoration [7.8] and a volume of collected papers on two-dimensional signal processing [7.9].

7.2. Two-Dimensional Transforms

Among the various transforms discussed in Chapter 2, Fourier transform undoubtedly has the widest application. The other transforms are mainly useful in image coding. Experience in the past few years has indicated that among the image-independent transforms (hence K–L transform is not included), the discrete cosine transform (DCT) and the slant transform give the best performance in image coding. The conventional way of computing DCT is using FFT [7.8, 11]. However, a recent fast algorithm promises to speed up by a factor of 6 [7.12].

To calculate two-dimensional transforms, such as Fourier and Hadamard transforms, on a computer whose core memory cannot hold the entire image, auxiliary memory such as disks are needed and matrix transposition is necessary. An efficient algorithm for matrix transposition was suggested by Eklundh in 1972 [7.13]. More recently, he has developed two new algorithms and some results on optimum strategy [7.14].

An interesting recent research area is the use of number-theoretic techniques in signal processing. Rader was the first person who suggested the use of number-theoretic transforms (e.g., Fermat-number transform) to do high-speed two-dimensional convolution [7.15–17]. An excellent introduction to this idea is the section written by Rader in [7.18]. By using these number-theoretic transforms, impressive computational savings can be obtained in a limited number of cases, where i) the sequences to be transformed are relatively short (fewer than around 250 samples), ii) considerable accuracy is needed, and iii) multiplication is more costly than addition. It is expected that Fermat-number transform is faster than the FFT method in doing two-dimensional convolution if the window size is less than 20×20 samples.

Winograd applied number-theoretic techniques to the calculation of the discrete Fourier transform (DFT) [7.19]. Compared to FFT, the number of multiplications is drastically reduced while the number of additions remains approximately the same. For example, for a sequence of 1024 samples, FFT requires 12288 multiplications and 26624 additions; while for a sequence of 1008 samples, the Winograd Fourier transform (WFT) requires 4212 multiplications and 25224 additions. Hardware for performing WFT has been constructed and incorporated in a high-speed signal processor [7.20]. Quantization errors (due to roundoff and coefficient quantization) for WFT in the fixed-point case were studied by Patterson and McClellan [7.21], who found that in general WFT requires one or two more bits for data representation to give an error similar to that of FFT.

7.3. Two-Dimensional Digital Filters

For developments up to 1975, the reader is referred to the review paper of Mersereau and Dudgeon [7.22], and a related one by Bose [7.23].

7.3.1. Nonrecursive Filters

In Chapter 3, four methods of designing nonrecursive two-dimensional filters were described: window technique, frequency sampling, conventional linear programming, and iterative linear programming. Since around 1975, two new good design techniques have emerged. The first is a method proposed by McClellan [7.24] and later generalized by

MERSEREAU et al. [7.25, 26]. In this method, one designs two-dimensional nonrecursive filters by applying a change of variables to optimum minimax one-dimensional filters. It is especially suitable for designing circularly symmetric low-pass and band-pass filters, and more generally filters with quadrilateral symmetry. The method is computationally easy and in many cases gives optimal designs.

The second method is the extension of the Remez multiple exchange algorithm to the design of two-dimensional nonrecursive filters [7.27]. This is probably the best general purpose design tool for nonrecursive filters with Chebyshev error norm (minimax).

7.3.2. Recursive Filters

One key problem in designing two-dimensional recursive filters is the test of stability. Many earlier results on stability were discussed in detail in Chapter 4. For recent developments on this and related problems, the reader is referred to the masterful review paper of JURY [7.28]. Recently, based on the pioneering work of PISTOR [7.29], DUDGEON [7.30] and EKSTROM and TWOGOOD [7.31], O'CONNOR and HUANG developed some very efficient methods of testing the stability of two-dimensional recursive filters [7.32–34]. Some related papers are [7.35–37].

In Chapter 4, it was conjectured that the double least squares inverse method would stabilize an unstable filter. It has been demonstrated since then that this conjecture is not true in general [7.38]. However, it is interesting to investigate under what conditions the conjecture is valid [7.39, 40].

Since around 1975, many new methods and extensions of earlier ones have emerged for the design of two-dimensional recursive filters. The approach of separable-sum approximation [7.41, 42] was pursued further by TAKAHASHI and TSUJI using Laguerre functions [7.43]. Several design methods based on optimization techniques have been developed [7.44–47]. In particular, the method of ALY and FAHMY [7.47] uses l_p approximations of magnitude and group delay, and guarantees stability through the use of a frequency transformation [7.48]. RAMAMOORTHY and BRUTON [7.48, 49] proposed a method based on the decomposition of an $n \times n$ multivariable positive real admittance matrix and using Koga's result that all such matrices can be realized as multivariable finite lumped passive networks. The method guarantees stability. Several design techniques impose relatively simple sufficient conditions for stability, which however are not necessary. These include the methods described in [7.50, 51]. Other design techniques are [7.52, 53].

All the design techniques mentioned above are for quarter-plane filters, i.e., the support of the impulse response of the filter is a quarter-plane. The design is more flexible if we can extend the support of the impulse response to a half-plane. The design of half-plane filters is in its infancy [7.54] and an important area for further research.

For analyses of quantization errors in two-dimensional digital filters, see [7.26, 7.55–59].

7.4. Image Restoration and Enhancement

The line of demarcation between restoration and enhancement is by no means clear. Roughly speaking, one talks about restoration when one has a definite mathematical model for the image degradations and aims to compensate for these degradations by suitable processing techniques. On the other hand, the aim of enhancement is usually more vague – one simply wants to improve the quality (usually subjective) of the images. In this sense, all the methods discussed in Chapter 5 lie in the category of image restoration. One should also keep in mind that in the literature image restoration usually refers to the compensation of *linear* degradations.

7.4.1. Image Restoration

In Sections 5.12–15, several spectrum extrapolation methods were described. In 1975, a new iterative method was proposed by PAPOULIS [7.60]. More recently, SABRI and STEENAART modified this method so that the total extrapolation process can be achieved by a single matrix operation [7.61].

In Chapters 1 and 5, several algebraic techniques of image restoration were described. Recently, YOULA published a generalized method of image restoration by alternating orthogonal projections [7.62]. HOU and ANDREWS proposed a technique of least squares image restoration using spline basis functions [7.63].

Among nonlinear restoration techniques, maximum entropy methods have drawn the most attention [7.64]. Recently, Bayesian methods were proposed by HUNT [7.65].

Specific degradations which have received attention include motion [7.66, 67] and atmospherical turbulence [7.68, 69].

An important technique for phase compensation was developed by KNOX and THOMPSON for the recovery of images from atmospherically degraded short-exposure photographs [7.70, 71]. This complements the earlier work of LABERYRIE [7.72], as reviewed in [7.73]. An alternative approach was studied by McGLAMERY [7.74]. A review on imaging through randomly fluctuating media was given by BATES et al. [7.75], see also [7.68]. O'CONNOR and HUANG [7.33] compared the methods of Knox-Thompson and McGlamery. Preliminary experimental results indicate that the Knox-Thompson method gives better restoration.

A posteriori estimation of the impulse response (point spread function) for specific degradations, such as linear motion blur and lens defocusing, was treated by CANNON [7.76]. For general linear shift-invariant degradations, two methods have been used: edge-gradient analysis and piecewise Fourier transformation [7.77, 78]. The latter method encountered difficulty in the determination of the phase of the frequency response of the degradation [7.77, 79]. The Knox-Thompson technique [7.70, 71] renewed hope in conquering this difficulty. However, two independent efforts [7.33, 80] have so far achieved only very limited success.

Systematic comparisons of the performances of different restoration techniques are badly needed. A step in the right direction was taken by CANNON et al. [7.81]. They compared three image restoration methods: Wiener (least squares), power-spectrum equalization [7.76], and maximum a posteriori probability estimation [7.65]. Two criteria were used: mean-square error, and subjective quality. The test images were degraded by either Gaussian blur or defocusing, at various signal-to-noise ratios. Their major finding is that all methods perform equally well on Gaussian-blurred images regardless of S/N, blur severity, or image type; however, in the case of defocusing, the MAP method excels, especially in a high S/N environment.

The aim of most restoration and enhancement task is to improve the *subjective* quality of images as viewed by human observers. The quantification of the human visual system is, however, an extremely difficult problem. Nonetheless, notable advances have been made [7.82–86].

7.4.2. Image Enhancement

Three important tasks in image enhancement are: contrast enhancement, noise reduction, and edge sharpening.

The most commonly used contrast enhancement technique is histogram modification. In many cases, the subjective contrast of an image is improved if one equalizes the histogram of the gray-levels of the picture elements (i.e., making the histogram uniform) [7.87]. FREI [7.88] argued that one should make the histogram of the subjective brightness of the picture elements uniform, or equivalently making the gray-level histogram hyperbola in shape. The reader is warned that although histogram modification may improve the subjective quality of an image, it may hurt the further machine analysis of the image (such as segmentation and classification).

An often used but seldom if ever published technique for enhancing the contrast of small details (such as contained in fingerprints) is variance equalization or local gain control [7.89]. The image here is normalized such that the sample variance in a small window (5×5 picture elements, say) is equal over the entire image.

An obvious way of noise reduction is to use Wiener (linear least squares) filters. An alternative is Kalman filters. HABIBI and NAHI were among the first who suggested the extension of Kalman filtering to two dimensions for image noise reduction [7.90–93]. More recently, important contributions have been made by JAIN and WOODS [7.94–97]. In particular, WOODS have proposed several ways of making the computation more efficient.

In order to reduce noise but keep the edges in the image sharp, one has to use linear shift-varying or nonlinear filters. The former include adaptive Wiener or Kalman filters. The latter include median filtering [7.98–100].

Edge sharpening can be done by raising the amplitudes of the high spatial frequencies of an image relative to those of the lower spatial frequencies. However, usually one gets better results if the high-emphasis filter is applied to the logarithm of the gray-levels of the image and the output exponentiated [7.82]. This method, called homomorphic filtering, tends to enhance the local contrast as well as sharpen the edges. For an insightful discussion on this and related issues, read [7.85].

References

7.1   A. ROSENFELD: Computer Graphics and Image Processing **2**, (2), 211–242 (1978)
7.2   A. ROSENFELD, A. C. KAK: *Digital Picture Processing* (Academic Press, New York, 1976)
7.3   R. GONZALEZ, P. A. WINTZ: *Digital Image Processing* (Addison-Wesley, Reading, Mass. 1977)
7.4   W. K. PRATT: *Digital Image Processing* (Wiley, New York 1978)
7.5   A. ROSENFELD (ed.): *Digital Picture Analysis*, Topics in Applied Physics, Vol. 11 (Springer, Berlin, Heidelberg, New York 1976)
7.6   J. K. AGGARWAL, R. O. DUDA, A. ROSENFELD (eds.): *Computer Methods in Image Analysis* (IEEE Press, New York 1977)
7.7   R. BERNSTEIN (ed.): *Digital Image Processing for Remote Sensing* (IEEE Press, New York 1978)
7.8   H. C. ANDREWS, B. R. HUNT: *Digital Image Restoration* (Prentice-Hall, Englewood Cliffs, NJ. 1977)

7.9   S.K.MITRA, M.P.EKSTROM (eds.): *Two-Dimensional Digital Signal Processing* (Dowden, Hutchinson, and Ross, Stroudsburg, Pa. 1978)

7.10 N.AHMED, T.NATARJAN, K.R.RAO: IEEE Trans. Computers, C-23, 90–93 (1974)

7.11 B.D.TSENG, W.C.MILLER: IEEE Trans. Computers, C-27, (10), 966–968 (1978)

7.12 W.H.CHEN, C.H.SMITH, S.C.FRALICK: IEEE Trans. Communications, COM-25, (9), 1004–1009 (1977)

7.13 J.O.EKLUNDH: IEEE Trans. Computers, C-21, 801–803 (1972)

7.14 J.O.EKLUNDH: Efficient Matrix Transposition with Limited High-Speed Storage, FOA Rpt., 12, (1), 1–19 (1978) National Defense Research Institute, Linkoping, Sweden

7.15 C.M.RADER: IEEE Trans. Circuits and Systems CAS-22, 575 (1975)

7.16 R.C.AGARWAL, C.S.BURRUS: Proc. IEEE, 63, 550–560 (1975)

7.17 I.S.REED, T.K.TRUONG, Y.S.KWOH, E.L.HALL: IEEE Trans. Computers, C-26, (9), 874–881 (1977)

7.18 L.R.RABINER, B.GOLD: *Theory and Application of Digital Signal Processing* (Prentice-Hall, Englewood Cliffs, NJ. 1975), 419–434

7.19 H.F.SILVERMAN: IEEE Trans. Acoustics, Speech, and Signal Processing ASSP-25, (2), 152–165 (1977)

7.20 A.PELED: "A Low-Cost Image Processing Facility Employing a New Hardware Realization of High-Speed Signal Processors", in *Advances in Digital Image Processing*, ed. by P.STUCKI (Plenum, New York 1979)

7.21 R.W.PATTERSON, J.H.MCCLELLAN: IEEE Trans. Acoustics, Speech, and Signal Processing 26, (5), 447–455 (1978)

7.22 R.M.MERSERAU, D.E.DUDGEON: Proc. IEEE 63, (4), 610–623 (1975)

7.23 N.K.BOSE: Proc. IEEE, 65, (6), 824–840 (1977)

7.24 J.H.MCCLELLAN: The Design of Two-Dimensional Digital Filters by Transformations. Proc. 7th Annual Princeton Conference on Information Science and Systems (1973) 247–251

7.25 R.M.MERSEREAU, W.F.G.MECKLENBRÄUKER, T.F.QUATIERI, JR.: IEEE Trans. Circuits and Systems, CAS-23, (7), 405–414 (1976)

7.26 W.F.G.MECKLENBRÄUKER, R.M.MERSEREAU: IEEE Trans. Circuits and Systems, CAS-23, (7), 414–422 (1976)

7.27 D.B.HARRIS, R.M.MERSEREAU: A Comparison of Iterative Methods for Optimal 2-D Filter Design, Proc. International Conference on Acoustics, Speech, and Signal Proc. (1977) p. 527

7.28 E.I.JURY: Proc. Ieee, 66, (9), 1018–1047 (1978)

7.29 P.PISTOR: IBM J. Res. Dev. 18, 59–71 (1974)

7.30 D.DUDGEON: IEEE Trans. Acoustics, Speech, and Signal Proc. ASSP, 25, 476–484 (1977)

7.31 M.EKSTROM, R.TWOGOOD: A Stability Test for Two-Dimensional Recursive Digital Filters Using the Complex Cepstrum. Proc. IEEE Intern. Conf. on Acoustics, Speech, and Signal Proc. (1977) pp. 535–538

7.32 B.T.O'CONNOR, T.S.HUANG: Stability of General Two-Dimensional Recursive Digital Filters. Tech. Rpt. TR-EE 77-36, School of Electrical Engineering, Purdue University (Oct. 1977)

7.33 B.T.O'CONNOR, T.S.HUANG: Techniques for Determining the Stability of Two-Dimensional Recursive Filters and Their Applications to Image Restoration. Tech. Rpt., TR-EE 78-18, School of Electrical Engineering, Purdue University (May 1978)

7.34 B.T.O'CONNOR, T.S.HUANG: IEEE Trans. Acoustics, Speech, and Signal Proc. ASSP-26, 550–560 (1978)

7.35 M.G.STRINTZIS: IEEE Trans. Circuits and Systems C-24, 432–437 (1977)

7.36 R.DECARLO, R.SAEKS, J.MURRAY: Proc. IEEE 65, 978–979 (1977)

7.37   G. A. SHAW: An Algorithm for Testing Stability of Two-Dimensional Digital Recursive Filters. Proc. IEEE Intern. Conf. on Acoustics, Speech, and Signal Proc. (1978)

7.38   Y. GENIN, Y. KAMP: Electron, Lett, 11, 330–331 (1975)

7.39   E. K. JURY, V. R. KOLVAENNU, B. D. O. ANDERSON: Proc. IEEE, 65, 887–898 (1977)

7.40   Y. V. GENIN, Y. G. KAMP: Proc. IEEE 65 (6), 873–886 (1977)

7.41   S. TREITEL, J. L. SHANKS: IEEE Trans. Geosci. Electron., 9, (1), 10–27 (1971)

7.42   S. CHAKRABARTI, N. K. BOSE, S. K. MITRA: J. Franklin Inst., 299, (1), 53–60 (1975)

7.43   S. TAKAHASHI, S. TSUJI: A Design Method of Two-Dimensional Recursive Digital Filter by Using Laguerre Functions, Trans. Inst. Electronics and Communication Engineers (Japan), 60, 521–528 (1977), (in Japanese)

7.44   D. E. DUDGEON: IEEE Trans. Acoustics, Speech, and Signal Proc. 23, 264–267 (1975)

7.45   G. A. MARIA, M. M. FAHMY: IEEE Trans. Acoustics, Speech, and Signal Proc. ASSP, 22, (1), 15–21 (1974)

7.46   G. A. MARIA, M. M. FAHMY: IEEE Trans. Circuits and Systems CAS-21, (3), 431–436 (1974)

7.47   S. A. H. ALY, M. M. FAHMY: IEEE Trans. Circuits and Systems CSA-25 (11), 908–916 (1978)

7.48   P. A. RAMAMOORTHY, L. T. BRUTON: Frequency Domain Approximation of Stable Multi-Dimensional Discrete Recursive Filters. Proc. Intern. Symp. on Circuits and Systems (April, 1977)

7.49   P. A. RAMAMOORTHY, L. T. BRUTON: Design of Stable Two-Dimensional Digital Recursive Filters and Applications in Image Processing. Proc. 20th Midwest Symp. on Circuits and Systems (August, 1977)

7.50   A. CHOTTEVA, G. A. JULLIEN: Designing Near Linear Phase Recursive Filters Using Linear Programming. Proc. IEEE Intern. Conf. on Acoustics, Speech, and Signal Processing (May, 1977) pp. 88–92

7.51   A. M. ALI: Design of Inherently Stable Two-Dimensional Recursive Filters Imitating the Behavior of One-Dimensional Analog Filters. Proc. IEEE Intern. Conf. on Acoustics, Speech, and Signal Proc. (April, 1978) pp. 765–768

7.52   M. LAL: Space-Domain Design of Two-Dimensional Recursive Digital Filters. Proc. IEEE Symp. Circuits and Systems, Newton, Mass. (March, 1975)

7.53   N. A. PENDERGRASS, S. K. MITRA, E. I. JURY: IEEE Trans. Circuits and Systems CAS-23, 26–35 (1976)

7.54   M. P. EKSTROM, J. W. WOODS: Some Results on the Design of Two-Dimensional Half-Plane Recursive Filters. Proc. IEEE Intern. Symp. on Circuits and Systems, Munich (April, 1976)

7.55   G. A. MARIA, M. M. FAHMY: IEEE Trans. Circuits and Systems CAS-22 (10), 826–830 (1975)

7.56   K. O. SHIPP, JR., J. K. AGGARWAL: IEEE Trans. Acoustics, Speech, and Signal Proc. ASSP-24 (4), 339–341 (1976)

7.57   M.-D. NI, J. K. AGGARWAL: IEEE Trans. Computers C-25 (7), 755–759 (1976)

7.58   T.-L. CHANG: IEEE Trans. Circuits and Systems CAS-24 (1), 15–19 (1977)

7.59   P. AGATHOKLIS, E. I. JURY, M. MAUSOUR: A Note on the Evaluation of Quantization Error in Two-Dimensional Digital Filters. Preprint (Oct., 1978)

7.60   A. PAPOULIS: IEEE Trans. Circuits and Systems CAS-22 (9), 735–742 (1975)

7.61   M. S. SABRI, W. STEENAART: IEEE Trans. Circuits and Systems CAS-25 (2), 74–78 (1978)

7.62   D. C. YOULA: IEEE Trans. Circuits and Systems CAS-25 (9), 694–702 (1978)

7.63   H. S. HOU, H. C. ANDREWS: IEEE Trans. Computers C-26 (9), 856–873 (1977)

7.64   B. R. FRIEDEN, W. SWINDELL: Restored Pictures of Ganymede, Moon of Jupiter, Science (March 26, 1976)

7.65   B.R.HUNT: IEEE Trans. Computers C-**26** (3), 219–229 (1977)
7.66   A.O.ABOUTALIB, L.M.SILVERMAN: IEEE Trans. Circuits and Systems CAS-**22** (3), 278–286 (1975)
7.67   A.O.ABOUTALIB, L.M.SILVERMAN, M.S.MURPHY: IEEE Trans. Automatic Control AC-**22** (3), 294–302 (1977)
7.68   J.W.STROHBEHN (ed.): *Laser Beam Propagation in the Atmosphere*, Topics in Applied Physics, Vol. 25 (Springer, Berlin, Heidelberg, New York 1978)
7.69   G.I.MARCHUK, G.A.MIKHAILOV, M.A.NAZARALIEV, R.A.DARBINJAN, B.A.KARGIN, B.S.ELEPOV: *The Monte Carlo Method in Atmospheric Optics*, Springer Series in Optical Science, Vol. 12 (Springer, Berlin, Heidelberg, New York 1979)
7.70   K.KNOX, B.THOMPSON: Astrophys. J. **193**, 145–148 (1974)
7.71   K.KNOX: J. Opt. Soc. Am. **66**, 1236–1239 (1976)
7.72   A.LABEYRIE: Astron. Astrophys. **6**, 85–87 (1970)
7.73   J.C.DAINTY (ed.): *Laser Speckle and Related Phenomena*, Topics in Applied Physics, Vol. 9 (Springer, Berlin, Heidelberg, New York 1975)
7.74   B.McGLAMERY: NASA Tech. Rpt. SP-**256**, 167–192 (1971)
7.75   R.BATES, M.McDONELL, P.GOUGH: Proc. IEEE **65**, 135–143 (1977)
7.76   M.CANNON: IEEE Trans. Acoustics, Speech, and Signal Proc. ASSP-**24** (1), 58–63 (1976)
7.77   T.S.HUANG, W.F.SCHREIBER, O.J.TRETIAK: Image Processing, Proc. IEEE (Nov., 1971)
7.78   M.P.EKSTROM: IEEE Trans. Computers C-**22**, 322–328 (1973)
7.79   A.FILIP: Estimating the Impulse Response of a Linear Shift-Invariant Image Degrading System. Ph. D. thesis, Dept. of Electrical Engineering, M.I.T. (1972)
7.80   J.B.MORTON: An Investigation into an a posteriori Method of Image Restoration. Ph. D. thesis, Dept. of Electrical Engineering, Univ. of Southern Calif. (1978)
7.81   T.M.CANNON, H.J.TRUSSELL, B.R.HUNT: Appl. Opt. **17** (21), 3384–3390 (1978)
7.82   T.G.STOCKHAM, JR.: Proc. IEEE **60**, 828–1309 (1972)
7.83   G.L.ANDERSON, A.N.NETRAVALI: IEEE Trans. Systems, Man, and Cybernetics SMC-**6** (12), 845–853 (1976)
7.84   T.HEUTEA, B.E.A.SALEH: IEEE Trans. Systems, Man, and Cybernetics SMC-**8** (12), 883–888 (1978)
7.85   W.F.SCHREIBER: Proc. IEEE **66** (12), 1640–1651 (1978)
7.86   D.E.TROXEL, W.F.SCHREIBER et al.: Image Enhancement/Coding Systems Using Pseudorandom Noise Processing, To appear in Proc. IEEE; preprint, Dec. 1, 1978
7.87   R.A.HUMMEL: Histogram Modification Techniques. Tech. Rpt. TR-329, Computer Science Center, Univ. of Maryland (Sept., 1974)
7.88   W.FREI: Image Enhancement by Hostogram Hyperbolization, Computer Graphics, and Image Processing **6**, 286–294 (1977)
7.89   J.HARRIS: Image Enhancement by Variance Equalization. Private communication (1970)
7.90   A.HABIBI: Proc. IEEE **60** (7), 878–883 (1972)
7.91   M.G.STRINTZIS: Proc. IEEE **64**, 1255–1257 (1976)
7.92   N.E.NAHI, T.ASSEFI: IEEE Trans. Computer C-**21** (7), 734–738 (1972)
7.93   N.E.NAHI, C.A.FRANCO: Recursive Image Enhancement Vector Processing, IEEE Trans. Commun. CO-**21** (4), 305–311 (1973)
7.94   A.K.JAIN, E.ANGEL: IEEE Trans. Computers C-**23** (5), 470–476 (1976)
7.95   J.W.WOODS, C.H.RADEWAN: Reduced Update Kalman Filter — A Two-Dimensional Recursive Processor. Proc. Johns Hopkins Conf. on Inf. Sci. and Syst. (March, 1976)
7.96   J.W.WOODS, C.H.RADEWAN: IEEE Trans. Information Theory IT-**23**, 473–482 (1977)

7.97   J. W. WOODS, V. K. INGLE, A. RADPOUR, H. KAUFMAN: Recursive Estimation with
       Non-Homogeneous Image Models. Proc. IEEE Conf. on Pattern Recognition and
       Image Processing, Chicago (May, 1978)
7.98   W. K. PRATT: Median Filtering, Semiannual Report Image Processing Inst., Univ. of
       Southern Calif. (Sept., 1975) pp. 116–123
7.99   B. R. FRIEDEN: Opt. Soc. Am. **66**, 280–283 (1976)
7.100 T. S. HUANG, G. J. YANG, G. Y. TANG: A Fast Two-Dimensional Median Filtering
       Algorithm. IEEE Trans. Acoustics, Speech, and Signal Proc. ASSP-**27** (2), 13–18
       (1979)

Further References with Titles

H. CHANG, J. K. AGGARWAL: Design of two-dimensional semicausal recursive filters. IEEE
    Trans. Circuits and Systems CAS-**25** (12), 1051–1059 (1978)
K. HIRANO, J. K. AGGARWAL: Design of two-dimensional recursive digital filters. IEEE
    Trans. Circuits and Systems CAS-**25** (12), 1066–1076 (1978)
P. DELSARTE, Y. V. GENIN, Y. G. KAMP: Planar least squares inverse polynomials: Part I –
    Algebraic properties. IEEE Trans. Circuits and Systems CAS-**26** (1), 59–66 (1979)
A. FETTWEIS: Suppression of parasitic oscillations in multidimensional wave digital filters.
    IEEE Trans. Circuits and Systems CAS-**25** (12), 1060–1065 (1978)
G. S. ROBINSON, J. J. REIS: Spectral extrapolation with spatial constraints, Proc. Society
    of Photo-Optical Instrumentation Engineers, Vol. 149, Applications of Digital Image
    Processing (1978)
J. A. CADZOW: An extrapolation procedure for band-limited signals. IEEE Trans. Acous-
    tics, Speech, and Signal Processing ASSP-**27** (1), 4–12 (1979)
A. K. JAIN: Partial differential equations and finite-difference methods in image processing,
    Part 1: Image representation, optimization theory and applications. IEEE Trans.
    Automatic Control AC-**23** (1), 65–91 (1976)
A. K. JAIN, J. R. JAIN: Partial differential equations and finite-difference methods in
    image processing, Part 2: Image restoration. IEEE Trans. Automatic Control AC-**23**
    (5), 817–834 (1978)
J. W. WOODS: Two-dimensional discrete Markov fields. IEEE Trans. Inform. Theory
    IT-**18**, 232–240 (1972)
E. WONG: Recursive causal linear filtering for two-dimensional random fields. IEEE
    Trans. Inform. Theory IT-**24** (1), 50–59 (1978)

Subject Index

Applied Physics

A monthly journal

Board of Editors
S. Amelinckx, Mol. **V.P. Chebotayev,** Novosibirsk
R. Gomer, Chicago, IL., **H. Ibach,** Jülich
V.S. Letokhov, Moskau, **H.K.V. Lotsch,** Heidelberg
H.J. Queisser, Stuttgart, **F.P. Schäfer,** Göttingen
A. Seeger, Stuttgart, **K. Shimoda,** Tokyo
T. Tamir, Brooklyn, NY, **W.T. Welford,** London
H.P.J. Wijn, Eindhoven

Coverage
application-oriented experimental and theoretical
physics:

Solid-State Physics   *Quantum Electronics*
Surface Sciences   *Laser Spectroscopy*
Solar Energy Physics   *Photophysical Chemistry*
Microwave Acoustics   *Optical Physics*
Electrophysics   *Integrated Optics*

Special Features
rapid publication (3-4 months)
no page charge for **concise** reports
prepublication of titles and abstracts
microfiche edition available as well

Languages
mostly English

Articles
original reports, and short communications review
and/or tutorial papers

Manuscripts
to Springer-Verlag (Attn. H. Lotsch), P.O. Box 105 280
D-6900 Heidelberg 1, F.R. Germany

Springer-Verlag
Berlin
Heidelberg
New York

Place North-American orders with:
Springer-Verlag New York Inc., 175 Fifth Avenue,
New York, N.Y. 100 10, USA

Acoustic Surface Waves

Editor: *A. A. Oliner*

1978. 198 figures, 16 tables. XI, 331 pages
(Topics in Applied Physics, Volume 24)
ISBN 3-540-08575-0

Contents:

A. A. Oliner: Introduction. – *G. W. Farnell:*
Types and Properties of Surface Waves. –
H. M. Gerard: Principles of Surface Wave
Filter Design. – *E. A. Ash:* Fundamentals
of Signal Processing Devices. –
A. A. Oliner: Waveguides for Surface
Waves. – *A. J. Slobodnik, Jr.:* Materials
and Their Influence on Performance. –
H. I. Smith: Fabrication Techniques for
Surface Wave Devices.

Digital Picture Analysis

Editor: *A. Rosenfeld*

1976. 114 figures, 47 tables.
XIII, 351 pages
(Topics in Applied Physics, Volume 11)
ISBN 3-540-07579-8

Contents:

A. Rosenfeld: Introduction. –
R. M. Haralick: Automatic Remote
Sensor Image Processing. – *C. A. A. Har-
low, S. J. Dwyer III., G. Lodwick:* On Radio-
graphic Image Analysis. – *R. L. McIlwain,
Jr.:* Image Processing in High Energy
Physics. – *K. Preston, Jr.:* Digital Picture
Analysis in Cytology. – *J. R. Ullmann:* Pic-
ture Analysis in Character Recognition

Springer-Verlag
Berlin
Heidelberg
New York

T. Pavlidis

Structural Pattern Recognition

1977. 173 figures, 13 tables.
XII, 302 pages
(Springer Series in Electrophysics,
Volume 1)
ISBN 3-540-08463-0

Contents:

Mathematical Techniques for Curve
Fitting. – Graphs and Grids. – Funda-
mentals of Picture Segmentation. –
Advanced Segmentation Techniques. –
Scene Analysis. – Analytical Description
of Region Boundaries. – Syntactic Ana-
lysis of Region Boundaries and Other
Curves. – Shape Description by Region
Analysis. – Classification Description
and Syntactic Analysis.

Nonlinear Methods of Spectral Analysis

Editor: *S. Haykin*

1979. 45 figures, 2 tables. Approx.
320 pages
(Topics in Applied Physics, Volume 34)
ISBN 3-540-09351-6

Contents:

S. Haykin: Introduction. – *S. Haykin,
St. Kesler:* Prediction: Error Filtering and
Maximum-Entropy Spectral Estima-
tion. – *T. Ulrych, M. Ooe:* Autoregressive
and Mixed Autoregressive-Moving
Average Models and Spectra. –
E. A. Robinson: Iterative Least-Squares
Procedure for ARMA Spectral Estima-
tion. – *J. Capon:* Maximum-Likelihood
Spectral Estimation. – *R. N. McDonough:*
Application of the Maximum-Likelihood
Method and the Maximum-Entropy
Method to Array Processing.